GEORGE GABRIEL STOKES

GEORGE GABRIEL STOKES

Life, Science and Faith

Edited by

MARK MCCARTNEY
ANDREW WHITAKER
ALASTAIR WOOD

OXFORD
UNIVERSITY PRESS

OXFORD
UNIVERSITY PRESS

Great Clarendon Street, Oxford, OX2 6DP,
United Kingdom

Oxford University Press is a department of the University of Oxford.
It furthers the University's objective of excellence in research, scholarship,
and education by publishing worldwide. Oxford is a registered trade mark of
Oxford University Press in the UK and in certain other countries

© Oxford University Press 2019

The moral rights of the authors have been asserted

First Edition published in 2019

Impression: 1

Published in the United States of America by Oxford University Press
198 Madison Avenue, New York, NY 10016, United States of America

British Library Cataloguing in Publication Data
Data available

Library of Congress Control Number: 2018963937

ISBN 978-0-19-882286-8

DOI: 10.1093/oso/9780198822868.001.0001

Printed and bound by
CPI Group (UK) Ltd, Croydon, CR0 4YY

FOREWORD

At a meeting in Bangalore in 1988, marking the birth centenary of the Nobel Laureate C. V. Raman, I was asked to give several additional lectures in place of overseas speakers who had cancelled. During one of those talks, I suddenly realised that underlying each of them was one or more contributions by Sir George Gabriel Stokes. Understanding divergent series, phenomena involving polarized light, fluid motion, refraction and diffraction by sound and of sound, Stokes's theorem (I didn't know then that he learned it from Kelvin), . . . : the list seemed endless.

My enthusiasm thus ignited, I acquired Stokes's collected works and explored the vast range and originality of his physics and mathematics (separately and in combination). Paul Dirac was certainly wrong in his uncharacteristically ungenerous assessment (reported by John Polkinghorne) dismissing Stokes as 'a second-rate Lucasian Professor'. On the contrary, in every subject he touched his contributions were definitive, and influenced all who followed. Perhaps Dirac failed to understand, as we do now, that discovering new laws of nature is not the only fundamental science: equally fundamental is discovering and understanding phenomena hidden in the laws we already know.

I resolved to learn more about Stokes's origins. So did Alastair Wood, who organized a visit with our families to the village of Skreen, on the west coast of Ireland south of Sligo, where Stokes spent his early years. His father had been a Church of Ireland minister, caring for a flock scattered over four fishing villages; inside the church in Skreen is a plaque celebrating his work there. As we drove up to the modern house replacing the one (now demolished) where Stokes was born, the woman living there—wife of the local doctor—approached us: 'You must be from Dublin, about the whale'. Over a generous and suddenly improvised lunch, we learned that a dead whale had been washed up on the beach, and she learned what a great scientist Stokes had been.

In this welcome celebration of the bicentenary of Stokes's birth, chapters by distinguished authors span aspects of his life, his beliefs, and of course the many fields of science in which he excelled.

Michael Berry
University of Bristol

PREFACE

On 13 August 1819 one of the greatest of Victorian scientists, Sir George Gabriel Stokes, was born youngest son of the Rector of Skreen on the wild and remote Atlantic coast of Ireland. He was to die, after fifty-three years in the most famous mathematical chair in Cambridge, on 1 February 1903. What happened in between and how was it achieved? This volume is our tribute to him on the two hundredth anniversary of his birth.

In his famous letter on semi-convergent series published in *Acta Mathematica* in 1902, in the last year of his life, we found the poetical sentence 'As [the parameter] passes through the critical value, the inferior term enters as it were into a mist, is hidden for a little from view, and comes out with its coefficient changed.' There are a number of remarkable features about this sentence. The first is that appears in a *mathematical* journal, albeit that as a letter it was not refereed and standards of rigour were different in these days. The second is that it came from a man who was apparently noted for his serious and taciturn nature. But how many times as a boy had he gazed from the Rectory schoolroom window at the clouds sweeping across the brow of Benbulben across Sligo Bay? What happens inside this mist has been explained over the past fifty years by R. B. Dingle, M. V. Berry and others, but there were other areas of physics where Stokes himself cleared away the mist. We ourselves felt that, like other great Victorians, the life and works of Stokes had disappeared too much into the mist and now was the time to restore them to public notice.

For every book there is a moment in time which can be identified as its genesis. In this case it occurred during the incident of the whale beached on Skreen strand described in the Foreword. So moved were the visitors (one of whom is an editor of the present volume) by the warmth and generosity of their welcome in the village that they went on to organize four summer schools in the tiny parish hall, which attracted Stokes scholars worldwide. These led on to other Stokes events. Independently, and about the same time, in the northeast corner of the island, the two other editors found themselves becoming interested in the lives of two other Victorians, James Clerk Maxwell and Lord Kelvin. This interest resulted in a pair of books, edited jointly with Raymond Flood and published by Oxford University Press. Kelvin, Maxwell and Stokes were all towering figures in Victorian natural philosophy, and so a third book on Stokes was a natural progression. Thus in 2017, through a series of unlikely events, these two streams merged to produce the present volume.

We hope that this volume will appeal to academics and students across a wide sweep of mathematics and physics, and their respective histories. Like many other disciplines, mathematical physics has become much more compartmentalised in the twenty-first. century than it was in the nineteenth. It is therefore difficult for one modern researcher to offer a comprehensive evaluation of every aspect of Stokes's work .We have invited contributions from several experts in the fields of mathematics and physics studied by Stokes, while asking them to remain accessible to other working scientists and to the broader public interested in Victorian Science. Chapter 2, written by a descendant of the first Gabriel Stokes, contains new material on the Stokes family and its Irish connections and will be of considerable interest to historians of science. The other chapters are written by historians of science and mathematics. We thank all of these authors who have submitted with grace and patience to our editorial requests. We also wish to thank Howard Emmens, Dan Taber, Lydia Vavavil and Katherine Ward for their guidance, advice and patience as this book moved through OUP from commissioning to final production.

Almost all books stand on the shoulders of others. Stokes's successor in Cambridge, Sir Joseph Larmor, published the *Memoir* (written by Stokes's daughter Isabella) *and Scientific Correspondence of the Late Sir George Gabriel Stokes* in 1907. This has been an important source for many of the essays. We also acknowledge the scholarship of David B. Wilson who in 1987 produced *Kelvin and Stokes: A Comparative Study in Victorian Physics* (Adam Hilger) and in 1990 he edited *The Correspondence between Sir George Gabriel Stokes and Sir William Thomson, Baron Kelvin of Largs* (2 volumes), published by Cambridge University Press. But as far as we know ours is the first volume to be dedicated entirely to G. G. Stokes.

Mark McCartney, Andrew Whitaker and Alastair Wood
January 2019

CONTENTS

CHAPTER 1

Biographical Introduction

ALASTAIR WOOD

Stokes's Place in Victorian Science

One of the greatest of Victorian scientists, Sir George Gabriel Stokes, was born the youngest son of an obscure parson in a remote Irish glebe house in 1819. He died a celebrated scientist and honoured public servant in Cambridge in 1903. Stokes's contributions to mathematics and physics have endured over a century through Stokes drift, Stokes's law, Stokes's theorem, Stokes parameters, the Stokes phenomenon, Stokes's conjecture and the Navier–Stokes equations. But while Stokes's discoveries are impressive, many believe that he made an equal contribution to scientific administration, providing sound advice and good judgement in a key century for British science. Some great researchers deliberately avoid administration, committees, and editorship of journals on the grounds that such activities would stifle their creativity. Stokes threw himself into these roles. Although a shy man of few words, he was regarded with affection by colleagues for his encouragement of their work and communication of ideas, especially through his remarkably extensive correspondence. Besides describing his outstanding professional career, this introductory chapter explores Stokes's lesser-known family background and early life, its effect on his character and personality, and his contributions outside the fields of mathematics and physics.

Social History and Family Background

George Gabriel Stokes was born in Skreen Rectory, County Sligo, on the north-west coast of Ireland, on 13 August 1819. His father, Rev. Gabriel Stokes, recorded as being of a taciturn nature, was aged 52 when George Gabriel was born. A study of the Stokes family tree shows a predilection for the names George and Gabriel, and we will refer to him by both to avoid confusion. His mother was beautiful, but stern, and the children stood in awe of her.[2] Coupled with his grandfather's known shyness, it is easy to imagine the source of Stokes's 'rich silences'. Stokes was a member of a well-established Anglo-Irish family of Trinity College Dublin academics and clergymen (short biographical details of these are given in Chapter 2). His grandfather, John Stokes, had been Regius Professor of Greek in Dublin

Wood, A., *Biographical Introduction*. In: *George Gabriel Stokes: Life, Science and Faith*, Mark McCartney, Andrew Whitaker, and Alastair Wood (Eds): Oxford University Press (2019). © Oxford University Press. DOI: 10.1093/oso/9780198822868.001.0001

Fig. 1.1 Skreen Rectory as it appeared in the early twentieth century. From Rev. James Greer, *The Windings of the Moy* (Dublin: Alex Thom & Co., 1924).

University. George Gabriel's three eldest brothers were to be ordained as Anglican priests and throughout his life he clung to the basic evangelical truths that he had learned from his father. This web of Anglican family connections advanced his early education and his career at Cambridge. His family background explains the positions that he was to adopt in later life, in particular the maintenance of the established status of the Anglican Church in England and Ireland (he was not successful in the latter), his earnest commitment to the Act of Union between Great Britain and Ireland and his dedication to the monarch, Queen Victoria, as Defender of the Faith.

The Stokes family had settled in Dublin from England some two hundred years earlier. The first of the family to be mentioned in Ireland was a Gabriel Stokes born in 1682, a mathematical instrument maker in Essex Street, Dublin, who became Deputy Surveyor General of Ireland. At that time the Test Act and the Penal Laws, which placed Catholics under major political and social disadvantage, were still in force. All the Stokes were Anglican in religious matters. The Church of Ireland was predominantly the church of the Anglo-Irish and English landowners, although in County Sligo the congregation would have been mainly agricultural settlers and artisans who had come over in the wake of Cromwell's army in the 1640s. The largest religious grouping were the Catholics, at that time mainly peasants and tenant farmers in a subsistence economy, their rights removed by the Penal Laws.

A third group consisted of Protestant Dissenters, organized, cohesive, and concentrated mostly in the northern counties of Ireland. It is into this group that Stokes's distinguished contemporary and friend William Thomson, elevated to the peerage in the New Year Honours List of 1892 as Lord Kelvin, was born, his family having settled from Scotland in 1641. (We will follow the convention of referring to Thomson by this title only after 1891.) The Dissenters were allies of the Church of Ireland in troubled times. Although refused legal toleration (they could not own land), they were prosperous and secure by comparison with the Catholics. But as Nonconformists they were denied entry to the older universities in England and Ireland, although not in Scotland where the established church was Presbyterian. This difficulty was later to raise its head in 1849 when Thomson was attempting to persuade Stokes to apply for the vacant chair of mathematics in Glasgow. Thomas Turton, Lucasian professor from 1822 to 1826, was an advocate of the exclusion of Nonconformists from Cambridge. In this he was supported by William Whewell, the Trinity philosophy don who eventually became Cambridge's Vice-Chancellor. Throughout his career Whewell played a large part in canvassing heads of colleges in support of professorial candidates. There was still a body of opinion which saw the main role of the University as a provider of priests to the established Church.

The social structure in nineteenth-century Ireland meant that Anglo-Irish families enjoyed a considerable advantage in every walk of life. In some respects they formed a separate community, disliked and largely avoided by the Irish, but mistrusted by the British government. Nonetheless they contained a small, closely knit scientific elite who made a disproportionately high contribution to nineteenth-century physics. This group included, among others, the astronomer William Parsons, 3rd Earl of Rosse (1800–67), the mathematician and physicist Sir William Rowan Hamilton (1805–65), the mathematician James MacCullagh (1809–47), the seismologist Robert Mallet (1810–81), the geometer George Salmon (1819–1904), the physicist Samuel Haughton (1821–97), the natural philosopher George Francis Fitzgerald (1851–1901) and the statistician Ysidro Edgeworth (1845–1926), nephew of the hydrographer Admiral Francis Beaufort (of wind-scale fame).

By the time of Stokes's birth in 1819, compromise had begun to replace the Penal Laws and relations had become easier. Although his father's cousin Whitley Stokes, a medical fellow of Trinity College Dublin, had a minor involvement with the United Irishmen in the Rebellion of 1798, the Stokes, in common with other Protestant families, were alarmed by the invasion of a French army and the violence that accompanied this movement. Thenceforth they took a pro-unionist stance, including when the Irish Parliament was abolished through the Act of Union of 1800. But in the case of George Gabriel Stokes his daughter, Mrs Laurence Humphry, records in her memoir[1] 'The late Queen's Jubilees were occasions which he thoroughly enjoyed, for like all Irishmen of his way of thinking, he was a very loyal subject.' There can be little doubt that Stokes regarded himself as British, but he was greatly influenced throughout his life by his Anglo-Irish upbringing.[1]

As to his immediate family (see Chapter 2), in 1798 Gabriel Stokes, son of John Stokes, Rector of Skreen and Vicar-General of Killala, married Elizabeth, the daughter of John Haughton, the Rector of Kilrea in County Londonderry. Haughton was a Methodist preacher who took holy orders in the Anglican Church in the 1760s. It was the Carlow branch of the Haughton family, an originally Quaker family of millers who turned Anglican in 1833, that produced the physicist and polymath Samuel Haughton mentioned in the list of Anglo-Irish scientists above. Skreen is a scattered village, really more of a dispersed 'townland' in the Irish sense, near the sea, about 12 miles south-west of the county town of Sligo: Killala lies across the county border in Mayo. The village of Skreen straddles the present day national road from Sligo to Ballina. Their first child, Sarah, died in infancy, but they produced seven further children, of whom George Gabriel was the youngest. All three of his surviving brothers became clergymen. The oldest, John Whitley (1800–83), who was already 20 in 1819 when George Gabriel was born, became Archdeacon of Armagh and Rector of Aughnacloy. William

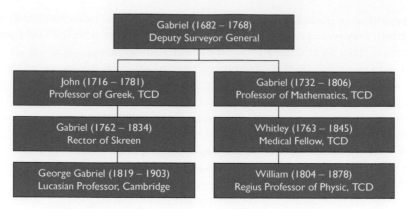

Fig. 1.2 The academic descendants of Gabriel Stokes (1682–1768).

Haughton (1802–84) was Fellow of Gonville and Caius College, Cambridge and later Rector of Denver in Norfolk. Henry George (1804–78) became Rector of Ardcolm in County Wexford. Of the two surviving daughters, Elizabeth Mary (1810–1904), who remained unmarried, was particularly close to George Gabriel. Sarah Ellen (1813–95) married her cousin, Hudleston Stokes, of the Madras civil service, and lived much of her life in India. With four clergymen in his immediate family and a mother who was a daughter of a rector, it is not hard to account for George Gabriel's lifelong piety.

Childhood in Skreen

The upbringing of Stokes appears to have rested mainly in the hands of his eldest sister Elizabeth and second eldest brother William, who trained him to be hardy and brave, and walked great distances with him while out shooting. Later in life Elizabeth wrote:

> Such a pleasant-faced, plump, very fair, rosy baby. His cap, for babies in those days always wore caps, had half fallen off his head, showing his pretty hair. He grew into a dear, good child, a little passionate, but with a strict sense of honour and truthfulness, I do not think he ever told a lie.

She also records:

> My mother found him very slow in learning to read: she was teaching him out of *Cobwebs to Catch Flies* in words of one syllable, a book which was probably too childish for his mind. One day he asked her if he might read the Psalms instead, and she thought she would try him; he got on at once and she had no further difficulty. I believe he was very well grounded in Latin grammar by my father before he went to school.

But perhaps the deep interest in religious questions, coupled with the stern, uncompromising evangelicalism of his parents was overdone. The Rev. H. P. Stokes (no relation), Vicar of St. Paul's Church, Cambridge, records Stokes telling him 'As a child, horrified by the idea of endless torments, I wished that there was no future state, lest I fall into them.' This fear of eternal punishment is more fully discussed in Chapter 10.

The Church of Ireland in Skreen remains in use today, although the rector now resides out-side the parish. Despite his taciturn nature, Gabriel Stokes appears to have been genuinely liked by his parishioners. A plaque in the Parish Church records:

> To the Memory of
> The REV. GABRIEL STOKES A.M. late Rector of this Parish
> This tablet was erected
> By the FRIENDS and PARISHIONERS of this venerable Man
> As a Memorial
> Of nearly twenty Years of affectionate intercourse
> In which as a Father he lived among his Flock
> His mild sincere and evangelical piety increasing
> In lustre and beauty to its earthly close.
> He was Born AD 1762 and died May 8th. 1834
> Aged 72 Years.
> "He walked with me in peace and equity,
> And did turn away many from iniquity."
> *Malachi Chap. II, Ver. 6.*

Church records show that in 1834 Gabriel Stokes had asked the Government for money to provide coffins for paupers and to support orphans. None was granted, but the expenses were met by the church locally.

It is a remarkable coincidence that George Gabriel Stokes was not the only son of a Rector of Skreen to become a famous mathematical physicist. The Rev. W. Crofton was Rector from 1847 to 1851, the next but one to follow Gabriel Stokes. His eldest son, Morgan William Crofton F.R.S. (1826–1915), was the first Professor of Natural Philosophy in what is now University College, Galway (1849–52), leaving to succeed J. J. Sylvester as Professor of Mathematics at the Royal Military Academy, Woolwich.

Schooldays in Dublin and Bristol

Stokes received his early education in Skreen from the Parish Clerk, George Coulter (1807–1902). Coulter's direct descendants still live in Skreen[2] today. Besides being a substantial tenant farmer and latterly village postmaster, Coulter was a successful schoolmaster. A Royal Commission Report of 1835 records him as having sixty-nine male pupils and seventeen female; average daily attendance was sixty-six. The subjects were reading, writing (see later remarks on Stokes's handwriting), arithmetic, and religion. Coulter recorded 'Master George' as 'working out for himself new ways of doing sums, far better than those given in Voster's arithmetic; and clever people were surprised by the questions which he used to solve by the arithmetical rule of False Position'. It is tempting to speculate that Master George occupied a stool in the former Parochial School, which still stands today and is used as the church hall. Although the building is Victorian, a splendid example of its type and very little altered, Parish records show that it was erected by the Rev. Edward Nangle, who succeeded to the Rectory in 1851. It is more likely that Coulter went up to the Rectory to coach the incumbent's children. Master George also read classics with his father, who by this time was ageing and in poor health; recall that he had been 52 when George Gabriel was born. A decision was taken in 1832 to send George Gabriel to Dr Wall's School in Dublin.

Thus in 1832, at 13 years of age, George Gabriel came to live with his father's oldest brother, John Whitley, in Dublin so that he could attend, as a day boarder, the school run by the Rev. R. H. Wall in Hume Street. This arrangement was no doubt motivated by the heavy expense of raising a large family on a rector's stipend. The curriculum of Dr Wall's establishment included English, French, Italian, astronomy, mathematics, 'Counting House' and Military College. As part of his training, Stokes was obliged to take lessons in riding. There is no evidence that Stokes ever enjoyed this activity, or pursued it in later life. His horse threw him, breaking his arm. Rather than returning to the house of his uncle John, where his mother was in residence, the boy had the good sense to go directly to the rooms of his father's medical cousin, yet another Gabriel Stokes, in Harcourt Street to have his arm set. Dr Gabriel Stokes was the brother of the celebrated Dr William Stokes, of Cheyne-Stokes respiration fame. Elizabeth records that Stokes was greatly intrigued when Röntgen rays, on which he worked towards the end of his life, showed this fracture most clearly.

At Dr Wall's School, Stokes attracted the attention of the mathematics master by his elegant solution of geometrical problems. But in May 1834, news reached him that his father had suddenly died from heart failure. He appears to have been more affected by his father's death than his reserve, and the age difference between father and son, would suggest. Stokes's daughter, Mrs Laurence Humphry, who looked after Stokes following the death of his wife in 1899, writes that:

> I remember that in 1901, when I was wearing a pair of small and unremarkable silver buckles he [Stokes], usually most unobservant of details in dress, looking earnestly at them requested me not to wear them again, as they were knee-buckles worn by his father on the morning he was found lying dead in 1834.

After Gabriel Stokes died in 1834, his widow and two daughters had to leave Skreen Rectory to make way for the new incumbent, but money was found to send George Gabriel to school in England. His second brother, William, had been Sixteenth Wrangler (equivalent to a first class honours award in today's system) in the Cambridge Mathematical Tripos of 1828, and had obtained a fellowship at Caius College. It was he who recommended Bristol College, whose Headmaster was Joseph Henry Jerrard, also a mathematician. An honorary fellow of Caius, Jerrard had been a contemporary of William Stokes at Cambridge. Thus both would have encountered the modernized syllabus for the Tripos examination, including the Continental version of the calculus, introduced by Herschel, Babbage, Peacock and Whewell (see Chapter 3) around 1820. Jerrard would not have been tardy in aligning his Bristol syllabus to take account of these changes;[4] meeting Cambridge-style mathematics two years later would thus have been less of a shock for the younger Stokes brother.

William and George Gabriel made the crossing from Waterford to Bristol in the packet *Killarney* in 1835. The boat nearly sank in the rough seas of St George's Channel, and William recounted how George Gabriel had taken off his greatcoat so as to be ready to swim. A much-quoted anecdote about Stokes seems to have originated during his two years in Bristol College:

> His habit, often remarked in later life, of answering with a plain yes or no, when something more elaborate was expected, is supposed to date from his transference from an Irish to an English school, when his brothers chaffed him and warned him that if he gave long Irish answers he would be laughed at by his school fellows.[1]

Stokes's mathematics teacher in Bristol was Francis Newman, brother of Cardinal Newman. As is discussed in chapter 3, in Stokes Newman had a very strong student. Further, Stokes appears to have had a great affection for Newman, whom he records as having 'a very pleasing countenance and kindly manners', and maintained a correspondence with him on mathematical topics when both had become famous. Stokes was presented in January 1837 with a prize for eminent proficiency in mathematics and, in June 1837, Dr Jerrard wrote to him:

> I have strongly advised your brother to enter you at Trinity [Cambridge, not Dublin], as I feel convinced that you will in all human probability succeed in obtaining a Fellowship at that College.

Early Days in Cambridge

Stokes entered Pembroke College, Cambridge as an undergraduate in 1837—the year that Queen Victoria, of the same age as Stokes, ascended to the throne. Stokes was to outlive her by two years. Distinguished graduates from Pembroke included the martyr Bishop Ridley, the poets Spenser and Gray, and the statesman William Pitt.

Although a mathematical prodigy at school, Stokes was beaten into second place in his first year at Pembroke by one John Sykes. At that time the teaching of mathematics in Cambridge by college lecturers, and even by professors of the university, was of variable quality and sometimes non-existent. To ensure the best possible classification in the the highly competitive Mathematical Tripos examinations, many students hired professional coaches from outside their college. From his second year onwards Stokes studied with the greatest of these private tutors, the 'wrangler maker' William Hopkins, who was known for taking on only the most promising students from first year. These private coaches formed a central locus through which the prevailing views of past wranglers, who had become the examiners and textbook writers of the present, were passed down to the wranglers of the future.[3] Some less scrupulous tutors taught weaker charges to memorize bookwork and to develop speed-writing with a quill pen, and trained them to recognize the (then) mostly predictable questions in the problems sections of the examinations. A less able student, well-drilled by his crammer, could sometimes beat a mathematical genius into second place, as happened to William Thomson in 1845.

Hopkins was a Derbyshire farmer who, following the death of his first wife, had entered Peterhouse at the age of 29—what we would now call a 'mature student'—and had graduated in 1827 as Seventh Wrangler (that is, placed seventh in mathematics in the whole university). Normally this would have gained him a college fellowship, but because he had meantime remarried he was disbarred by the strict celibacy rules of the time. Hopkins had a distinguished career as a physical geologist, but his invaluable contribution to Victorian science was to successfully coach almost two hundred undergraduates. The most distinguished among them were (besides Stokes) William Thomson, James Clerk Maxwell, Peter Guthrie Tait, Francis Galton, Isaac Todhunter, and Edward John Routh. Though less known than Maxwell, Routh was Senior Wrangler in 1854, when Maxwell was Second Wrangler. But in the same year they were joint first in the examination for the Smith's Prize (see Chapter 3). It was Routh who tutored John William Strutt, later Lord Rayleigh, another research collaborator of Stokes, to match his own senior wranglership.

Fig. 1.3 George Gabriel Stokes in youth and old age. From (a) George Gabriel Stokes, *Mathematical and Physical Papers*, ed. Joseph Larmor, vol. V (Cambridge: Cambridge University Press, 1905), frontispiece; (b) *Memoir and Scientific Correspondence of the Late Sir George Gabriel Stokes*, ed. Joseph Larmor, vol. I (Cambridge: Cambridge University Press, 1907), facing p. 16,

So effective were his studies with Hopkins that Stokes was Senior Wrangler and First Smith's Prizeman in 1841. It had been a good year for the relatively small Pembroke College with five out of the thirty-seven wranglers, including, in third place, Stokes's first-year adversary, John Sykes. Given his results, Stokes's election to a Fellowship at Pembroke was a formality. Unlike his fellow tutees, the adventurous Thomson and Clerk Maxwell, the more cautious Stokes was to remain in Cambridge until his death in 1903, a total of 66 years.

Lucasian Professor

Stokes did not have long to wait for his chair at Cambridge. His predecessor in the Lucasian Chair, Joshua King, retired on grounds of ill health in 1849. He had been active in university administration through syllabus and examination development in his earlier years as Fellow of Queens' College, but, after suffering a stroke in 1843, lectured and published very little. On 23 October 1849 the College Masters, including the influential Master of Trinity William Whewell, elected Stokes unopposed to the Lucasian Chair of Mathematics. There is no doubt that the lecturing and research interests of Stokes matched more closely their desire to introduce the newer science subjects. The next university calendar announced a new lecture course in hydrostatics, pneumatics, and the physical theory of light.

Two days later we find Thomson, who had seen the appointment in *The Times*, writing to Stokes for the first time as

My Dear Professor,

I am very glad for the sake of mathematics as well as for your own that you have got the chair as you will now have every inducement to go on as you have been doing, and we shall feel much surer of you than when you only had your fellowship to connect you with Cambridge. No wonder you have just discovered a theorem!

Set up by a statute of King Charles II in 1664, the achievements of many of the nineteen holders of the Lucasian professorship have made it one of the most famous chairs of mathematics in the world. It has been held largely by applied mathematicians and theoretical physicists. The second professor was Sir Isaac Newton (1669–1702) and the eleventh was the originator of modern computing, Charles Babbage (1828–39).

Since Stokes (1849–1903) there have been only six incumbents: the last major champion of the 'luminiferous ether' as a basis for the theory of light, Sir Joseph Larmor, (1903–32); the Nobel Laureate in Physics Paul Dirac (1932–69); the fluid dynamicist and admirer of Stokes Sir James Lighthill (1969–80); the cosmologist Stephen Hawking (1980–2009); Brian Green (2009–15); and Michael Cates (2015–).

Although prestigious, the chair was relatively poorly endowed from the fluctuating agricultural income of estates in Bedfordshire, and to augment his earnings Stokes also taught at the Government School of Mines in London during the period 1854–60. It was not a new problem. A previous Lucasian professor George Airy (1826–8) complained that in resigning his Trinity fellowship to accept the chair he was taking a drop of one third in his income. When Airy was appointed in 1827 as the Plumian professor and Director of the University Observatory, the Lucasian chair fell vacant. Whewell warned a possible candidate, John Herschel, that 'so far as income is concerned, the Lucasian is rather a starving matter'.[4]

In keeping with the recent traditions of the Lucasian chair, Stokes was far from being a pure mathematician. While his obituary notice in *The Times* invites the reader 'to dwell upon his masterly treatment of some of the most abstruse problems of pure mathematics', his output of mathematical papers (only seven) was dwarfed by his output in the physical sciences. In the Royal Society's *Catalogue of Scientific Papers* Stokes is credited with 138 publications, only one of which was jointly authored (with Hopkinson). To these must be added three others discovered by David B. Wilson and listed in the index to volume 2 of his *The Correspondence between Sir George Gabriel Stokes and Sir William Thomson*.[6]

Even then, his mathematical results arose mainly from the needs of the physical problems which he and others studied. In this, Stokes was a man of his time, together with Cambridge and Scottish colleagues Airy, Maxwell, Tait and Thomson creating mathematical theories to meet the needs of Victorian industry and of communications within the expanding British Empire. This growth in what we would now call 'applied mathematics' was reflected in changes to the undergraduate syllabus in their respective universities. In Cambridge this movement was commenced by George Airy and driven on by Stokes himself and his former student James Clerk Maxwell (later the first Director of the Cavendish Laboratory). The next two Directors, Lord Rayleigh and J. J. Thomson, had also been taught by Stokes. During the 1870s Stokes introduced more questions on heat, electricity and magnetism (essential for the fast growing telegraph industry) into the Smith's Prize examination at Cambridge. He had been preceded by Thomson in Glasgow University and by Tait in Edinburgh University who had made similar alterations during the 1860s to their natural philosophy degrees.

Table of Lucasian Professors of Mathematics

Isaac Barrow	1663–9
Isaac Newton	1669–1702
William Whiston	1702–10
Nicholas Saunderson	1711–39
John Colson	1739–60
Edward Waring	1760–98
Isaac Milner	1798–1820
Robert Woodhouse	1820–2
Thomas Turton	1822–6
George Airy	1826–8
Charles Babbage	1828–39
Joshua King	1839–49
George Gabriel Stokes	**1849–1903**
Joseph Larmor	1903–32
Paul Dirac	1932–69
James Lighthill	1969–80
Stephen Hawking	1980–2009
Brian Green	2009–15
Michael Cates	2015–

The mathematics of Stokes will be fully discussed in Chapter 7 of this volume: a brief outline of its applicable nature is given here. Stokes's paper on periodic series concerned conditions for the expansion of a given function in what we now know as a Fourier series. In the course of this work he made use of what we now know as the Riemann–Lebesgue lemma some seven years before Riemann. Stokes is also credited with having had the idea of uniform convergence of a series, albeit in a restricted case.

His major contribution to mathematics was the asymptotic expansion of integrals and of solutions of differential equations. By the time of Stokes the theory of convergent series was well established, but asymptotic series belong to the class of divergent series which was largely ignored by pure mathematicians. But if such series are suitably truncated they may still provide useful information from a 'resummation' of the divergent tail, as will be explained in Chapter 7. Typically for Stokes, this asymptotic problem arose from the optical research of his predecessor Airy on caustics behind rainbows. Airy had obtained a function representing the intensity of light of a particular colour in terms of an integral containing a parameter m which measured angular distance across the caustic. Stokes was able to approximate the location of the zeros of the Airy function by two different exponential terms, one for large positive, the other for large negative values of m. He was the first to recognize what we know today as Stokes's Phenomenon, whereby these exponential representations could be connected by continuing them round the complex plane. This discovery is fundamental to the modern subjects of superasymptotics and hyperasymptotics. He subsequently applied this idea to other special functions of mathematical physics and to differential equations.

It is perhaps strange that this major contribution by a Lucasian professor received no mention in the book *Divergent Series* by a later Cambridge professor, the distinguished mathematical analyst G. H. Hardy (Sadleirian chair 1931–42), published two years after Hardy's death in 1947.[5] But Stokes wrote in the discursive style of the mathematical physicist rather than the definition-lemma-theorem-proof style of the rigorous pure mathematician. In a separate 1918 paper, however, Hardy did recognize the contribution of Stokes's earlier work to the concept of uniform convergence.

In other areas of asymptotic analysis Stokes employed the saddle-point method for integrals in the complex plane a full decade before Riemann, to whom it is usually attributed. He also realized the link between his asymptotic methods for integrals and Kelvin's method of stationary phase. The well-known theorem in vector calculus which bears his name, and is fundamental to modern differential geometry, is however not due to Stokes, the result having been communicated to him in the postscript of a letter from Thomson in July 1850. The proof was set by Stokes as a problem in the Smith's Prize examination at Cambridge in 1854. It is said that James Clerk Maxwell was the only student to successfully attempt the question.

Although appointed to the Lucasian chair for his outstanding research, Stokes showed a concern in advance of his time for the welfare of his students, stating that he was 'prepared privately to be consulted by and to assist any of the mathematical students of the university'. His predecessor but one, Charles Babbage (Lucasian Professor 1828–39) enjoyed independent means, and—even though the university statutes required him to give lectures—was rarely present in Cambridge and never once addressed classes, being occupied in London with his 'calculating engine'. Stokes immediately advertised that 'the present professor intends to commence a lecture course in Hydrostatics', which he was still delivering fifty-three years later, in the last year of his life. Stokes's manuscript notes still exist in the University Library in Cambridge, although he eventually became one of the first people in Britain to make regular use of a typewriter.

The Correspondence

Stokes was a prolific correspondent, using the penny post as a modern scientist might use e-mail. He communicated endlessly with the leading scientific figures of his day, his exchanges with Thomson alone comprising 656 letters which have survived to the present day.[6] Aside from correspondence dealing with the refereeing of papers during his thirty-one years as Secretary of the Royal Society, he corresponded with most of the leading figures in British and Irish mathematics and physics, such as the astronomers John Herschel and Thomas Romney Robinson, the mathematician Arthur Cayley, and the physicists Peter Guthrie Tait and James Clerk Maxwell. Sometimes he wrote to the same person twice in the same day, although this may have arisen from the need to catch the post, which departed at a fixed time each day. Often we find a brief letter announcing his result sent in the morning, with the detailed reasoning following in the evening. Many of his correspondents, Thomson included, found his writing very difficult to read. Stokes acquired his first typewriter, which used only upper-case letters, in 1878. The second, used from 1886, also had only capital letters, though the third, used from 1890, possessed both upper and lower cases. But these early machines were not without their technical problems: we find a letter of 17 March 1896 on Röntgen rays beginning:

> My dear Kelvin,
> The pull wire of the letter that lies between d & f got broken, and I have sent it to London to be repaired.

Stokes was a shy and self-effacing character, known for his reticence, and although not anti-social he certainly had no reputation for small talk. He was interested only in mathematics, physics, and religion. There is a story in Cambridge about a dinner party[4] where Stokes once found himself seated next to a young American woman. The guests were amazed to find Stokes engaged in animated conversation and even smiling. When asked later about this rare event, the young woman replied that she had merely asked Sir George whether he preferred algebra or geometry.

Stokes had the custom of working long into the night, fortified only by strong tea from his trusty 'Brown Jenny' teapot. His careful and diligent approach is typified by the two letters to Kelvin of 13 March 1896.

> I have a good deal to write you, and have not time to catch this post with it. I write just to say that I do not by any means think that Lord Blythswood's experiment proves the reflexibility of the X-rays.

He continues in a longer letter, started the same day, to give detailed reasoning for his assertion, but breaks off midway to restart the letter with

> March 14. So far last night, or rather the commencement of the small hours of the morning. I resume before breakfast.

Other letters were written in snatches over a longer period. In December 1898 a single letter on the discontinuity in the motion of a liquid contained by two pistons in a cylinder was written over a seven-day period, although it is true that the Stokes family were visiting for Christmas. After one page of this letter, started on the 20, we find

> But now the clock has struck XII, and it is time to go to bed.
> Dec 21. I resume . , ,

The letter was eventually completed and sent off on 26 December, but the amazing thing is that Stokes had sent four other letters to Kelvin in the interim! This controversy continued by return of post throughout the festive season until 7 January 1899. Their correspondence was not always of such frequency and in the years 1853 and 1891 no letters seem to have been exchanged. It could be that these letters have failed to survive, but this seems unlikely in the first case since Thomson in his letter of 20 February 1854 begins

> It is a long time since I have either seen you or heard from you, and I want you to write to me about yourself & what you have been doing since ever so long. Have you made any more revolutions in science?

and proceeds to describe what he himself has been doing the previous summer.

Research in Continuum Mechanics

Stokes enjoyed his most active period from 1845 to 1860, concentrating his activities on his very significant contributions in light and fluid dynamics, thereafter calmly and competently managing the running of British science. Almost all of Stokes's 138 published papers (only one was jointly authored) appear in the five volumes of his collected *Mathematical and Physical Papers*.[7] The first three volumes were edited by Stokes personally, the last two by his successor in the Lucasian chair at Cambridge, Sir Joseph Larmor, another Irishman, whose *Memoir and Scientific Correspondence of the Late Sir George Gabriel Stokes, Bart.*,[1] published in 1907, is also an important source for later scholars. Larmor performed the similar service of editing the last three volumes of the *Mathematical and Physical Papers* of Kelvin, also born in Ireland and a fellow believer in the luminiferous ether.

While publishing their results independently, the 656 letters mentioned above show that Stokes served as a sounding board for Thomson's ideas, keeping him abreast of research that had already been done in the areas he chose to enter. When Thomson was carried away by some new enthusiasms, it was the quiet and cautious Stokes who brought him back down to earth, pointing out flaws in his arguments. Lord Kelvin, as Thomson had become in 1892, acknowledged that 'Stokes gave generously and freely of his treasures to all who were fortunate enough to have opportunity of receiving from him'.[4] The different temperaments of the two men are well illustrated by an anecdote of J. J. Thomson,[4] Director of the Cavendish Laboratory in Cambridge.

> When Kelvin was speaking, Stokes would remain silent until Kelvin seemed at any rate to pause. On the other hand, when Stokes was speaking, Kelvin would butt in after almost every sentence with some idea that had just occurred to him and which he could not suppress.

But when Stokes spoke, he did so with authority. William Thomson acknowledged this in his Baltimore Lecture of 1884 when he stated 'I always consult my great authority, Stokes, whenever I get a chance'.

Stokes commenced his research at the age of 23 in the area of hydrodynamics (a subject suggested to him by his former tutor Hopkins), both experimental and theoretical. Stokes's handwritten undergraduate notes of the lectures given by the Plumian professor James Challis still exist. Stokes was later to question the correctness of Challis's model and their disagreements led to a lengthy exchange of papers in the *Cambridge Philosophical Magazine*. During this period Stokes put forward the concept of 'internal friction' of an incompressible fluid. The fundamental equations for the motion of incompressible fluids were first published in 1822 by the French civil engineer Claude Navier—though to modern physicists his analysis is based on an unacceptable notion of intermolecular forces. Using his concept of internal friction in fluids, it was Stokes who in 1845 put the derivation of these equations on a firm footing. Thus it is by both names, Navier–Stokes, that these equations are known throughout the world today. They are used to describe the wake behind a boat or the turbulence behind a modern aircraft, and are employed on a daily basis by aeronautical engineers, ship designers, hydraulic engineers, and meteorologists. While simple examples, such as steady flow in a straight channel, can be solved exactly and some more complicated cases admit an approximate numerical solution by large-scale computer packages, the mathematical problem of the existence and uniqueness of a general solution to the Navier–Stokes equations remains unsolved today and is one of the Millennium Prize problems for which a reward was offered by the Clay Foundation in 2000. Stokes's work was independent of that of Poisson and Saint-Venant appearing in the French literature at the same time, but his methods could also be applied to other continuous media such as elastic solids. He applied the same models to the transmission of sound.

Stokes's other big contribution to fluid mechanics was in the theory of water waves, a subject that had been studied by his predecessor George Airy and the earlier Cambridge mathematician George Green. This research had been encouraged by the experimental results of the engineer and naval architect John Scott Russell, who had observed a solitary wave on the Scottish Union Canal in 1834 and had followed it for some miles on horseback. These were not explained by the existing linear theory. Stokes developed a 'weakly nonlinear' approximation which showed that the crests of sinusoidal waves propagating on the surface could be sharpened and the troughs flattened. This is more fully described in Chapter 6. At the same time there was a small increase in the wave speed. He also found that in this model the water particles do not travel in closed orbits, as predicted by linear theory, but that there is a small average drift velocity in the direction of propagation, the so-called Stokes drift. It is still used today in the modelling of the dispersal of floating rubbish on our ocean currents.

All of Stokes's important work on fluids was published over an eight-year period between 1842 and 1850. Over the next thirty years, particularly during the early years up to his becoming involved as Secretary of the Royal Society in 1854, his publications were mainly in optics. But after his marriage in 1859 there was a renewal of Stokes's interest in fluids, namely oscillatory water waves. His family visited his father-in-law Dr Robinson every summer and took regular trips to resorts on the north coast of Ireland, most frequently to Portstewart, whence they made excursions to the Giant's Causeway. Although primarily a theoretician, Stokes was not afraid to experiment: he measured the waves breaking in the Land Cave there, and also on the sloping sandy beaches at Portstewart. Typical of his letters to Thomson about this time is one written from the Observatory, Armagh on 15 September 1880. Stokes had made a mathematical conjecture that the angle at the crest of the wave of greatest height should be 120 degrees.

> You ask if I have done anything more about the greatest possible wave. I cannot say that I have, at least anything to mention mathematically. For it is not a very mathematical process taking off my shoes and stockings, tucking my trousers as high as I could, and wading out into the sea to get in a line with the crests of some small waves that were breaking on a sandy beach.... I feel pretty well satisfied that the limiting form is one presenting an edge of 120 degrees.

This celebrated Stokes Conjecture on the wave of greatest height has been verified mathematically only in recent years.

He wrote on the same topic a week later and Thomson replied from his yacht the *Lalla Rookh* at sea in the Clyde. In another letter Thomson writes 'Will you not come and have a sail with us and see and *feel* waves? We would take you away out to the west of Scilly for a day or two if that would suit best.'

Despite his seaside paddling, Stokes's interest in ocean waves was a serious one undertaken in consequence of his membership of the Meteorological Council. Stokes was aware that long waves radiating from distant storms travelled faster than short waves from the same source. In situations where unusually high seas were observed in the absence of a local wind, Stokes was able to analyse records of the direction and period of the waves to predict the location and direction of travel of the storm which had given birth to them. Larmor[1] describes his fascinating correspondence with the Admiralty Experimental station at Torquay and various sea captains, most notably Captain William Watson of the SS *Algeria*, on observations in ships' logs:

> We had waves on our homeward voyage on the Campania which cannot have been less than 60 feet high from hollow to crest.... We were on the Banks of Newfoundland, depth perhaps 40 fathoms at the time. But now!! I was told by several officers of the Campania that in both her and her sister ship the Lucania they had seen unbroken wave-crests right ahead of them and quite near, in line with the crowsnest on the foremast. This is 90 feet above the sea,...

The estimate given by Watson is not far short of the theoretical height eventually established almost a century later. Stokes also advised the Council on meteorological instruments, and a sunlight recorder designed by him was in use in the station at Valentia until recent times.

Research in Optics

Stokes's major advance in the 1850s was in the wave theory of light, by then well established at Cambridge, examining mathematically the properties of the 'luminiferous ether', which he treated as a sensibly incompressible elastic medium. 'Ether' was the old-fashioned name given to the medium filling all space which was thought to carry light waves as vibrations analogous to sound waves. It predated the theory of light as an electromagnetic phenomenon, introduced by his former student, the Scottish physicist James Clerk Maxwell, who was appointed to the Cavendish chair of experimental physics at Cambridge in 1871. After Hertz's 1887 experiment showed that electromagnetic waves could be generated by an electric circuit, the concept of the ether was attacked by another Irish physicist (and correspondent of Stokes) George Francis Fitzgerald, the leading follower of Maxwell. Together with other correspondents these men formed an 'invisible college' of researchers.[8]

His friend Thomson was also an enthusiast of the ether. To begin with he used the model of the ether as an elastic solid, developed by George Green at Cambridge in the 1840s, but switched to the compromise of a very viscous liquid, likened to 'glue water' or 'jelly', developed by Stokes. While Stokes in his later years had begun to doubt the validity of the ether model, Kelvin was still elaborating the ether model at the turn of the century. His experiments with wax and pitch, both at the extreme of viscous liquids, are still to be seen in the laboratories in Glasgow.

The concept of the ether enabled Stokes to obtain major results on the mathematical theory of diffraction, which he confirmed by experiment, on polarization of light, introducing the idea of Stokes parameters, and on fluorescence, which led him into the field of spectrum analysis. The spectrum of light passed through a prism gives us the characteristics of its source; the process of examining this spectrum quantitatively is known as spectroscopy. When light is shone on or through a material, its spectrum is modified because the material absorbs light of certain preferred wavelengths. Stokes pointed out in a letter to Thomson, some years before the publications of Kirchhoff and Bunsen, the possibility of determining the material composition of distant objects by analysing this spectrum.[8] He also applied spectral analysis to very small objects, discovering that haemoglobin exists in two states, only one of which carries oxygen. This enabled him to understand the role of haemoglobin in delivering oxygen through the blood.

His last major paper on light was his study of the dynamical theory of double refraction, presented in 1862, although later in life he delivered in the University of Aberdeen the comprehensive series of Burnett Lectures on Light (1884–7), published in London in three volumes, which some have classed as the textbook which he never wrote. From 1896 onwards, he was involved in the early investigation of Röntgen rays, now known as X-rays, in cooperation with the industrial chemist Sir William Crookes who had already assisted Rayleigh and Ramsay with the spectroscopy in their isolation of the element argon.

Stokes also enjoyed a certain involvement in industrial applications. Through his position in the School of Mines in London, probably initiated for financial reasons due to the poor salary of the Lucasian chair as noted above, Stokes continued his researches in the principles of geodesy (a link with his surveyor great-grandfather) and helped in the establishment of the

Indian Geodetic Survey. He pioneered the concept of gravitational variation around the Earth, which underlies methods still used today to prospect for minerals, oil and gas.

In other areas he also acted, over a period of many years, as consultant to the lens-maker Howard Grubb, who ran a successful and internationally known optical works in Rathmines, now a suburb of Dublin. It was Grubb who supplied the mirror for the Great Leviathan, the largest telescope in the world for seventy years, constructed by William Parsons, third Earl of Rosse, in his grounds at Birr Castle and completed in 1845. Stokes was a correspondent of both men and visited Birr to advise the Earl.

Stokes was also an adviser to Trinity House on lighthouse illuminants. His collected works include a paper on a differential equation relating to the breaking of railway bridges and, following the Tay Bridge disaster, he served on a Board of Trade committee to report on wind pressure on railway structures (see Chapter 9). But unlike his contemporary Kelvin in Glasgow, there is no evidence that Stokes ever made any money from his advice and consultancy so freely given. The Stokes family as a whole were not interested in material things and somewhat unworldly.

The Victorian Scientific Establishment—Stokes and the Royal Society

The second half of Stokes's life was increasingly taken up with scientific and academic administration. A major reason for this change was that in 1851 he, along with Thomson, had been elected a fellow of the Royal Society and shortly afterwards, in 1854, became a Secretary of the Society, where for thirty-one years he performed an important role in advising authors of research papers about possible improvements and related work. Besides editing their publications, Stokes also administered the distribution of government funds to support individual research projects. From 1885 to 1890 he was President of the Royal Society.

Many famous scientists tried out their half-formed ideas on Stokes who, like Thomson, was extremely active in the British Association for the Advancement of Science and, in 1869, its president. His close colleagues regretted his taking on these administrative duties and P. G. Tait (co-author with Thomson of the celebrated textbook *Treatise on Natural Philosophy*) even went so far as to write a letter to *Nature* protesting at 'the spectacle of a genius like that of Stokes' wasted on drudgery [and] exhausting labour'. Thomson wrote to him in 1859 of 'the importance to science of getting you out of London and Cambridge, those great juggernauts under which so much potential energy for original investigation is crushed' while attempting to persuade Stokes to apply for the vacant Professorship of Astronomy in Glasgow. But even although the religious tests, which had impeded him in 1849, had been removed, Stokes did not apply. It was perhaps unrealistic of Thomson to expect Stokes to leave such a prestigious chair in Cambridge, even although his administrative activities severely limited his time for personal research. Thomson was to make one more attempt in November 1884 to lighten the load on Stokes by urging him to apply for the Cavendish professorship in Cambridge, on the grounds that

> the income of the experimental physics chair is decidedly more than you have in the Lucasian and I thought possibly the difference might amount to even a money compensation for giving up the Royal Society work. Thus I thought of the whole thing rather as freeing you from fatiguing or possibly irksome work.

The reply of Stokes was typically altruistic:

> I feel that those who would be under me would be knowing more about the subjects than I do myself. Also it is hardly fair to block the way of promotion to younger men who might reasonably be expected to rise in their profession.

The Religious Scientist

Following the publication of *The Origin of Species* by Charles Darwin in 1859, the latter part of the nineteenth century was a time of conflict in British science between the supporters of Creation as set out in the Bible and followers of the theory of evolution. It is hard for us today to comprehend the importance of religious issues either to the Victorian scientists or to the man in the street who, while excited by new scientific inventions and discoveries, saw a certain conflict with the role of God as Creator. Stokes and Thomson were firmly in the Creation camp, although for different reasons, and neither of them would have taken literally the time periods given in the Genesis account of Creation. Thomson's scientific work on the age of the sun and the solar system appeared to show that there had been insufficient geological time for evolution to have taken place at the pace suggested by Darwin. In response to Thomson's results Darwin subsequently modified his views on the rate of evolution, but the irony was that Thomson's estimate of the age of the Earth, based on its rate of cooling, did not take account of the contribution of the heat of radioactivity, which was yet to be discovered. Stokes believed that the Bible account was true in the sense of an ongoing creation of organic life through a process controlled by God. Despite this difference of opinion, Darwin and Stokes bore each other no ill will and worked well together for several years as joint Secretaries of the Royal Society. Darwin was present at a High Table for Stokes's Jubilee Dinner in 1899.

On the other hand, the physicist John Tyndall and Thomas Henry Huxley were leading supporters of evolution within the Royal Society and the British Association. Both these strong agnostics took more extreme views than Darwin. After Tyndall's provocative address to the Association in 1868, Stokes was moved to strongly refute it in his presidential address the following year. Matters were made worse by Tyndall's own presidential address to the Association in Belfast in 1874. The disagreement was still rumbling on in 1887 when Stokes was elected MP for Cambridge University. Huxley perceived a conflict of interest and wrote anonymously in *Nature* that, as President of the Royal Society, Stokes should not simultaneously be a Member of Parliament. But there was a precedent in the form of his predecessor in the Lucasian Professorship, Isaac Newton, who had successfully combined the holding of both offices. Thomson wrote quickly to support Stokes:

> We were *very much* displeased with that article in Nature. I think on the contrary that your agreeing to be a member was most patriotic and public-spirited.

Huxley also felt that the President of the Royal Society should not be President of the Victoria Institute (see below).[1]

Stokes's initial reluctance to admit women to his lectures may also be attributed to his literal interpretation of the Holy Scriptures, particularly the Letters of the Apostle Paul. With the foundation of Girton and Newnham Colleges in the second half of the nineteenth century women were admitted to lectures and allowed to take the Tripos examinations, although they could not be granted degrees. But it seems that Stokes abandoned this dogmatic position after representation from his daughter Isabella[1] who felt her father might be denying a future Mrs Somerville, a famous early Victorian science writer.

A profoundly religious man, Stokes had always been interested in the relationship between science and religion. From 1886 to 1903 he was President of the Victoria Institute (see Chapter 10), whose aims were 'To examine, from the point of view of science, such questions as may have arisen from an apparent conflict between scientific results and religious truths; to enquire whether the scientific results are or are not well founded'. In 1891 and 1893 Stokes delivered the Gifford Lectures in Natural Theology in the University of Edinburgh. The same period saw the publication of Stokes's principal work on theology, *Conditional Immortality: A Help to Sceptics* (1897). He held extensive discussions with his vicar in Cambridge, Rev. H. P. Stokes, on the validity of this principle.

Marriage and Family Life

In 1859 Stokes vacated his fellowship at Pembroke, as he was compelled to do by the regulations at that time, on his marriage to Mary Susannah (Fig. 1.4), daughter of Dr Thomas Romney Robinson, Astronomer at Armagh and a correspondent of Stokes. Given the intensity of his work, the scarcity of women in the sciences, and his lack of social skills, Stokes must have had difficulty encountering a suitable wife. Their daughter Isabella reports in the *Memoir*[1] that the couple first met briefly at a Meeting of the British Association. Stokes was much taken with her and hoped to meet again, but he had mistaken her name. Then Fate took a hand and there was a second encounter at Birr Castle, when both Stokes and Dr Robinson were inspecting the Great Telescope. There is mathematical folklore that Stokes proposed marriage inside the tube of the Telescope, but Isabella makes clear that that this 'is absolute fiction' (see Chapter 2).

Fig. 1.4 Lady Mary Susannah Stokes. G. G. Stokes wrote many letters to his young love, showing a combination of affection, his love of science and mathematics, and even doubt at his fitness for married life. Image courtesy of the Stokes family.

Stokes's regular visits to his brother's parish of Aughnacloy, only 15 miles from Armagh, permitted the relationship to develop. Stokes announced his engagement in a letter to his friend Thomson in September 1856, adding that they planned to marry in July 1857. One of the delightful features of the Stokes–Kelvin correspondence is the ability of both men to include personal snippets amongst the mathematical physics.

But it was to be a long engagement. Worried by his inability to support a wife solely on his professorial income, in 1857 Stokes had written to Mary that his position would be ameliorated if College fellowships remained open to married dons. Fortunately he was able to obtain other sources of income and the marriage went ahead. Following a change in regulations, he was subsequently able to resume his fellowship and for the last year of his life served as Master of Pembroke. Shortly after their marriage the couple moved to Lensfield Cottage, a happy and charming home, in which Stokes had a 'simple study' and conducted experiments 'in a narrow passage behind the pantry, with simple and homely apparatus'.

Financial problems aside, it says a lot for Mary's patience and forbearance that the marriage ever took place. Prior to their wedding day Stokes, who, as we have already remarked, was a tireless writer of letters, had carried on an extensive (one letter ran to fifty-five pages) and frank correspondence with his fiancée, sometimes doubting his ability to show affection. In another letter, the theme of which will be familiar to all spouses of research physicists, he states that he has been up until 3 a.m. wrestling with a mathematical problem and fears that she will not permit this after their marriage! Based on other remarks in this highly personal correspondence, David Wilson suggests that 'Stokes himself may have welcomed what others regretted—his abandonment of the lonely rigours of mathematical physics for domestic life and the collegiality of scientific administration'.[6]

George Gabriel and Mary had five children: Arthur Romney (1858–1916) who was a schoolmaster in Shrewsbury; Susanna Elizabeth (1859–63) who died of the scarlet fever which also infected her father; Isabella Lucy, who married Dr Laurence Humphry and died in 1934; William George Gabriel (1863–93) who trained as a medical doctor but died in tragic circumstances in County Durham (see Chapter 2); and another daughter Nora Susanna who died in infancy in 1868. Arthur Romney, who inherited the baronetcy on the death of his father in 1903, died without male issue and the baronetcy was extinguished in 1916. All the direct descendants of George Gabriel Stokes come from Isabella Lucy or from Arthur Romney's two surviving daughters.

As might be expected from the previous section, the family were regular worshippers at St. Paul's Church, Cambridge, where George Gabriel was churchwarden for thirty-four years. Unlike his brothers, he never took holy orders. He is recorded by a plaque in the church listing his distinctions. As noted above, he held the last of his close discussions on his Theory of Conditional Immortality with his Vicar, Rev. H. P. Stokes, just a few days before his death.

Parliament and Public Life

When the Liberal party split over Gladstone's Home Rule measures for Ireland, the anti-Home Rule members formed the Liberal Unionist party. Seventy-eight Liberal Unionists were returned to Westminster to become a minority party in the Conservative-led government of Lord Salisbury. This coalition contained among other smaller parties G. G. Stokes, who had been returned as an Independent for the Cambridge University seat at a by-election in 1887. We find Thomson, a supporter of the Liberal Unionists[9] in Scotland, writing to his friend in March 1888

I hope you have been enjoying Parliament. It must be very satisfactory, and pleasant, to see all going so well.... I am now feeling quite hopeful that I might live to see the last of government by party.

It is not certain that Stokes did enjoy Parliament. He spoke only thrice in five years, on topics close to his interests: university representation on Oxford and Cambridge town councils; in support of two officials of the British Museum (of which he was a trustee) who worked for the special Irish Commission; and in favour of the Free Education Act to enable ten shillings to be paid to every child attending school during 40 weeks of the year. Stokes did not stand again at the 1892 election, finding the long hours uncongenial. By contrast, Thomson had become Baron Kelvin of Largs in the New Year honours list of 1892 in recognition of his 'most valued service to science and progress in this country'. Unlike Stokes in the Commons, Kelvin enjoyed attending sittings of the House of Lords and spoke on fourteen occasions.

Stokes was made a baronet by Queen Victoria in 1889. In contrast to Thomson, whose wealth and ability to support the lifestyle expected of a peer had been factors in his elevation, Stokes anguished long on this financial question before accepting this hereditary honour. It was difficult enough for him in a university professorship to maintain appearances, but how would his elder son, Arthur, a schoolmaster in Shrewsbury, be able to shoulder the expenses of the hereditary title? In those days a baronet was expected to enjoy a certain lifestyle.

Stokes was awarded the Copley Medal of the Royal Society in 1893, and in 1899 was given a Professorial Jubilee (fifty years as Lucasian Professor) by the University of Cambridge (see Fig. 1.5). I am grateful to Michael Sandford[10] for the original table plan, passed down to him by

Fig. 1.5 Attendees at Stokes's Professorial Jubilee held in Cambridge in 1899. Stokes is seated at the front directly below the vertex of the doorway. Image courtesy of the Master and Fellows of Pembroke College, Cambridge.

Rev. Hudleston Stokes who was present at this dinner for 230 eminent guests. The two high tables contained, besides the usual senior government, church, and legal figures, the University Chancellor and Vice-Chancellor, the College Masters and many of the scientific correspondents of Stokes mentioned in this volume. These included Lord Kelvin, the Earl of Rosse, Lord Rayleigh, Sir Frederick Abel, Sir William Crookes, Professor Darwin, Professor Michelson, and his second cousin Sir William Stokes from the medical side of the family. Foreign professors seated at the high tables were Darboux, Mittag-Leffler, Voigt, Arrhenius, Riecke, Schuster, Becquerel, Rénard and Rucker. The French delegation brought with them the Arago Medal of the Institute of France: Stokes also held the Helmholtz Medal of the Berlin Academy. Present at other tables were Professors J. J. Thomson, Reynolds, Fitzgerald and Dr Routh.

Stokes died at Lensfield House (the home of his son-in-law Dr Humphry) at 1 a.m. on Sunday 1 February 1903. Although obsessed with his scientific work, Stokes had excited feelings of warmth and admiration among his contemporaries for his honesty and integrity. Some colleagues felt that he could have done more in the field of physics in later life, but Stokes himself seemed to find fulfilment in his role as a gatekeeper and arbiter of the Victorian scientific establishment. Rayleigh and Kelvin both published obituaries of Stokes, Kelvin observing at the time that his heart was in the grave with Stokes. Sadly, the gravestone of Stokes in Cambridge's Mill Road cemetery has now vanished from view, but he is commemorated by a bronze roundel (designed by Sir Hamo Thorneycroft) in the north choir aisle of Westminster Abbey, a plaque in St. Paul's Church, Cambridge and by a stone memorial at his birthplace in Skreen, County Sligo (Fig. 1.6).

Fig. 1.6 The author (right) and the author of Chapter 7 at the unveiling of the memorial to Stokes at Skreen in 1995. Stokes's birthplace, the old rectory, is now demolished and has been replaced with the new rectory which can be seen in the top left of the picture.

We conclude with an extract from Stokes's obituary in *The Times* of 3 February 1903:

> We may enumerate his scientific papers, we may expatiate upon his work in optics or hydro-dynamics, we may dwell upon his masterly treatment of some of the most abstruse problems of pure mathematics, yet only a select body of experts can readily understand how great he was in these various directions, while possibly not all experts understand how much greater was the man than all his works.... Sir George Stokes was as remarkable for the simplicity and singleness of aim, for freedom from all personal ambitions and petty jealousies, as the breadth and depth of his intellectual equipment. He was a model of what every man should be who aspires to be a high priest in the temple of nature.

• •

ACKNOWLEDGEMENTS

The author would like to thank various relatives of Stokes—Michael Sandford, the late Professor Anne Crookshank, Dr Michael Purser (both of Trinity College Dublin), Ms Teresa Stokes and Dr Nick Stokes of Canberra—for their help in the early stages of this book. He also acknowledges the late Dr John Dougherty, sometime Stokes lecturer in the University of Cambridge.

CHAPTER 2

The Stokes Family in Ireland and Cambridge

MICHAEL C. W. SANDFORD

Introduction to The Stokes Family

In this chapter we usually refer to George Gabriel Stokes simply as Stokes, but give the Christian names and dates of his relatives where needed to avoid confusion.

The Stokes family was quite prominent in Ireland. Stokes descends from a well-known Dublin engineer, Gabriel Stokes, Deputy Surveyor General for Ireland who had been born 1682 to one John Stokes. We first detail Stokes's place in the Stokes family and identify his maternal ancestral lines. Then using the account by his daughter, Isabella, writing as Mrs Laurence Humphry her *Notes and Reflections* in the 1907 memoir edited by Joseph Larmor,[1] we explore the family life and character of Stokes. We conclude this chapter with an appendix containing biographical notes on a selection of the relatives of Stokes who were eminent, particularly in academic and medical fields, or as church ministers or colonial administrators. None in the family seems to have acquired very great wealth or property, establishing their positions through their own intellectual efforts albeit sometimes receiving support from relatives.

For my own part, I remember when studying physics at school in the 1950s learning about Stokes's work and being told by my own grandmother, born Mary Eileen Stokes (1885–1967), that she remembered around 1900 meeting in Cambridge her famous 'uncle' Professor George (actually her first cousin twice removed). For many years I did not understand the correct relationship. When eventually I saw my grandmother's notes on her descent from Gabriel (1682–1768), then with the aid of the entry for the family in Burke's *Irish Family Records*[2] the blood line of Stokes became clear (see Fig. 2.1). However, the family's origins prior to the 1650s are still uncertain. Perhaps DNA will one day provide clues to clearly identify this particular Stokes family among the many.

Like many in the Victorian age Stokes carried out extensive correspondence, with both family and scientific colleagues. Fortunately Stokes kept many of his papers, which unsorted had been consigned to numerous packing cases; much of this is now preserved in the Cambridge University Library (17,700 letters) and was catalogued by Wilson,[3] who in a comparative study

Sandford, M. C. W., *The Stokes Family in Ireland and Cambridge*. In: *George Gabriel Stokes: Life, Science and Faith*, Mark McCartney, Andrew Whitaker, and Alastair Wood (Eds): Oxford University Press (2019). © Oxford University Press. DOI: 10.1093/oso/9780198822868.001.0002

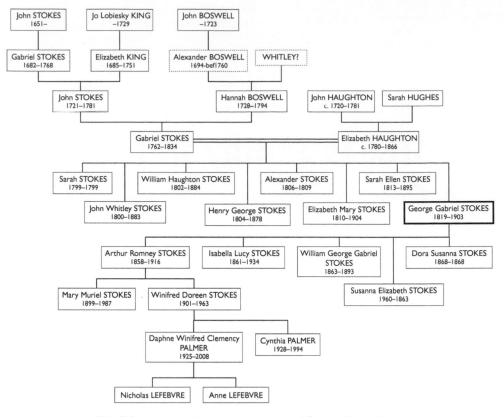

Fig. 2.1 Ancestors, siblings and descendants of George Gabriel Stokes.

of Kelvin and Stokes included a chapter on Stokes as a Victorian Correspondent.[4] Wilson recorded 1,060 letters from seventy-three relatives. This vast archive has not yet been systematically accessed for a detailed study of the Stokes family.

George Gabriel Stokes: Ancestry and Cousins

The origin of this Dublin Stokes family is not certain further back than Stokes's great-great-grandfather, John Stokes, who is recorded as a 'taylor within the White Lyon', a seventeenth-century district of Dublin. It was John's son, Gabriel (1682–1768) who was granted arms in 1721 (Fig. 2.2). As Dublin Protestants these Stokes were very likely of English origin.

A major genealogical study of the family was carried out during the years 1978–86 by Alan Geoffrey Stokes (1912–96), who was a great-great-grandson of William Stokes (1793–1864).[5] Alan's grandfather, the civil engineer William Stokes (1846–1902), had emigrated to Queensland in 1874. There were few family records held by this Australian branch. However, with the aid of the published Irish pedigrees[2] Alan was able to contact members of the Stokes Family in Ireland, England, and Canada and he carried out much research in Irish and English archives. He provides an overview of previous research and also reviews the efforts that had been made to link the Irish Stokes family to Stokes families in England, in particular those recorded

Fig. 2.2 Arms granted to Gabriel Stokes in 1721. Image courtesy of Teresa Stokes.

in Kent and also *c*.1300 at Sende, Wiltshire, and in Gloucestershire. He concluded no definite proof of a link existed despite sharing the blazon *Gules demi-lion rampant double-queued Argent.*

More recently, a series of chance connections gave me the opportunity to transcribe a manuscript entitled 'The Stokes Family Book', written between 1897 and 1901 by Hudleston Stokes (1831–1904), son of Henry Stokes (1793–1868). This has provided complementary information about the family, including newspaper cuttings and Cambridge University's arrangements for the 1899 Jubilee of Stokes's Lucasian professorship described in Chapter 1.

There are about 30,000 people named Stokes in the UK, making it the 328th most common surname.[6] 'Stoke' occurs in many place names and many unrelated families doubtless adopted Stokes as their surname. One longer-term hope for finding connections between Stokes families is through testing the Y-DNA of males. The Y-DNA is passed on from father to son with only minor mutations over the generations. So far, one male line descendant of Gabriel Stokes (1682–1768) has been tested and has a partial match with a Stokes in the USA, who has speculatively traced his line back to Birmingham in the eighteenth century but with no obvious connection to the Dublin Stokes. Testing of the Y-DNA of men in other principal descendant lines from Gabriel would be very useful to confirm the genealogy described here and to throw light on the origins of this family of Irish Stokes. It would also be desirable for more English and Irish Stokes, especially those with records extending back to the seventeenth century or earlier, to be tested in the hope of establishing a match which might then point the direction for future archival researches.

The most complete family tree of these Irish Stokes and their descendants around the world is currently maintained on line by my sixth cousin Teresa Stokes, a great-great-granddaughter of Henry Stokes (1808–87).[7] Figure 2.3 shows all the descendants from Gabriel Stokes (1682–1786) as far as his great-grandchildren, the generation of our Stokes, with the exception of his siblings who are shown in Figure 2.1. Stokes's first and second cousins are marked. In this chapter, when referring to relatives in later generations we give both their birth and death years where known, and also the relationship to the appropriate sibling or cousin of Stokes. This enables their relationship to Stokes himself to be easily calculated, and also enables the relative to be located in more extensive family trees such as that maintained by Teresa Stokes.

We note that for the fourteen Stokes daughters shown in Figure 2.3 only two marriages are known: those of Sarah Ellen (1813–95) and of Mary (born about 1771) to Francis Amyot Professor of French at Trinity College Dublin (TCD). This could be the result of bias by the family genealogists who have concentrated on men bearing the Stokes surname. At any rate no descending lines are known from these daughters.

It will be noted that in these family trees the names Gabriel and Whitley frequently appear as first names—the latest count is twenty-eight Gabriels and fifteen Whitleys. It was not recorded from where the name Whitley came into the family. However, it occurs as a surname in Ireland, particularly in Ulster. The earliest two bearers of the name Whitley Stokes in Figure 2.3 are the eldest children of the brothers John (1721–81) and Gabriel Stokes (1732–1806). These brothers had each married a Boswell, Hannah and Sarah respectively, presumably sisters or cousins. So it is possible that the Whitley surname occurs in the Boswell ancestry. Support for this hypothesis is provided by the occurrence of two Whitleys in the family tree of Boswells of Ballycurry, Co. Wicklow.[8] A Whitley Boswell was a son of the Alexander Boswell who could have been father of the sisters. Also Alexander had a grandson, the Dublin inventor John Whitley Boswell (c.1767–1841) who had entered TCD as a pensioner under a Dr Stokes on 9 June 1784 and received his BA in 1788. His numerous patents include an award on 20 May 1802 'for a method of building or fabricating ships or vessels for navigation'.[9] In reference 2 it is stated that Hannah Boswell was of Ballycurry. As a junior fellow at TCD in 1746, John Stokes (1721–81) would have known the John Boswell (d.1749), one of the family from Ballycurry, who had become a fellow of TCD in 1745. It is possible that this led to his introduction to his future wife Hannah Boswell. The Boswells of Ballycurry descend from Sir Robert de Bosville (d. c.1092) who arrived with William the Conqueror. Thus, even if the origins of the Stokes family are uncertain, it seems that Stokes could claim Norman ancestry through his grandmother, Hannah Boswell.

Stokes's parents were Gabriel Stokes (1762–1835), who had been a scholar at TCD, and Elizabeth Haughton (1781–1866). Stokes recalled in 1857 that it was while his father was curate of St Anne's, Dublin that he fell in love with Elizabeth, who was then very young. A marriage licence was issued in Dublin in 1798[10] and the marriage is recorded as having taken place on 17 February 1798. Gabriel was afterwards vicar of Carnalway in Co. Kildare (in 1806[11]) and of Ardfinnan in Co. Tipperary, before becoming rector of Skreen in Co. Sligo and vicar general of Killala and Achnory.

Elizabeth Haughton (1781–1866) was the daughter of John Haughton, a weaver from Chinley End in the Peak District of Derbyshire. As early as 1741, John Haughton had become one of John Wesley's first itinerant preachers.[12] In 1749 Ireland was a single circuit in the Methodist church, and Haughton was one of the two preachers appointed to cover the whole of the island. In 1750 he was with Wesley during riots in Cork. In 1760 he was again sent to Ireland, but shortly afterwards left the Methodists and sought ordination as a clergyman in the Anglican Church.

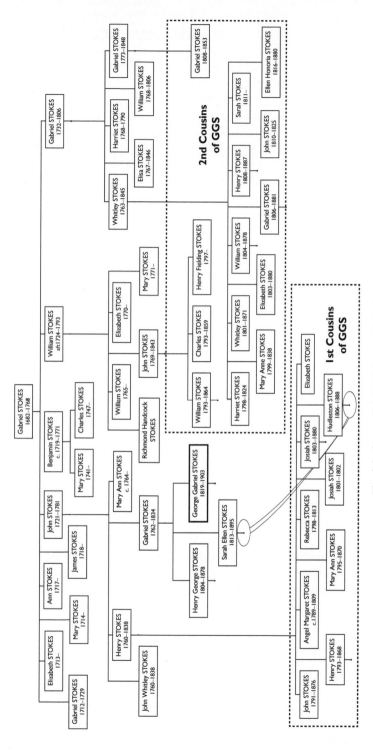

Fig. 2.3 George Gabriel Stokes and his Stokes cousins.

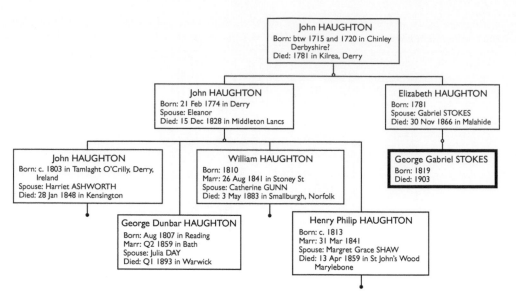

Fig. 2.4 The Haughton relatives of Stokes include four first cousins.

After ten years as a curate, in 1777 he was installed as rector of Kilrea where he was also a local magistrate. Despite joining the established Church he remained a strong supporter of the Methodist movement all his life, and in 1778 Wesley visited John in Kilrea and preached there. John Haughton died in 1781, the year that Elizabeth was born. Based on the assumption that he had been a young man of 21 to 26 when he joined Wesley, we estimate his birth year to be between 1715 and 1720. This would make him rather old, between 61 and 66, when Elizabeth was born.

Elizabeth Haughton had a brother, John Haughton (1774–1828), who entered TCD in 1790, receiving his BA in 1794, and became a minister. In 1803 this brother must have been still living or connected with Tamlaght O'Crilly, a parish adjacent to Kilrea in County Londonderry, because his eldest son was born there. John then came to England, becoming in 1808 vicar at St Giles, Reading. This is where his sister, then Elizabeth Stokes, on a visit from Ireland had her third daughter baptized in 1810. John had obtained in 1809 an MA at Christ's College, Cambridge. John's four sons (see Fig. 2.4), first cousins to Stokes, were all Oxford or Cambridge educated and were ordained in the Church of England. It is interesting to note the rise of this Haughton family in just two generations, from the apparently humble origins of a weaver in Derbyshire, to produce these four well-educated grandsons as well as the illustrious Stokes and his siblings. Presumably the association with John Wesley's work had unlocked the potential of the weaver.

Personal and Biographical

The 111 pages which form Section I of Volume 1 of the 1907 memoir edited by Larmor were the responsibility of Stokes's daughter Isabella Humphry (1861–1934), who, in addition to writing her own reminiscences and selecting letters written by Stokes, collected material from both family and friends. They provide a wealth of detail about the character and family life of Stokes. As the major primary source on his personal life it has been used for many biographies of

Stokes, including Chapter 1 of the present volume. In this section we select quotations from Isabella's recollections and from the letters.

The main details of Stokes's childhood in Skreen, a remote parish on the Atlantic coast in County Sligo (see Chapter 1), were recorded by Isabella based on letters from her aunt Elizabeth Mary Stokes (1811–1904) who had been eight years old when Stokes was born. Although during the summer Isabella often visited her aunt at Malahide, a seaside village (at that time) north of Dublin, she regretted that she never had the opportunity to visit Skreen with her father and potentially learn directly from him more about his childhood.

The next period of family life covered in Isabella's *Recollections and Letters* deals with Stokes's courtship and marriage (7 July 1857) to Mary Robinson, who was born in April 1823. Her father was Rev. Dr Thomas Romney Robinson (1792–1882),[13] an eminent scientist, born in Ireland and educated at TCD where he was a Fellow with interests ranging across physics, mathematics, chemistry, and natural history. Although he was not an astronomer, he was appointed to the Armagh Observatory in 1823.

Portions of twenty-five letters which Stokes wrote to Mary during the latter part of their courtship were selected for publication by Isabella. They cover the period from 5 January to 8 June 1857. She writes:

> His suit was successful; but on one occasion a letter of fifty-five pages about his scientific preoccupations gave room for misunderstanding, and it seemed that the engagement might terminate. The passages printed below are chosen from many letters that were written at this period. At first sight it seemed as though even these were of too intimate a nature for publication; but on thinking the matter over and taking advice from friends the letters were so unlike ordinary love letters, so dignified and impersonal in their expression, that, written, as he said, to explain his character, they must be of legitimate interest to others as containing the only self-revelation that he apparently ever consciously made. They are remarkable also from the curious place which he assigned to his original investigations; it almost seems as if he considered them the height of dissipation, and everything else a duty. He evidently thought that his correspondent had not been unjust in thinking his nature deficient in warmth at this period, and that he was conscious of a too overwhelming absorption in his investigations and experiments. Nor is this surprising, when we consider that this period coincided with the development of some of his most striking discoveries. As she felt this anxiety when about to sever herself from her old home, she was wise and true in expressing it at the risk of pain to them both. He never after-wards heard of a broken engagement without pain, holding that if not two, anyhow one person usually suffered acutely. Even if he hardly knew people, he grieved at such news.
>
> His marriage was a singularly happy one. He first met Miss Robinson, daughter of the Rev. Thomas Romney Robinson, Astronomer of the Armagh Observatory, at a Meeting of the British Association [hereafter denoted by B.A.], and had some difficulty in re-discovering her, as he imagined her to be a daughter of Sir David Brewster. It was but a cursory first interview, but he was so much charmed by her appearance and her manner that he cherished the hope that this might be the lady of his affection. They next stayed together at Lord Rosse's: report said that he proposed to her in the tube of the Great Telescope, but this is absolute fiction.

In a letter to Mary on 24 January 1857 Stokes writes:

> To reassure you after the great pain I fear I have caused you, I will tell you one thing which will show how I regarded you about two and a half years ago and regard you still. I have never told

it to you yet. When the clouds seemed to clear off in that wonderful way, and I saw how I was on the point of sinking into an old bachelor (I mean no disrespect to the genus) and felt how much better married life, if carried out as I looked forward to, would be, I felt that perhaps my marriage with you would be even the turning-point of my salvation.

Two and half years takes us back to July 1854, well before the B.A. meeting held in Liverpool in the autumn of 1854, so Stokes and Mary first met either at the B.A. in Hull 1853, or much more likely at the Belfast meeting of 1852. Stokes was already then a vice-president of the B.A.'s Section A committee on mathematics and physics. In his presidential address to the assembly, Colonel Edward Sabine referring to Stokes noted 'the gratification that all who cultivate science in this part of the United Kingdom must feel at the rising eminence of their highly accomplished fellow-countryman'. He explained that because of the interest in Stokes's work for several branches of the sciences, instead of being presented in one of the parallel discipline meetings it had been selected for one of two evening general meetings which all could attend. So on Friday 3 September 1852 at 8 p.m. Stokes gave his lecture, *Recent Discoveries in the Properties of Light*, i.e. fluorescence,[14] to the whole Association (and presumably guests) in the May Street Church, Belfast.

It was clearly a major occasion for Stokes. His sister Elizabeth attended and recorded:

> We fixed upon Friday for our journey to Belfast that I might go to George's Lecture, and I had the great pleasure, of seeing the very high estimation in which he is held. He spoke for about two hours and was listened to with the deepest attention, and such stillness, except a burst of applause now and again. He was most perfectly at his ease, and spoke so distinctly that I did not lose a word. When he had concluded Colonel Sabine rose and said he was sure he was only fulfilling the wishes of the ladies and gentlemen present in conveying to him their sincere thanks for his kindness in coming forward, and for the very clear explanation he had given of his discovery; that he felt sure there were but few present who could follow the subject in all its details; yet he was also sure there were none who had not derived pleasure and profit, and that many would look back with delight to their presence there that evening, as they watched the onward progress of him whose present discovery was but a first step, of him who, if God is pleased to spare his life, promises to be one of the first scientific men of his age or of any other; that his countrymen have good reason to be proud of him, and so on. Had Colonel Sabine been his father I think he could not have taken more deep interest in him than he appeared to do.

We can speculate that Mary Robinson was likely to have been present with her father for this lecture, and this led to their first meeting either after the lecture or at one of the B.A.'s two evening soirées. In the confusion of praise that he was receiving one can readily understand how Stokes mistook her parentage, as recorded by Isabella (see above).

Certainly the praise of the attractive 33-year-old professor was echoed by Mary's father Romney Robinson, who extolled Stokes in a speech concluding the meeting:

> Of the communications brought before Section A, most certainly that of Professor Stokes, which the public has heard here in a more popular form, holds the highest place, from the singular light which it has thrown on that class of phenomena first brought into notice by Herschel and Brewster. I consider that this may be termed the third epoch in the history of light—the first, being that in which Newton discovered the decomposition of the solar ray by means of the prism; the second, the discovery of the polarisation of light by Malus—and I regard this as not inferior in importance to the other two. Already does it open to the eye that

views its development a course of which we can scarcely contemplate the end, and which is not more promising in its theoretical results than it is in the host of practical applications of which it is susceptible.

Eventually Stokes did rediscover Mary. Things progressed, and, as noted in Chapter 1, in September 1856 they became engaged. However money was still a problem, for when married Stokes would have to vacate his Pembroke College fellowship. The insufficient endowment of the Lucasian chair at Cambridge made it necessary for him to supplement his income from other sources. For much of the 1850s he was a lecturer at the School of Mines, then located in Jermyn Street, London. He also received an income as a Secretary of the Royal Society. But he had continuing doubts about income and the difficulty of juggling jobs in London and in Cambridge.

Some of these doubts may well have been expressed in the notorious fifty-five-page letter from Stokes to Mary which was not reproduced by Isabella. We can deduce it was posted on the morning of Saturday 17 January 1857, for later that day he writes in a follow-up:

> As no mails go out on Sunday you will not I fear get this for a day and a half after the 55-pager. I am afraid the latter, though it will probably be interesting to you (if I may judge how I felt about similar letters to me), may have given you some pain. That gave me occasional pain, being no longer pent up, can now give me no more, and I feel so happy now that there is nothing to interfere with my love. The ghosts looked always wrong; but now how silly and contemptible they appear as well!…
>
> The 55-pager explained to you my motives for the first time, and it is better you should love me as what I am than as something else. My letter this morning was for a letter something like Hofmann's methylethylamylophenylammonium for a word: I guess you never got a 55-pager before.

On 3 February 1857 he writes:

> Even before I went to the Observatory last August, I thought that as to the time of our marriage I ought not to put it off merely for the sake of saving more money, when I had what prudent and experienced people considered enough (I thought it might be deemed necessary or rather advisable to save money for two or three years in case it should turn out that you had not any, but it was not so deemed either by your friends or mine, but the contrary), but that I ought to put it off a little if my marriage would throw difficulties in the way of University reform so far as relates to the endowment of my own Professorship. Everybody allows that in the abstract the Professorships ought to be endowed, but where is the money to come from?…
>
> Now, my notion is this. Talk may go on for years and come to nothing, and I don't know whether I ought to put off my marriage for mere talk….
>
> If I were called into residence and my Fellowship were added to the Professorship, my income would be just about what it is independently of my Fellowship which I should have to give up if married at Easter, but our position would be far, far pleasanter. I should be in a fixed and highly respectable position instead of being like a 'bookseller's hack' as Airy expressed it to me, 'a scientific hack' as one of my intimate friends said to me I was, I should do one thing well (at least I hope so) instead of having so many dissimilar things to attend to that I feel as if I were doing them all badly. I should have (probably) much more leisure for researches, which would then become part of my business, to keep up the reputation of the Chair. I should have pleasant intellectual society. And you, instead of being a nobody buried in big smoky noisy London, would become all at once a full-blown Frau Professorin, with extremely pleasant

society at the distance of a short walk from you, living in what I almost consider the country, able to drop in to those lady friends with whom you might become intimate instead of having to take a cab for two or three miles to call on a friend.

I wish you would speak seriously to your father about this: I should like to know what he thinks....

I have run on and written a pretty good letter after all, and I will now bid you good-bye. P.S. I must explain that my 'castles' meant merely bright anticipations of steady mutual affection. I did not by using that word imply that I looked on them as fancies not to be realized. Your character has been uninterruptedly sunshiny in my eyes.

Then on 7 February:

As to the time of our marriage, I shall proceed on the supposition that it is to be Easter Tuesday unless something occurs to prevent it. I don't think this at all likely as far as changes are concerned.

Easter Tuesday, the date being planned for the wedding, was 14 April, but the ceremony was evidently put back to the summer vacation.

On 11 May 1857 Stokes wrote:

If we are married at the time we are at present thinking of, and go to Switzerland as we talked of, I think I will bring a couple of quartz prisms, a quartz lens, and a piece of uranium glass with me, to observe the spectrum on top of the Rigi or Faulhorn.

What scientific observations Stokes actually performed on his honeymoon we do not know, Mary's journal recorded, 'George is so fond of lightning'; later, 'he puts his head under all the waterspouts he can find'; then, 'he flew about, now up, now down, trying to find a better path; he quite enjoys dangerous places and looks so happy where his neck might be broken'.

Finally, all was set, and from London on 12 June 1857 Stokes wrote:

To-day has been rather a busy day with me, seeing people on scientific matters, correcting press and so forth. I felt rather fagged, as if I wanted relaxation, and thought if I could only get some talk with you and hear some of your music it would freshen me again and set me up. I ordered a suit for a wedding, as well as 400 cards and envelopes.

From London on 18 June 1857:

The preparations for our marriage are made; the day is named: but even now refuse me if you wish it....I cannot feel as if you would draw back, but you must take your choice. 1 a.m. on Sunday morning...then it is right that you should even now draw back, nor heed though I should go to the grave a thinking machine unenlivened and uncheered and unwarmed by the happiness of domestic affection. But I will not dwell on this for I do not believe it can be the case: you mentioned it merely as passing thoughts which troubled you, and which you told lovingly and conscientiously to me, and I don't love you the less for having told them out or even for having had them.

Till your yesterday's letter there was nothing to prevent the feeling of full satisfaction with which I could make you my bride, and there is nothing still except your suspicion. The happiness of deep affection outweighs in my mind the happiness of the scientific leisure which I give up, but the happiness of the scientific leisure may outweigh mere milk and water affection. I feel prepared to make my promises provided you feel prepared to believe in them; but a great love on the one side requires a great trust on the other, and you must trust me for love as well.

They were duly married on 7 July 1867 in Armagh Cathedral by Stokes's elder brother John Whitley Stokes (1800–63), the archdeacon of the Cathedral. Their first Cambridge residence, in 1858, was in tiny rooms over a nursery gardener's some way out on the Trumpington Road. Then after a brief sojourn in London lodgings, while he lectured at the School of Mines, they took a small, then isolated, house in Cambridge, Lensfield Cottage.

We can follow Stokes's household through the Census. In 1851 he is recorded as a fellow and professor of mathematics at Pembroke College. The 1861 census return for the Stokes household at Lensfield Cottage was filled in and signed in clear precise handwriting by Stokes himself. It is a welcome contrast to the messy and sometimes hard-to-decipher hand of many enumerators. The household comprised George Gabriel Stokes, Mary Susanna Stokes, Arthur Romney Stokes, and Susanna Elizabeth Stokes. Visitors were Frances Waller, age 38, born Ireland and Maria Edgeworth, age 21, born Ireland. Servants were Anne Leslie, nurse, age 36, born Canada; Sarah Watson, cook, age 25, born Cambridge; Elizabeth Marks, housemaid, age 25, born Ireland; and Sarah Taylor, nurserymaid, age 15, born Midd[lese]x.

In 1871, the Lensfield Cottage census-taker records three children, with Isabella and William replacing the daughter Susanna who had died. Maria Edgeworth is still a visitor (Occupation: shareholder). Alice Leslie still heads the servant list of four which now includes a husband and wife as butler and cook.

The visitor Maria Edgeworth (1839–93) recorded in the 1861 and 1871 census was a half-niece of Mary's stepmother, Lucy Jane Robinson née Edgeworth (1805–98). These visits to Stokes's household and Cambridge society proved fruitful for they resulted in her marriage in Cambridge to Rev. John Sanderson during the second quarter of 1871.

In 1881, there were still four servants, headed by the faithful Anne Leslie. By 1891 all the children have left the household. Lucy Jane Robinson (the 86-year-old widowed stepmother of Mary) and three servants make up the household. Lucy Robinson died in Cambridge in the first quarter of 1898 and is buried in the Mill Road cemetery.

Stokes's wife, Mary, died on 30 December 1899, and although Isabella records that Stokes then came to live with her husband and her, to judge from his letter headings he continued to use Lensfield Cottage for a while: until at least 4 January 1901. However, in the 1901 census, which was recorded on 31 March, Lensfield Cottage is shown between 9 and 10 Union Road, but renamed as Stokes Lea and is occupied by a clergyman's family. Stokes is recorded with his daughter Isabella Humphry and her husband at Lensfield House, which was located at the north-west corner of Lensfield, where Panton Road and Lensfield Road meet. This household is shown headed by Laurence Humphry, 44-year-old Doctor of Medicine; William C. Robinson, a 74-year-old retired Indian Company man and uncle to Isabella was visiting; and there were three servants.

Lensfield Cottage (Fig. 2.5) in Union Road had been built during the initial Newtown development of land belonging to a former Downing fellow, the London lawyer John Lens:[15] a four-acre field now bounded by Lensfield Road to the north, Hills Road to the east, and Panton Street to the west, with its southern boundary lying not on but just to the north of Union Road. Originally there were two large residences, each with ample grounds. One was 'Lensfield' in the north-west corner, built by William Wilkins in about 1811 for his own use, with grounds that stretched eastwards along Lensfield Road towards the junction with Hills Road at Hyde Park Corner. This survived until the present Chemistry Laboratory was built in 1953. Land in the eastern part of John Lens's holding was originally the site of another large house, then

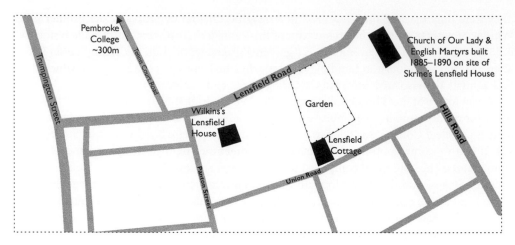

Fig. 2.5 Sketch map showing locations of Lensfield Cottage and the two Lensfield Houses.

named Lensfield House, built in 1810 for a prominent and well-connected local banker, Julian Skrine. That house was demolished in mid-1885 for the building of the Church of Our Lady of the Assumption and English Martyrs which today stands on the corner of Hills Road and Lensfield Road.

As recorded by Isabella, Lensfield Cottage on the southern boundary was isolated in 1858, and indeed it can be inferred from the 1861 census that Union Road had not then been developed around Lensfield Cottage. The Cottage seems to have had access to, if not possession of, some of Wilkins's land to form a good-sized garden some 70 metres wide and 100 metres deep on which the Scott Polar Laboratory now stands. If there were a suitable back gate onto Lensfield Road, Stokes would have had quick access to the back of Pembroke College via Tennis Court Road. Of this large garden Isabella records:

> He never worked in the garden, but used to stroll in it, picking off the dead flowers from the geraniums, one of his favourite plants. He loved brilliant colours intensely; and my mother and I used to be amused at the vehement colours he wished us to select when we had patterns of dress stuffs.

Here we quote from a selection of Isabella's recollections in which she describes Stokes's character and the family life at Lensfield Cottage. In several passages she gives a far friendlier picture of his character than might be assumed by someone experiencing his widely known taciturn nature.

On shyness/silence:

> He rejoiced in silent companionship, often taking one of us on his long quick walks, perhaps not saying a single word for miles. Then some day his interest would be aroused by something heard or seen, and he would have a sudden fit of eloquence.
>
> He often regretted his habit of silence; sometimes when asked why he had not taken pity on an uncomfortably shy person, he would answer that he wished to speak, but that he could think of nothing worth saying. Though silent he was very sympathetic, and almost everyone felt keenly attracted to him; even children, whom it might have been thought that his abstractedness would have daunted, went to him at once. His little grand-daughter was quite devoted

to him, and would coax him to take her to see the ducks, tortoises, and gold-fish in the Botanical Gardens.

Had he not married, he might have had a lonely old age; for people did not often visit him socially, even his old pupils.... Perhaps his silence was most painful to him when some foreigner came from a distance on purpose to meet and talk with him. It might have been imagined that polite and sociable Frenchmen, interested in the same subjects, and coming with the express object of conversing with him, would have vanquished any degree of reserve; but they were not always successful. Yet he very much enjoyed talking, and hearing conversation. It seemed on the whole easier to him to converse with women than with men, and during the last few years of his life, when less busy, he showed much pleasure in the society of ladies. He would often sit at tea-time amused by the chat, and then would suddenly launch into the conversation. The absence of obligation to talk inclined him to do so.

He did not often take these keen fancies for people, seldomer against them; but if there were anything droll in manner, if they poured out torrents of words, or gave an impression of special vanity or egoism, he would be very much entertained, as appeared in little amused twitches of his mouth, and reined-in smiles, with a certain twinkling of the eyes. We were often amused by the details he noticed in people staying in the house, a keen observation of which they were quite unaware because of his silence and apparent absence of mind.

Although of a very quiet and silent disposition, he by no means liked being alone; he would often bring his work into the drawing-room in the evening, and had a folding-table kept close to the door of that room so that he could work in family surroundings. He did not like the talk to stop on his account; indeed his power of concentration prevented its being a worry to him; it just seemed to reach him as a cheerful and soothing buzz. Still it was interesting to note that one never knew when he was listening; and the most unexpected subjects occasionally arrested his attention, when he would launch suddenly into the conversation in the intervals of his work.

He had a great love of parties and public functions of all kinds, and rarely refused invitations. I used laughingly to tell him that whether it were a wedding or a funeral did not make the slightest difference. This keen sympathy with human joys and griefs was characteristic of him; it was felt even towards those with whom he was but slightly acquainted, and he sometimes expressed a wish that he had the power of conveying his sympathy in words.

When friends were ill he would often go to see them and sit with them quite a long time. His fondness for frequent church-going on Sunday seemed to have its root in this same turn for companionship and good-fellowship, especially when linked by union of action or idea. But we used to wish that after the labours of the week he would not go three or even four times...

On exercise (walking and swimming):

In his undergraduate days sports were not the fashion with reading men, who took 'grinds' or country walks instead. This habit he maintained in youth, and until long past middle life long walks were the custom, both summer and winter, at a pace of nearly four miles an hour. At eighty-three years of age he still went the Grantchester 'Grind,' of three or four miles, and other equally long walks as his afternoon exercise.

Professor G. D. Living, the chemist who first met Stokes in 1850, records in his Appreciation that Stokes was a frequent and strong swimmer and was an early member of a new Swimming Club at Grantchester which started in the 1850s. As will be noted in Chapter 6 this swimming ability aided Stokes in his observations of water waves off the north-western coast of Ireland.

On anxiety about speaking:

He always had a difficulty in lecturing, and felt considerable anxiety about it. Even his professorial lectures, to which it might have been thought he would have become accustomed by long use, worried him, and he was anxious as to whether he was getting on too quickly, whether his class were following his lectures and his experiments, and was evidently always afraid of sinking his level by the staleness of custom. On our walks, when it was noticeable that he was deeply pondering and must not be disturbed, after he emerged from his long fit of abstraction he would not infrequently say that he had been thinking about his lectures and deciding on his course, and how he should present his subject attractively; and afterwards he would sometimes show that he was depressed about his lectures and thought them a failure. On the other hand, he would be quite gay when he had a specially nice class, and would thoroughly enjoy his course. But any single lecture or course of special or popular lectures was a thorn in his side, and worried and tired him more than anything else. It seemed that he thought that other people had some extremely high standard, would expect so much, and be so likely to be disappointed.

On female students:

When it was becoming customary for the University Professors to admit lady students to their lectures someone wrote to him asking for permission for the ladies to attend his lectures also. He said that he had almost decided to refuse; but I begged that he would take them, and asked him how he would feel if a Mrs Somerville had asked him to teach her, and he would not. At last he promised to admit them, and he became much interested in his lady students, and always knew how much they understood. He was much amused by one. After the first lecture he said, 'She frowns'—after the second, 'She is frowning horribly!'—after the third, 'Her forehead is one mass of corrugations; she won't be there next time!' and she was not. But some of the ladies got on splendidly, and he was much pleased when a Newnham lady who had attended his lectures brought him some original work which he approved.

On his papers in a mess:

He had two really wicked characteristics, that he would never allow anyone to help him with his work, not even permitting invitations to be answered for him, and that he kept every single thing he received by post, even advertisements. His study was enough to drive any housemaid 'wild.' He used gradually to acquire tables from the rest of the house, until there were as many tables as the room would hold, with narrow passages between, through which to squeeze if you could. On these tables papers were piled a foot or more deep. It may be imagined that, keeping everything, he could find nothing. One remembers the hunts there used to be before he went to London, every person in the house sometimes enlisted. It always began by his hunting alone and refusing all help rather fiercely. Then gradually, as the quest grew more desperate, the rejected suitors fell quietly into the ranks. Sometimes it was grim earnest, and the necessity was urgent; as for instance when one of the Gifford Lectures was missing, and it was nearly time to catch the only train that would connect with the mail, if he were to arrive in Edinburgh in time to deliver it. At last it was discovered, but only in the nick of time, in the round basket of an aged relative, which was called ever after 'the magpie's nest,' and came in for first search on subsequent occasions. When the study and the inner study had reached a state of repletion, my mother would wait for the Royal Society day, Thursday, for then he was often away for two nights or so, don apron and sleeves, and fall upon those rooms, when clothes-baskets full of unnecessary matter would be removed. Then for a long time afterwards we were considered the cause of the disappearance of every missing object or paper. We often

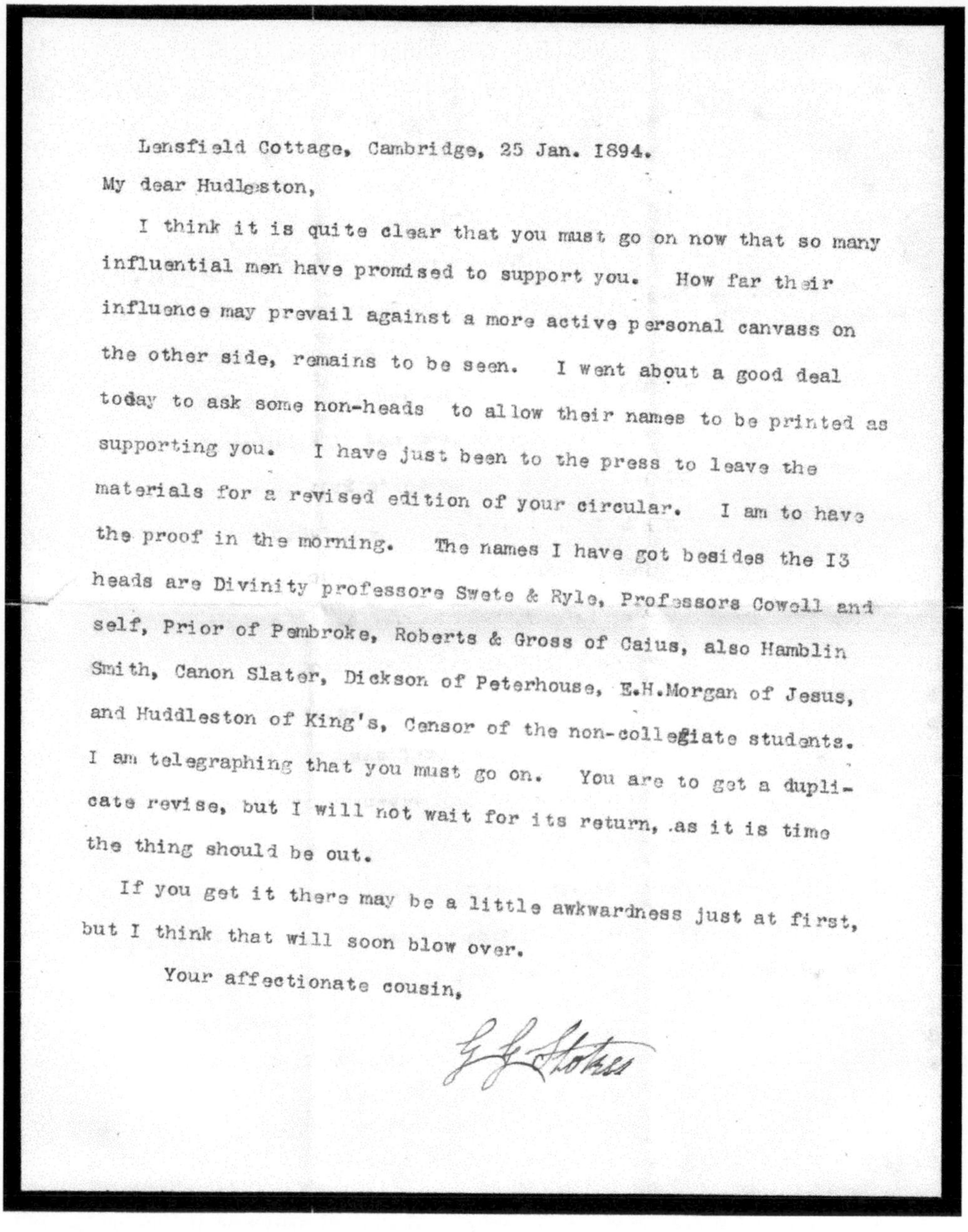

Lensfield Cottage, Cambridge, 25 Jan. 1894.

My dear Hudleston,

I think it is quite clear that you must go on now that so many influential men have promised to support you. How far their influence may prevail against a more active personal canvass on the other side, remains to be seen. I went about a good deal today to ask some non-heads to allow their names to be printed as supporting you. I have just been to the press to leave the materials for a revised edition of your circular. I am to have the proof in the morning. The names I have got besides the 13 heads are Divinity professors Swete & Ryle, Professors Cowell and self, Prior of Pembroke, Roberts & Gross of Caius, also Hamblin Smith, Canon Slater, Dickson of Peterhouse, E.H.Morgan of Jesus, and Huddleston of King's, Censor of the non-collegiate students. I am telegraphing that you must go on. You are to get a duplicate revise, but I will not wait for its return, .as it is time the thing should be out.

If you get it there may be a little awkwardness just at first, but I think that will soon blow over.

Your affectionate cousin,

G. G. Stokes

Fig. 2.6 Letter to Rev Hudleston Stokes (1831–1904), son of Henry Stokes (1793–1868), black-edged since Stokes was still in mourning for his son William George Gabriel, who had died five months previously.

presented him with letter-cases and other domestic inducements to tidiness, but they were usually found empty, or with things quite unimportant inside.

Stokes is recorded as being most kind to his relations. When Hudleston Stokes (1831–1904) (Fig. 2.7) arrived at Gonville and Caius in October 1850 he found that William Haughton Stokes (Fig. 2.1), a senior fellow there, only occasionally invited him to a breakfast. In contrast,

Stokes in Pembroke 'was most kind in frequently inviting me to his rooms, and in taking me for long walks during which he soared into mathematical heights to which my dull understanding strove in vain to follow'. After classical studies Hudleston became a vicar and was often in correspondence with Stokes. Forty years later Stokes lobbied the University Senate to secure support for the appointment of Hudleston to the vicarage of Stapleton, Shrewsbury. Stokes had 250 copies of a circular printed for distribution to residents who could vote, and reported back in four letters to Hudleston the progress and eventual successful outcome of the vote. Hudleston kept those letters and one is reproduced in Figure 2.6.

Stokes died at Lensfield House on 1 February 1903. The cause of death—certified by his son-in-law—was oesophageal cancer, from which he had suffered for six months and with pleurisy for the last four days. A certified nurse was in attendance. After having seen him as a fine family man through the writings of Isabella one just hopes he was as comfortable as possible in his last days.

As mentioned in Chapter 1 he had an elaborate funeral in Cambridge. Several memorials were erected including at Westminster Abbey. The memorial brass in St Paul's Church, Cambridge records that 'for 34 years he was a churchwarden of this parish and for the last 13 years of his life read the Holy Scriptures on the Lord's Day with singular reverence and discernment from the lectern of this church'.

Descendants of George Gabriel Stokes and Mary

In 1858 Arthur Romney Stokes, the first child of Stokes and Mary, was born. He was educated at Charterhouse and King's College, Cambridge. He taught first at Hereford School. In 1889 he married his first cousin once removed, Mary Winifred Garbett, granddaughter of Henry George Stokes (1804–78). In September 1889 he became master of the Fourth Form at Shrewsbury School and held the appointment for more than twenty years.[16] At Shrewsbury he took an active part in all school pursuits: boating, football, fives, and especially in rifle practice of the Corps to which he gave a cup for competition. In later years he had been unable to do more than occasional school work due to illness. A man of considerable originality of thought, he had travelled more than most schoolmasters. Arthur and Mary had daughters only, so the baronetcy which Arthur inherited from Stokes in 1903 died out on Arthur's death in 1916. Lady Stokes and her two girls all eventually settled in South Africa. Winifred, the younger daughter, had married Clement Victor Palmer in London and they settled on a farm at White River in Eastern Transvaal. This marriage produced two daughters: the elder, Daphne Palmer marrying Richard Lefebvre has living children and grandchildren in South Africa and Queensland.

Stokes and Mary's second child, Isabella Lucy, married Dr Laurence Humphry (1857–1920), who was the son of a London barrister. Humphry had studied medicine at Cambridge, and subsequently practised in the City. He also held posts at St Bartholomew's, the London Chest Hospital, and Addenbrooke's. He taught medicine in Cambridge and during the Great War was a lieutenant colonel in the Royal Army Medical Corps. He and Isabella had no children.

The third child was William George Gabriel Stokes (1863–93). He went to Marlborough, and entered Pembroke College, Cambridge aged eighteen in October 1881. His Cambridge records show Pens. 1881; Scholar 1882; B.A. (Class. Trip., Part I, 1st Class) 1884; M.B. and B.Chir. 1890; M.A. 1892. At St Thomas' Hospital he achieved the qualifications M.R.C.S., L.R.C.P. 1889. The census of 1891 shows him living at St Thomas' Hospital, London, with occupation given as surgeon. But two years later at Coundon, near Bishop Auckland, he was dead, aged only thirty. It has been said that an accidental overdose was the cause. However, my sixth cousin Teresa Stokes, great-great-grandaughter of Henry Stokes (1808–87), has tracked down the newspaper accounts of the inquest.[17] In 1893 William had gone to Bishop Auckland to take up a medical partnership with Dr Mark Wardle. After staying with Wardle for his first week he took lodgings in nearby Coundon on Tuesday 22 August, and despite apparently having been in good spirits was discovered by his housekeeper on the Friday morning insensible with all the symptoms of opium poisoning. Despite Wardle's continuous efforts for sixteen hours William died the following night. Stokes and Lady Stokes had been summoned from Ireland by telegram and they had arrived in time for the inquest. A note had been found saying, 'August 24. '93. my memory is going fast. my nerves are gone. I feel I am going mad. I can't sleep; I mean to sleep sound tonight. May God and my family forgive me. Wm. G.G. Stokes.'

On the outside of the envelope were the different times and amount of morphia taken:

10.40 - 8grs. morphia hydrochlorate.
1.20 I am getting drowsy, but have not slept.
2.45 - Have got no sleep. I am tired of this.

Following direction from the coroner, the jury pronounced a verdict of suicide by taking an overdose of morphia while temporarily insane. It was something the family seems to have kept quiet about, for suicide was then a criminal offence. After this family tragedy Isabella writes of her mother: 'In later years she went very little into society with him (Stokes), never having recovered from the shock of their younger son's death.'

The two youngest daughters of Stokes and Mary had died in infancy. Only Arthur Romney produced direct descendants to continue Stokes's line.

Appendix: Other Notable Descendants of the Stokes Family of Dublin

This appendix contains summaries of the major Stokes lines, starting with Gabriel Stokes (1682–1768) and including brief outlines of the lives of the more interesting or eminent ones not already addressed in the preceding sections. Figure 2.7 provides a simplified family tree showing the lines of descent and highlighting the persons included in this section. No fewer than eight of them appear in the *Oxford Dictionary of National Biography*. The three lines are treated in the following order:

1. Gabriel Stokes (1732–1808), whose line contains the notable Irish medical men;
2. John Stokes (1721–81), many of whose descendants (including Stokes) moved to England;
3. William Stokes (?–1793), whose line was linked in through the work of Alan Geoffrey Stokes (1912–96).

Gabriel Stokes (1682–1768)[18] attended the King's Hospital School, Dublin. On 31 July 1696 he was apprenticed to Joseph Moland, a notable surveyor who also taught navigation at the school. The earliest recorded example of Gabriel's work is a map of a plot of land in Christchurch Lane 'surveyed by Abr Carter and traced by Gab Stoakes', and dated 28 May 1700.[19] The earliest

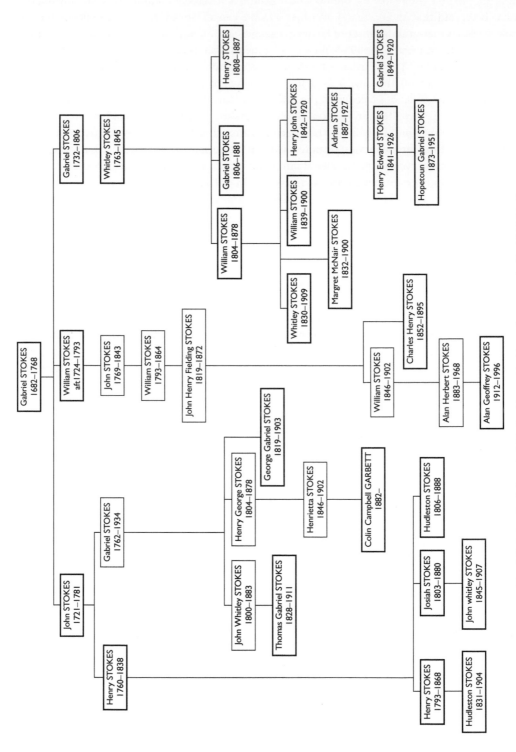

Fig. 2.7 Stokes family lines. Individuals addressed in this section are marked by highlighted boxes.

surviving example of his own surveying is a signed map dated 23 October 1714 of a four-acre plot at St Harold's Cross, Dublin. By around this time he had also become proficient as a mathematical instrument maker. In 1715, he repaired the quadrant belonging to TCD and obtained from the Surveyor General, Thomas Burgh, an endorsement of his work:

I always observed he had been particularly careful in amending what was amiss; and it appeared to me when the work was finished & the observations made with the instrument afterwards, that he had perfected his work and adjusted the Quadrant with skill and exactness.[20]

He operated from premises called 'The Dial' in Essex Street. Surviving instruments bearing his name and date of manufacture include a Gunter slide rule (1719, in the Whipple Museum, Cambridge), a horary quadrant (1738, in the National Maritime Museum), and surveying compasses (e.g. 1717, in the National Museum, Dublin). There is a portable sundial (1742) in the Havant Museum (see Fig. 2.8).

His arms (Fig. 2.2) were awarded by the Ulster King of Arms on 15 November 1721, and included as a crest a surveyor's forestaff, 'in recognition of his skill in his profession'.

In addition to his surveying and instrument manufacturing interests, in 1735 Gabriel published 'A scheme for effectually supplying every part of the city of Dublin with pipe water without any charge of water engines, or any water forcers, by a close adherence only to the natural laws of Gravitation, and the principles, rules and experiments of Hydrostaticks'. Invoking the great Sir Isaac Newton, he backed this up with a pamphlet, 'The Mathematical Cabinet of the Hydrostatical Ballance unlocked: or an Easy Key to all its uses'. An alternative water scheme was actually implemented. He is credited with the planning of the construction of Pigeon House Quay, a long mole on the South side of Dublin Bay.

In 1748 the Surveyor General, Jones-Neville, appointed Gabriel as Deputy Surveyor General of Ireland. While in that office, Gabriel set up an examination and certification system to identify competent surveyors, who could be recommended to Irish land owners.

Gabriel had married Elizabeth King, a merchant's daughter. Three major lines descend from their sons:

Fig. 2.8 Portable Sundial made by Gabriel Stokes, 1742.

Rev. Gabriel Stokes (1732–1806)[21] heads the line on the right in Figure 2.7. He was born in Dublin and educated at TCD where he graduated under the tutelage of his brother, John (1721–81), who was then a senior fellow. Soon after gaining a junior fellowship in 1755 he secured the College living of Ardtrea where he served for fourteen years. He married Sarah Boswell, likely the sister of Hannah Boswell who had married his brother John, and raised a family. He was in charge of the grammar school in Waterford, and became Chancellor of Waterford Cathedral. He then secured the living of Dysart-Martin in the diocese of Derry. He published an *Essay on Primate Newcome's Harmony of the Gospels* and also edited *Iphigenia in Aulis*. His death was caused by overexertion in helping to put out a fire.

Dr Whitley Stokes (1763–1845) was the eldest son of Rev. Gabriel Stokes (1732–1806) and entered TCD in 1779 where he won a scholarship. After graduating (B.A. 1783), he became a fellow in 1788 and was appointed King's Professor of the practice of medicine in 1793. As a known sympathizer of the United Irishmen Movement he was summoned before Lord Clare, the Vice-Chancellor, at a visitation in April 1798 held for the purpose of purging the College of all those involved in the build-up of what was to become the 1798 rebellion. He admitted having been a member of the Movement before 1792. However, he had not played a part more recently, except for providing medical care to an insurgent who was sick, and having furnished information to Lord Moira about the atrocities and torture inflicted on the people of the south of Ireland. Lord Clare declared himself 'gratified to find that the rumour of an eminent member of the university having been connected with a treasonable association was entirely refuted; but, nevertheless, as he had been drawn into with persons who were inimically disposed to the government of the country, he thought it his duty to prevent him from becoming a governing member of the university for a period of three years.'[22]

In 1805 Whitley was restored as a senior fellow, and in 1816 he was appointed lecturer in natural history. In 1830 he became Regius Professor of Physic to the university, a post he held until 1842 when he was succeeded by his son. He was instrumental in founding the College Botanical Gardens and in establishing the Zoological Gardens in Dublin. As a physician he had a large practice and was known for his treatment of fever during the severe epidemics of 1817 and 1827.

William Stokes (1804–78)[23, 24] was the son of Dr Whitley Stokes (1763–1845). He was educated in classics and mathematics by John Walker, a Fellow of TCD, and in science by his father. He was then sent to Edinburgh University to study medicine, graduating in 1825. In that year, while still a student, he published the first book in English on the use of the stethoscope, the invention of which had been announced in France in 1819. Thus he became known as the principal promoter of its use in Britain.

After qualifying he became a physician at the Meath hospital, where with his colleague, Dr Graves, he helped reform clinical teaching in Dublin. He became famous as a medical teacher but was equally renowned for his compassionate treatment of the poor of the city, notably in the great typhus epidemic of 1826, during which he caught the disease.

His major 1827 publication, *A Treatise on the Diagnosis and Treatment of Diseases of the Chest*, was considered to be highly original and 'a model of medical exposition'. Quoting the work of Dr Cheyne twenty-eight years earlier he described what became known as Cheyne–Stokes respiration.

Diseases of the Heart and Aorta (1854) was again based on original observations and description, in the field of cardiology. Together with Adams he first described the characteristic bradycardia and cerebral anaemia known as Stokes–Adams syndrome.

In 1843 William had assumed the duties of Regius Professor of Medicine at Trinity College, and on his father's death in 1845 he was appointed to the chair. Among his many innovations he was responsible for establishing a Diploma of Public Health at Dublin, the first in the British Isles. Involved with the great Irish epidemics during his early years in Dublin, he reported the first case of Asiatic cholera in 1832. He was also responsible for founding the Dublin Pathological Society.

In 1861 he was elected Physician to the Queen in Ireland, and became an FRS. He was President of the Royal College of Physicians of Ireland. According to Sir George Paget, who was Regius Professor of Physic at Cambridge, William was not only the most prominent physician in Ireland but also the greatest physician of his time in Europe. His works were translated into several European languages, and in 1876 he was awarded the Prussian Order of Merit for his medical writings.

He had married Mary Black of Glasgow, and it was their custom to hold an open evening each Saturday at their home in York Street. This circle of friends had a strong influence on Dublin society for many years. His interest in art, archaeology, and Irish history was recognized when he was elected President of the Royal Irish Academy in 1874. He was a friend of the painter Sir Frederick Burton, and of George Petrie whose *Life* he wrote.[25]

Whitley Stokes (1830–1909)[26], son of William Stokes (1804–78), went to TCD, then to the Inner Temple where he was called to the Bar in 1855 and practised law for six years. He held legal posts in Madras and Calcutta for twenty years during which he published a number of works on Indian law. He was made Companion of the Order of the Star of India (CSI) in 1877 and of the Indian Empire (CIE) in 1879.

Although he led a busy life as a lawyer he is best known for his Celtic studies. In his father's house he had met, amongst others, John O'Donovan and Eugene O'Curry, who were responsible for the marked growth in the study of Irish language, history, folklore, and poetry in the second half of the nineteenth century. From his early twenties, Whitley devoted himself to the words and forms of the Irish language. His first publication, *Irish Glosses from an old M.S. in T.C.D.*, appeared as a paper in the *Transactions of the Philological Society of London* (1859). For his first book, *A Medieval Tract on Latin Declensions*, he received the gold medal of the Royal Irish Academy.

With John Strachan he published *Paleo-Hibernicus*, 1,200 pages of old Irish glosses from pre-1000 AD manuscripts from the Continent and from Ireland. Other texts edited and translated by him include *Fis Adamnain, Togail Troi, The Tripartite Life of St Patrick*, and *Feilire of Aengus*. Beside these and many other Irish texts and glosses he published editions and translations of Cornish and Breton texts, many papers on grammatical subjects, and critical reviews of the work of other scholars in the field.

Following the centenary of his death a collection of essays on him has been published.[27]

Sir William Stokes (1838–1900),[28] another son of William Stokes (1804–78), after the Royal School, Armagh went to TCD where he graduated in medicine and surgery in 1863. He received his professional training in Dublin in the TCD School of Physic, in the Carmichael School, and at the Meath and Richmond hospitals. In 1861 he was awarded the gold medal of the Pathological Society of Dublin, and became the society's president in 1881. He was admitted a licentiate of the Royal College of Surgeons in Ireland in 1862 and became a fellow in 1874. After time in Paris, Vienna, Berlin, and Prague, where through his father's reputation he became friends with the most renowned teachers, he settled in practice in Dublin. From 1864 to 1868 he was surgeon to the Meath Hospital and then became surgeon to the House of Industry Hospitals.

It was here he performed the greater part of his operative work, which included a new method for the amputation of the leg at the knee, known as the Gritti–Stokes amputation. He was for some time lecturer in surgery in the Carmichael School of Medicine, and in 1872 he was elected Professor of Surgery at the Royal College of Surgeons of Ireland where he served as president in 1886–7. He received a knighthood in 1886 and in 1892 was appointed Surgeon-in-ordinary to the Queen in Ireland.

He served in the administration and teaching of his profession as a member of the Council of the Royal College of Surgeons in Ireland, for whom he was an examiner. He was also an examiner in surgery at the University of Oxford and at the Queen's University in Dublin.

Early in 1900 William was in South Africa as a consulting surgeon to the British Army engaged in the Boer war. He fell ill with pleurisy and died.

A man of great versatility, a good surgeon and a first-rate teacher, he was also an orator, a master of English composition, and a cultivated musician with a fine tenor voice which was often heard in private society in Dublin.

Margaret M'Nair Stokes (1832–1900)[29] grew up in her father's house in close contact with antiquaries James Tood, George Petrie, William Reeves, Sir Samuel Ferguson, and Edwin Wyndham-Quin (3rd Earl of Dunraven), which was a major influence on her life's work.

Margaret's published works commenced with in 1861 the illustration and illumination (based on The Book of Kells) of Ferguson's poem *The Cromlech of Howth*. She edited Dunraven's *Notes on Irish Architecture*, and published a work on Irish language inscriptions collected by Petrie, as well as many other works on art and architecture. Finally she researched the early Irish missionaries, which involved journeys in France and Italy tracing their footsteps.

She was an honorary member of both the Royal Irish Academy and the Royal Antiquaries of Ireland.

Adrian Stokes (1887–1927)[30] was the youngest son of Henry John Stokes (1842–1920). At TCD he obtained honours in anatomy. In 1911 he was awarded a medical travelling prize and the Banks medal by TCD. After spending six months at the Rockefeller Institute for Medical Research in New York, he was appointed assistant to the professor of pathology in Dublin, where he remained until the outbreak of war. He went to France as a lieutenant in the Royal Medical Cops in 1914. He had to deal with the ravages of the troops by tetanus. Packing the sidecar of his old motorcycle with anti-tetanic serum, he went round the field dressing stations, saving many lives. Thus the first mobile laboratory of the British Expeditionary Force was established. He invented the method of giving oxygen continuously through a nasal catheter to victims of gassing, a method later used in civilian practice. He also dealt with typhoid, cerebro-spinal fever, gas gangrene, trench nephritis, dysentery, and wound infections. In 1916, when jaundice appeared in Ypres, he proved by animal experiments that the disease was conveyed by rats, and was able to curtail the epidemic. For his war services he received the DSO and OBE and, for his work among the civil population, the Belgian Order of the Crown.

In 1919 he returned to TCD as Professor of Bacteriology and Preventive Medicine. In 1922 he became the William Dunn Professor of Pathology in London University, working in the Pathological Department of Guy's Hospital. In June 1927 he was sent to Lagos by the Rockefeller Commission of New York, and conducted decisive experiments showing that yellow fever could be transmitted to monkeys, paving the way for further progress using animal experiments. Unfortunately he himself developed yellow fever, from which he died.

Henry Stokes (1808–87) after an apprenticeship became county surveyor for Co. Kerry (1834–76). He maintained grant roads in the 1830s and 1840s and was an inspector under the

Landed Property Improvement Act in the 1860s and 1870s. Two of his sons and one grandson made their mark in the Indian Civil Service (ICS) and all three received knighthoods. The eldest son, **Sir Henry Edward Stokes** (1841–1916), came fourth in the entrance examination for the ICS. A younger son, **Sir Gabriel Stokes** (1849-1920), served in Madras and as Collector of Tanjore before becoming acting Governor of Madras in 1906. Henry Edward's son, **Sir Hopetoun Gabriel Stokes** (1873–1951), after a degree at Oxford also joined the ICS, eventually serving on the Executive Council of the Governor of Bombay.

John Stokes (1721–81) heads another descent line in Figure 2.7. He entered TCD aged 14, and became a fellow in 1746. On marriage he secured a good college living as Rector of Raynochy and Sharne (which appears with a variety of spellings and is located just north of Letterkenny in present day Donegal). He was Professor of Mathematics (1862–4), and then became Regius Professor of Greek. He is known for his work with Leland in editing Demosthenes.

John's three sons were all educated at TCD and became clergymen. The youngest, Gabriel Stokes (1762–1834), father of George Gabriel, we have covered in an earlier section. We note that Gabriel's grandson **Thomas Gabriel Stokes** (1828–1911), educated at TCD was one of the early family genealogists, and a great-grandson, **Sir Colin Campbell Garbett** (1882–1972), was made a Knight Commander of the Order of the Indian Empire (KCIE) following a career in the Indian Civil Service with military and civil duties in Mesopotamia, and in the Punjab where he held high office.

Henry Stokes (1760–1838) graduated at TCD in 1783 and became a tutor to the Cavendish family of Waterpark. This family had a branch at Doveridge, Derbyshire, and through this connection he secured the Rectorship of Doveridge in 1785. He married a daughter of the Rev. Josiah Marshall, another TCD graduate and rector of Fahan, Co. Donegal. The large family they raised has formed one of the principal Stokes branches in England. Two of Henry's sons entered the East India Civil Service, aided by their uncle John Hudleston MP, a director of the East India Company.

Josiah Stokes (1803–80) through his mother's connections undertook mercantile employment in Bombay. Returning to England he developed in Liverpool a prosperous merchant company (Stokes, Bates & Co.) which traded with South America. In 1867 he had a paddle steamer built on the Clyde for carrying goods on the River Paraguay. His wealth from the business was passed on, and my own great-grandfather **John Whitley Stokes** (1846–1907), after being educated at Cambridge and a spell in his father's business, was able to live a comfortable life as a country gentleman and amateur painter on the border with Wales.

Henry Stokes (1793–1868), married a wealthy widow when very young and lived off the income from her estates as a country gentleman. His son **Rev. Hudleston Stokes** (1831–1904) deserves special note for his manuscript 'The Stokes Family Book' (mentioned above) which includes several mentions of his interactions with Stokes. He also preserved documents describing the arrangements for Stokes's 1899 jubilee celebrations, which he attended.

We turn now to the third line in Figure 2.7, that of **William Stokes** (after 1724–1793). He appears not to have been recorded by those studying the family until more recently, when Alan Geoffrey Stokes from Australia traced his own Stokes line back through a series of Irish engineers and surveyors. He arrived at William and found a record of him dying in 1793 as the Clerk of the Board of Works. He identified him as a son of Gabriel Stokes (1682–1768), possibly based on an extract of his genealogical details made from his will by the antiquary Sir William Betham,[31] which apparently names his father only as Stokes with no first name (this was often Betham's custom when the father was not specifically named in the will). Alan speculated that the branch was little known by cousins in other lines because there had been no male residents in Ireland or England for many years. As was noted above, testing of the Y-DNA of the descendants of this Stokes branch could provide valuable confirmation of the genealogy.

The line from William leads down through John Stokes (1769–1864) who having started as an artificer and supervisor became the Chief Engineer of the Grand Canal. Next is William Stokes (1793–1864), a land surveyor, then John Henry Fielding Stokes (1819–72) who worked on the railways and became the Engineer-in-Chief of the Irish North Western Railway. One of John's sons, William (1845–1902) was an engineer who worked with his father. He went to Liverpool and then in 1874 sought his fortune in the gold mines of Queensland. He heads several Stokes families in Australia, his grandson being Alan Geoffrey Stokes (1912–1996), the family genealogist mentioned above.

Charles Henry Stokes (1852–95), son of William Stokes (1793–1864), is probably the most notorious of the Stokes family, otherwise so replete with respectable academics, vicars, civil servants, and merchants. He started working life as a clerk at a shipping office in Liverpool, and then trained as a lay preacher in the Church Missionary Society (CMS) and sailed to Zanzibar in 1878. He organized the transportation of supplies to the mission stations inland. He reportedly got on well with the natives. His wife, a CMS nurse, died in childbirth in 1884. He then married an unbaptized daughter of an African chief and consequently was dismissed from the CMS, so he set himself up as an ivory trader and porterage contractor. From 1894 he worked as an administrator for the German East African colonies. But he also continued his transport and trading businesses. Hoping to earn enough to be able to retire back to Ireland he set off into the Belgian Congo with a caravan of three thousand porters to buy ivory. But Belgium had decided that all ivory was the property of the Belgian Crown. After being arrested and subjected to a summary trial, Charles was immediately executed. This was viewed as murder and there was a huge furore only concluded after Queen Victoria's personal intervention with her cousin Leopold, King of the Belgians, when Belgium finally paid £8,000 compensation to both the British and German governments. Charles also had children by two native concubines given to him by the King of Buganda. Several Stokes lines exist today descending from these unions.

ACKNOWLEDGEMENTS

Sincere thanks are due to the following living descendants who have helped me over the last nineteen years with important contributions to the study of the Stokes family reported here: Meyricke Serjeantson, Alan Nicholas Stokes, Nicholas Lefebvre, and Teresa Stokes. Also I thank Alan Boswell for recent help with the Boswell connection.

CHAPTER 3

'Stokes of Pembroke S.W. & a very good one'

The Mathematical Education of George Gabriel Stokes

JUNE BARROW-GREEN

M y title comes from the diary of Joseph Romilly, Registrary of the University of Cambridge.[1] Romilly, who was writing on 22 January 1841, was remarking on the fact that Stokes had not only triumphed in the gruelling Cambridge Mathematical Tripos examination, winning the coveted title of Senior Wrangler (Fig. 3.1), but that he had excelled in doing so in a year when the papers were notoriously difficult.[2] It was a notable achievement but it was a prize hard won after several years of preparation, and not only years spent at Cambridge. When Stokes arrived at Pembroke College, he had spent the previous two years at Bristol College, a school which prided itself on its success in preparing its students for Oxford and Cambridge.

In this chapter I shall follow Stokes on his path to the senior wranglership, tracing his mathematical journey from its beginnings in Ireland to its close at the end of his final year of undergraduate study.

Early Education

Stokes began his education at home, being tutored in Latin by his father and learning arithmetic from the local parish clerk, George Coulter.[3] Even at a young age Stokes showed evidence of mathematical talent. Coulter, who taught using Voster's *Arithmetic*,[4] a popular text specifically 'adapted to the Commerce of Ireland', was delighted by the fact that the young Stokes quickly worked out for himself 'new ways of doing sums'[5] far better than those given by Voster. Coulter also reported that 'clever people' were astonished by Stokes's facility with the rule of false position.[6]

Barrow-Green, J., *'Stokes of Pembroke S.W. & a very good one': The Mathematical Education of George Gabriel Stokes*. In: *George Gabriel Stokes: Life, Science and Faith*, Mark McCartney, Andrew Whitaker, and Alastair Wood (Eds): Oxford University Press (2019). © Oxford University Press. DOI: 10.1093/oso/ 9780198822868.001.0003

Fig. 3.1 Stokes as Senior Wrangler in 1841. Courtesy of the Master and Fellows of Trinity College, Cambridge.

In 1832, aged 13, Stokes was sent to Dublin where he lived with his uncle, John Stokes, and attended the famous school of Dr Richard Henry Wall. Dr Wall's school, officially known as the Seminary for General Education, was located on Hume Street (the building is now part of a hospital). Among other pupils who at some time attended the school were the engineer William Thomas Mulvaney and the mathematician George Johnston Allman. The course of instruction included mathematics, arithmetic, and book-keeping, as well as weekly lectures on mechanics.[7] Stokes's mathematical ability once again attracted attention: this time it was his solutions to geometrical problems which caught a master's eye.[8]

Stokes's father died in 1834 and soon after the decision was made to send Stokes to school in England. The school chosen was Bristol College, the choice being made on the recommendation of Stokes's second brother, William Haughton, who was a friend of the Irish-born principal of the College, Joseph Henry Jerrard. Both of them had studied at Trinity College Dublin before going up to Caius College, Cambridge in 1824. William graduated as Sixteenth Wrangler, coming rather higher on the Mathematical Tripos list of 1828 than Jerrard who was 27th Senior Optime that year, although Jerrard was placed in the first class in the Classical Tripos.[9] Humphry described Jerrard as 'a mathematician of some note'[10] but evidence for her claim has proved hard to find. Graduating as 27th Senior Optime in the Mathematical Tripos would not warrant such an epithet. It is possible that Humphry confused Joseph with his elder brother, George Birch Jerrard, who was educated at Trinity College Dublin and who was well known in his day for his publications on the theory of equations.[11] It seems less likely that she confused him with his younger brother, Frederick William Hill Jerrard, who was Eighth Wrangler at Cambridge in 1833 and two months later came to the College as the first professor of mathematics and natural philosophy,[12] since the youngest Jerrard displayed no interest in mathematics after leaving the College, making his career in the church.

Bristol College

In 1829 a group of the 'more intelligent citizens' of Bristol, conscious of the lack of an institution of higher education in their city, proposed 'the erection of a college, with an efficient staff of masters and lectures, theological instruction according to the doctrines of the Church of England being also provided for such pupils as might desire to avail themselves of it'.[13] An *Outline of the Plan of Education to be pursued in the Bristol College*,[14] published by the Council of the College in 1830, provides information about the structure of the College, professors to be appointed, etc. The prospectus also makes clear that the teaching of mathematics was considered essential—something that was not the case in all schools at the time, Eton and Harrow providing conspicuous examples[15]—as it was part of a 'liberal education', the notion so strongly advocated by William Whewell.[16] And the course of study of mathematics at Trinity College, Cambridge, was to provide the model. As the *Outline* stated:

> Experience has proved, that a close application to the exact sciences is the best discipline for the mind, and the most suitable preparation for its advancement in the schools of philosophy. The Mathematics are therefore justly held to be an essential part of every liberal education.... The Mathematics will be taught in separate classes....It is intended however to adopt, with no more alteration than can be avoided, the plan at present pursued in Trinity College, Cambridge. It is expected, that the student will have been grounded in the elements of Geometry and Algebra, while in the junior classes. He will then proceed to Plane Trigonometry, to the higher parts of Algebra, and having become acquainted with the Differential and Integral Calculus, to the theory of Curves, and successively to Statics and Dynamics, Conic Sections, and the first three sections of Newton's *Principia*. Thus far he may advance in the first and second years : in the third, he will be occupied with the principles of Hydrostatics and Optics, and with the remainder of the first book of the *Principia*, as well as with Spherical Trigonometry and Physical Astronomy.[17]

When the College opened its doors on 17 January 1831 it had about thirty pupils, considerably fewer than originally envisaged. Owing to the constitution of the College with respect to religion—theological instruction was voluntary—the Bishop of Bristol had been openly hostile to the plan, which meant that the founders had been unable to raise as much capital as they had hoped. Feelings had run high with a heated discussion having taken place within the pages of *The Bristol Mercury*, and comparisons being made with 'that moral pest-house—the London University'[18] Nevertheless, despite the opposition, the College quickly achieved success with many of its pupils later attesting to its value.[19]

But despite its academic success, the College was to be short-lived. The problem once again lay in its religious tolerance. Sons of Nonconformists were allowed to attend the school without having to participate in religious instruction, and this did not sit well with a section of the local clergy. By 1840 the latter had garnered enough support, including the Bishop of Gloucester and Bristol,[20] to open a rival school. Bristol College was unable to withstand the competition and closed at Christmas 1841. As one commentator put it: 'its promoters were a generation before their contemporaries, and the institution was of too liberal a character for the age.'[21]

The first mathematics tutor of the College, who was also the Vice-Principal, was George Butterton, replacement for the original appointee Charles Smith who had been taken ill. Both Butterton and Smith were Eighth Wranglers, the former in 1827 and the latter in 1828. By March 1833, the number of students had increased to the extent that it was found necessary to establish a professorship of mathematics and natural philosophy in the senior department and it was then that Frederick William Hill Jerrard was elected to the post.

Thus by the time Stokes entered Bristol College in 1835, there were two members of staff who were high-ranking wranglers, both well placed to implement the course of Cambridge-style mathematics detailed in the prospectus, with one of them, the youngest Jerrard, having left Cambridge only two years before. There is no record of how Stokes fared under their instruction but one teacher at the College whom Stokes did recall was the writer Francis Newman, younger brother of John Henry (later Cardinal) Newman, and 'a man of charming character'.[22] He had distinguished himself at Oxford, gaining a double first in classics and mathematics in 1826, and joined the College in 1834 as classical tutor to lecture for Joseph Henry Jerrard who was unwell. However, it was his mathematics that made his mark with Stokes. Humphry recalled that Stokes considered he 'owed much' to Newman's teaching, and he 'subsequently corresponded with him on mathematical subjects when both had become famous'.[23] Although today Francis Newman is remembered as a social and religious reformer, his first publication was a pamphlet on Taylor series,[24] and he went on to write numerous books and articles on mathematics, including a book on elementary geometry[25] which was published in 1841, the year after he left the College.[26]

Stokes's ability for mathematics was readily apparent to those around him at the College. According to Lord Rayleigh, there was 'a tradition that he did many propositions of Euclid, as problems, without looking at the book'.[27] At the end of his first year Stokes was awarded a 'Grand Prize' which included a prize in mathematics and a certificate of honour in English prose composition.[28] At the end of his second and final year, in which he was examined in, elementary dynamics, mechanics, the differential and integral calculus, and trigonometry, with oral testing on the last three, he did not win the first mathematics prize but was awarded a prize for 'Extraordinary Progress in Mathematics', which possibly indicates that when he arrived at the College he was not as well prepared in mathematics as some of his fellow students.[29]

His final examinations over, the question now was where should Stokes go to next. His father and all three brothers had studied at Trinity College Dublin (TCD), but they had not been to a school which explicitly prepared students for Oxford or Cambridge. In fact Stokes's brother William had entered TCD at the age of 17 before going up to Cambridge five years later. But such a pattern of study was not unusual. The level of mathematics taught at TCD and other universities, such as Edinburgh University, Glasgow University, or University College London, was much lower than at Cambridge, and in the nineteenth century several mathematics students at Cambridge, including high-ranking wranglers such as William Thomson and James Clerk Maxwell, had already studied for a degree elsewhere.

On the final day of Stokes's examination, Joseph Jerrard wrote to Stokes: 'I have strongly advised your brother [William] to enter you at Trinity, as I feel convinced that you will in all human probability succeed in obtaining a Fellowship at that College'.[30] William, who in 1836 had become Dean of Caius College, counselled Stokes to take care when choosing his college.[31] However, when Stokes went up to Cambridge he did not go to Trinity but to Pembroke.

Why Stokes went to Pembroke and not to Trinity as advised by Jerrard is not known. Stokes himself seems never to have given a reason. In 1901, when in his early eighties he was asked to recall his student days, he gave no clue, only writing: 'I entered Pembroke College, Cambridge in 1837. In those days boys coming to the University had not in general read so far in mathematics as is the custom at present; and I had not begun the differential calculus when I entered the College, and had only recently read analytical sections.'[32] It appears, however, that Stokes's memory must have been playing tricks on him since, as noted above, the subjects in his final examination at Bristol College included the differential and integral calculus.

Cambridge

At the beginning of the nineteenth century, the normal route to a Bachelor of Arts degree at Cambridge was through the Senate-House Examination, popularly known as the Tripos. The examination was primarily in mathematics but included other subjects, such as logic, philosophy and theology. It began to be referred to as the *Mathematical* Tripos only in 1824, when the Classical Tripos was examined for the first time, although students could enter the Classical Tripos only if they had already obtained honours in the Mathematical Tripos—a situation which pertained until 1850. As the century progressed the examination took on an ever-increasing significance. There was a shift from oral to written examinations, with success in the final examination being paramount, and a concomitant rise in private tutoring without which such success was virtually impossible. A high place in the order of merit garnered national recognition and was a passport to the career of one's choice.

Mathematics was the core of study at Cambridge not because it was preparation for a career as a mathematician but because it provided a fundamental part of a liberal education. The reason for studying Euclid's *Elements* was not simply to learn geometry. It was a training of the mind. That being said, knowledge of Euclid provided (at least some) access to the single most important text a Cambridge mathematics student had to study: Isaac Newton's notoriously difficult *Principia*. Written primarily in the language of geometry, the *Principia* provided the most certain demonstration of man's knowledge of the natural world. It is no wonder that it played a dominating role in the university's undergraduate course.

Undergraduates generally began their studies under the direction of a college lecturer whose duties were to guide the reading of the students and to prepare them for the rigours of the college and Senate-House examinations. When Stokes arrived at Pembroke, the college mathematical lecturer was John Mills (Fifth Wrangler in 1831) so presumably it was Mills who acted as Stokes's first point of call. Little is known about the particular nature of Stokes's studies in his first year except that he came second in the college examinations, pipped by a certain John Sykes. Sykes's father came from Cambridge so perhaps Sykes had been even better prepared in Cambridge-style mathematics than Stokes. The following year Stokes came out top with Sykes coming only third.[33] In his second and third years, Stokes was privately tutored by the famous mathematical coach[34] William Hopkins (of whom more below). Sykes it seems was not tutored by Hopkins although he would have certainly received private tuition.[35]

As well as the tuition provided by the college, there were also lectures delivered by the professors. Not all students attended the lectures of the mathematics professors, and not all of the mathematics professors lectured. While Stokes was an undergraduate neither of the Lucasian professors—Charles Babbage, who held the chair from 1828 to 1839, and Joshua King, who held the chair from 1839 to 1849 (when he was succeeded by Stokes)—lectured, although both of them examined for the Smith's Prize,[36] King having previously examined several times for the Tripos. George Peacock, Lowndean professor of astronomy and geometry, advertised lectures on 'Science of astronomy and practical methods of observation; use of Instruments. Geometry, and general principles of Mathematical Reasoning' and it is likely that Stokes attended them.[37]

Stokes certainly attended the lectures of the Plumian professor of astronomy and experimental philosophy, James Challis, today best remembered for failing to identify the planet Neptune. Challis offered 'Laws of Hydrodynamics, Pneumatics, and Optics with special reference to the Mathematical Theories of Light and Sound. Explanations exhibited experimentally; explanations given of Principles employed in Mathematical Reasoning'.[38]

As well as Euclid's *Elements* and Newton's *Principia*, there were several mathematical textbooks which Cambridge students were expected to study, many of which had been written by former Cambridge wranglers and were designed specifically for students of the university. Amongst the most prolific and influential of writers who produced books in this category was John Hymers (Second Wrangler in 1826), a Fellow of St John's, who successfully combined his college career with private tutoring, and who examined for the Tripos in 1833 and 1834. With a reputation for being 'profoundly versed in mathematics',[39] Hymers had 'a vast acquaintance with the mathematics of the Continent'[40] which was evident in many of his books. For example, the second edition of his *Integral Equations* (1835) introduced English students to the newly discovered topic of elliptic functions, while his *Treatise on Conic Sections and the Application of Algebra to Geometry* (1837) became the standard textbook on analytic geometry.

The 1816 translation of Lacroix's introductory textbook on the differential and integral calculus by Babbage, Herschel, and Peacock, which was an important stimulus for the introduction of analytical methods into Cambridge, was followed by a number of new books that treated their subjects from an analytical perspective. Among these were the textbooks of William Whewell on mechanics (1819) and dynamics (1823). Whewell's *Treatise on Dynamics* is also notable for being one of the first books to treat the subject from a deliberately pedagogical perspective, as well as for promoting a problem-solving approach.[41]

George Biddell Airy's *Mathematical Tracts* was another staple of an undergraduate's diet. Originally published in 1826 while Airy was Lucasian professor, the second edition of 1831, which would have been studied by Stokes, included a new section on the wave theory of light. It provided an analytical approach to problems of physical astronomy, the shape of the Earth, and to its precession and nutation. And Airy did not shy from explaining why he considered other texts unsuitable: the standard text on Lagrange's calculus of variations by Lacroix was 'singularly confused and unintelligible' (a judgement with which later authors concurred). A more advanced text espousing a similar analytical approach which appeared a few years later was John Pratt's *The Mathematical Principles of Mechanical Philosophy and their Application to the Theory of Universal Gravitation* (1836). Pratt (Third Wrangler in 1833), a former student of Hopkins, provided a particularly clear account of the shape of the earth. Another book Stokes could have read is Mary Somerville's *Mechanism of the Heavens* (1831), Somerville's interpretation of Laplace's *Mécanique Céleste*. It was promoted by both Whewell and Peacock, the latter writing to the author in 1832 that he 'had little doubt that it will immediately become an essential work to those of our students who aspire to the highest places in our examinations'.[42]

William Hopkins—'The Senior-Wrangler Maker'

Stokes's coach, William Hopkins, went up to Cambridge in 1822 as an undergraduate at the age of twenty-nine, a change in family circumstances allowing him to abandon his inherited occupation of gentleman farmer. Graduating as Seventh Wrangler in 1827—a strong year in which Augustus De Morgan (1806–71) came fourth—he would have been a candidate for a college fellowship but as a married man he was ineligible. The same year he was appointed to a lecture-ship in Peterhouse and Esquire Bedell to the University (a partly administrative and partly ceremonial post). In 1839 he was one of only two candidates for the Lucasian professorship, the chair having become vacant on the resignation of Charles Babbage. But it was a contest he was doomed to lose. The other candidate was Joshua King, Senior Wrangler in 1819, President of Queens' College and one of the eight electors. King won the vote by a majority of seven to one.

Hopkins was the first of the Cambridge coaches to make a permanent living from private tutoring. He rapidly developed a reputation as an outstanding teacher and his results were remarkable. Between 1828 and 1849 he 'personally trained almost 50% of the top ten wranglers, 67% of the top three, and 77% of senior wranglers', which amounted to 108 in the top ten, 44 in the first three, and 17 senior wranglers, and earned him the sobriquet 'senior wrangler maker'.[43]

As Hopkins' reputation grew, he was able to pick and choose his students. By taking students in their second year, or occasionally their third, he had time to assess their abilities and select the most promising before taking them into his tutelage. Stokes was typical in this respect. He began studying with Hopkins in his fourth term and stayed with him until his final examinations. Hopkins taught in small classes, putting students of equal ability together. As Warwick has described, 'this meant that the class could move ahead at the fastest possible pace, the students learning from and competing against each other'.[44] Or as one of his obituarists wrote:

> The secret of his success as a teacher was the happy faculty he had of drawing out the thoughts of his pupils and make them instruct each other, while he took care that the subjects under discussion were treated in a philosophical manner so that mere preparation for the senate-house examination was subordinate to sound scientific training.[45]

A first-hand account of how Hopkins ran his classes was given by William Thomson (later Lord Kelvin). Thomson, who started with Hopkins in a class of about five students in the year after Stokes graduated, wrote to his father James Thomson (professor of mathematics at the University of Glasgow):

> What we have had already approximates very much to the plan which you pursue with your class. He [Hopkins] asked us all questions on various points in the differential calculus, in the order of his manuscript, which he has given us to transcribe, and gave us exercises on the different subjects discussed, which we are to bring with us tomorrow. He says he can never be quite satisfied that a man has got correct ideas on any mathematical subject till he has questioned him viva voce. I can judge very little yet of the other men whom I meet with him, but I hope they are not extremely formidable.[46]

Although the number in the class was relatively small, the competitive nature of the Senate-House Examination was clearly in evidence.

Of the atmosphere of the class, Francis Galton, a student with Hopkins from 1841, painted a rather convivial picture:

> 'Hopkins to use a Cantab expression is a regular brick; tells funny stories connected with different problems and is no way Donnish; he rattles us on at a splendid pace and makes mathematics anything but a dry subject by entering thoroughly into its metaphysics. I never enjoyed anything so much before.... [He charges] only £72 per annum instead of £100 as currently reported: this will make a jolly difference to my finances.[47]

But this letter was written when Galton was in only his first year of study with Hopkins and as time progressed the unremitting strain of competition began to tell on him. In the end his health broke down completely and he never graduated.[48] The brilliant Robert Leslie Ellis, Senior Wrangler the year before Stokes, unusually studied with Hopkins only in his final year and only so that 'his reading should be arranged and put in a form suitable for the Cambridge examinations'. Although Ellis detested the system which he described as 'the crushing down of the mind and body for a worthless end',[49] he knew that if he was to have any chance of being a high wrangler he had to go to Hopkins.

In contrast to Ellis, the system appears to have suited Stokes rather well. He kept the notes he made while reading with Hopkins, although they are rather difficult to read.[50] There are sets of notes for several different courses including differential and integral calculus, mechanics, dynamics, optics, hydrostatics, hydrodynamics, sound and light, and calculus of variations.[51] From these it is possible to get a good idea of Hopkins' style of teaching. The standard theory is given, examples are worked through and others are left for the student to complete. Hopkins was also particularly well placed to teach on topics covered in Airy's *Tracts* since he would have attended Airy's lectures while he was an undergraduate.

None of Hopkins' own notes appear to have survived although he did publish a textbook, *Elements of Trigonometry* (1833). A feature of this text is his use of history of mathematics both to elucidate and to generate interest in the mathematics discussed. (Another textbook, on the differential calculus, was promised but never materialized.[52]) An insight into Hopkins' views on mathematics and what informed his teaching can be gleaned from a pamphlet he wrote in 1841 in response to proposals to exclude from the Tripos applications of mathematics to astronomy, optics, and hydrodynamics, which would have rendered the retention of partial differential equations and higher geometry pointless:

> It is only when the student approaches the great theories, as Physical Astronomy and Physical Optics, that he can fully appreciate the real importance and value of pure mathematical science, as the only instrument of investigation by which man could possibly have attained a knowledge of so much of what is perfect and beautiful in the structure of the material universe, and in the laws which govern it. It is then that he can form an adequate conception of the genius that has been developed in the framing of those theories, and can feel himself under those salutary influences which must be ever exercised on the mind of youth by the contemplation of the workings of lofty genius, in whatever department of science or literature it may have been called into action.[53]

Thus for Hopkins mathematics was important because it was the means by which the secrets of the universe could be revealed. In short, Hopkins 'regarded the Newtonian system of the world and the wave theory of light as the crowning achievements of the mathematical investigation of nature'.[54] The proposals would also have meant that there was little point in retaining partial differential equations and higher geometry. Little wonder Hopkins was opposed.

Hopkins' pamphlet also provided a response to a recent attack by George Peacock on private tutoring. Peacock had made no bones about his views:

> The rapid growth of the private tuition in late years, which is due to various causes, is an evil of the most alarming magnitude, not merely as a great and ruinous increase of the expenses of academical education, but as threatening to supersede the system of public instruction, both in the colleges and in the university. [55]

He considered even the best coaching to be a system 'of forced culture' which although it 'may accelerate the maturity of the fruit' was 'inconsistent with the healthy and permanent product-iveness of the tree'. His solution was to ban private tutoring altogether, although he did concede it would cause financial hardship to some talented men, and, without explicitly mentioning him by name, he did acknowledge Hopkins' 'great skill and pre-eminent success as an instructor of youth'.[56]

Since a student aspiring to high honours had to engage a private tutor, an important factor in the debate, aside from the threat to the university, was the cost of private tutoring. In 1839 Ellis paid Hopkins £42 for a year's worth of coaching, a fee which indicates that he was being coached six days a week.[57] The fact that Galton was charged £72 per annum indicates that he was having additional coaching in the long vacation.[58] In 1852 Hopkins himself estimated that to a good student the cost of private tutoring for three years was approximately £150. [59,60] Stokes was fortunate that Hopkins' fees were not an issue.

Hopkins had a lasting influence on Stokes, and not only a pedagogical one. While Stokes was still an undergraduate, it was Hopkins who advised him to study hydrodynamics,[61] the subject in which he began his research.

Examinations

During the early decades of the nineteenth century, the Senate-House Examination underwent a number of reforms. At the beginning of the century it lasted for three days, but it gradually got extended so that by the time Stokes came to sit it in 1841 it was six days, having been extended from five in 1839, with the papers becoming progressively more difficult. The ques-tions were of two types: bookwork and problems. The former required students to reproduce standard definitions, theorems, and proofs, while the latter tested students' ability to apply what they had learnt to increasingly technical and challenging problems. These were not prob-lems to be found in the back of textbooks but problems constructed specifically for the exam-ination, and it was not unusual for the examiners to base questions on their own research.[62] Importantly, it was the problems that effectively determined the order of merit.

There were two examinations every day apart from Sunday, two-and-a-half hours in the morning and three hours in the afternoon, making a total of thirty-three hours examination altogether. For the first two papers, students were not allowed to use the differential calculus. Every undergraduate had to take the first four papers and a failure to pass resulted in the stu-dent being 'plucked'; i.e. not allowed to continue his studies.[63]

The papers were set by two Moderators and two Examiners. They undertook essentially the same tasks, the only difference being that the Moderators were responsible for the 'papers of original problems', i.e. for the more difficult ones.[64] Usually at least one of the Moderators con-tinued as either a Moderator or an Examiner in the following year, thereby ensuring continuity. In Stokes's year, the Moderators were Alexander Thurtell (Fourth Wrangler in 1829) and Edwin Steventon (Third Wrangler in 1830), and the Examiners were Henry Wilkinson Cookson (Seventh Wrangler in 1832) and Edward Brumell (Third Wrangler in 1837), none of whom made a distinguished career in mathematics. Stokes himself would be a Moderator in 1846–8 and an Examiner in 1849.

The preparation for the examination was a punishing experience and it is little wonder that the health of students was sometimes compromised. The American Charles Bristed,[65] who studied for the Tripos between 1841 and 1844 and wrote a book chronicling his experience, declared:

> Indeed a man must be healthy as well as strong—"in condition" altogether to stand the work. For in the eight hours a-day which form the ordinary amount of a reading man's study, he gets through as much work as a German does in twelve; and nothing that our students go through can compare with the fatigue of a Cambridge examination. If a man's health is seriously affected, he gives up honors at once, unless he be a genius like my friend E[llis], who "can't help being first."[66]

Ellis suffered from poor health throughout his time at Cambridge and indeed for most of his life. Stokes was more robust. He told his daughter that 'he never read more than eight hours a day, even before an examination' and that 'he had never been reduced to binding his head up with a wet towel'.[67]

The examination produced its own casualties. In 1842 C. T. Simpson, who was Second Wrangler to Cayley, 'almost broke down from over exertion…and found himself actually obliged to carry a supply of ether and other stimulants into the examination in case of accidents'.[68] While some students were unable to stay the course altogether:

> A singular case of *funk* occurred at this examination [1843]. The man who would have been second (also a Johnian) took fright when four of the six days were over, and fairly ran away— not only from the examination, but out of Cambridge, and was not discovered by his friends or family till some time after.[69]

The 'man' in question was Thomas Minchin Goodeve who had been expected to be second but had ended up as ninth, his days of absence from the Tripos thus proving not too calamitous.[70] Indeed, as Bristed himself observed, the papers of the last two days affected the places of only the best ten or fifteen students.

When it came to Stokes's turn, he had to tackle 175 questions over the course of the twelve papers.[71] There were the standard questions which required the reproduction and proof of propositions from Euclid's *Elements* and from Newton's *Principia*. For example, one of the questions required the statement and proof of 'the tenth lemma of Newton's first section'.

The lemma states:

> *The spaces which a body describes when urged by any finite force, whether that force is determinate and immutable, or is continually increased or continually decreased, are at the very beginning of the motion in the squared ratio of the times.*

In other words, the position varies as the square of the time, providing the initial acceleration is non-zero.[72]

Students were thus not only expected to be able to prove lemmas from the *Principia*, they were also expected to remember how they were numbered! The majority of the other questions were on algebra, the calculus, mechanics, dynamics, astronomy, hydrostatics, and optics, with only a few on heat, electricity, and magnetism.

The results were announced in the Senate-House on 22 January, ten days after the final examination. The following day the degrees were conferred by the Vice-Chancellor at a ceremony in the Senate House (Fig. 3.2). Although no description exists of Stokes's ceremony, Ellis recorded his experience in the previous year:

PRESENTATION OF THE SENIOR WRANGLER TO THE VICE CHANCELLOR.

Cambridge, January 1842

Fig. 3.2 Arthur Cayley, Senior Wrangler 1842, being presented to the Vice-Chancellor, William Whewell, at the Senate House to have his degree conferred. Andrew Warwick (*Masters of Theory*, p. 207) considers it probable that the figure on the left with his right hand on the table is William Hopkins. From V. A. Huber, *The English Universities*, tr. F. Newman, vol. 1 (London: William Pickering, 1843), frontispiece.

> When all was ready [William Hopkins] and the other Esquire Bedell made a line with their maces, and Burcham led me up [the Senate House]. Instantly my good friends of Trinity and elsewhere, two or three hundred men, began cheering most vehemently, and I reached the Vice [Chancellor's] chair surrounded by waving handkerchiefs and most head rending shouts. Burcham nervous. I felt his hand tremble as he pronounced the customary words *"vobis presento hunc juvenem."* Then I took the oaths of allegiance and supremacy and I knelt before the Vice [Chancellor] who pattered over the *"Auctoritate mihi & c"* and then shaking hands wished me joy. I walked slowly and stiffly down the Senate House – more cheering.[73]

By the end of the proceedings, Ellis had turned so pale that he was made to sit down and a young woman from the crowd offered him some smelling salts! Stokes, one assumes, was made of sterner stuff.

As was common practice, the results of the examination were widely reported, both in the national press—*The Times* of London regularly printed a full list of the successful students—and in the local press in which often a 'local hero' was celebrated. It so happened that in 1841 the results caused a particular stir due to the high number of failures among classics students. The *Hull Packet* took up the story:

> The number of names on the printed list, published previously to the examination, contains 145 names; that on the return list is 117. It must not, however, be taken for granted, that all whose names were on the first list went into the Senate House, and that, consequently, 28 had

the misfortune of being 'plucked'. I believe that 25 is the exact number of these unhappy ones,—amongst them are some of the best classics in the University, who are debarred from the privilege of going in for the classical tripos and the medal. For the latter honour, even a Junior Optime cannot go in. Trinity [College] is like a little town in a roar this morning, on the disagreeable subject of a pair of their best classical scholars (of the house too) being plucked for *plus* and *minus*.[74]

The failures were the subject of correspondence in *The Times*[75] and Bristed too remarked on them, noting that 'the mathematical examination was very difficult and made great havoc amongst the classics'.[76] Later that same year, whether or not prompted by the 'havoc', Peacock voiced his concerns:

> The problems which are proposed in the senate-house are very generally of too high an order of difficulty, and are not such as naturally present themselves as direct exemplifications of principles and methods and require for their solution a peculiar tact and skill, which the best instructed and most accomplished student will not always be able to bring to bear upon them. It is not unusual to see a paper of questions proposed for solution in the space of three hours, which the best mathematician in Europe would hesitate to complete in a day.[77]

Given that the statutes explicitly stated that the papers should not contain more questions than well-prepared students could answer within the time allowed, the difficulty of the papers was clearly a legitimate cause for concern.[78]

Putting Stokes's triumph in the context of the uproar over the stiffness of the examination gives Romilly's remark about Stokes being 'a very good' senior wrangler extra resonance, especially since Romilly was a Trinity man. But Stokes's trials were not over with his victory in the Senate-House Examination.

Shortly after the results had been declared, the top wranglers knuckled down again to compete for the Smith's Prize.[79] This took the form of further examination papers, each one of which was sat over the course of a day and was set by a different examiner. Unlike the Tripos, the questions were usually geared towards evincing an original or creative approach.[80] In general only the most distinguished wranglers sat the examination, so the numbers entering were usually small, and it was not unknown for the number of candidates to be the same as the number of prizes.

In 1841 the Smith's Prize examiners were the three mathematics professors: Peacock, Challis, and King. Each paper consisted of around twenty-five questions ranging over a variety of subjects from pure mathematics to the construction of astronomical instruments, and often included a discursive element. For example, one question on Peacock's paper asked for a solution to the functional equation:

$$\left(\varphi x\right)^2 = \varphi\left(2x\right)+2,$$

another asked for an explanation of the construction of the 'Huyghenian eye-piece', while Peacock's final question asked for short dissertations on half a dozen different topics including the theory of parallels and the theory of the rainbow. Detailed knowledge of classic texts was assumed. One of King's questions gave a description of a problem in Newton's *Principia* and then referred to different solutions by Newton in different editions of the *Principia* and to three further solutions by Lagrange in his *Calcul des Fonctions*, none of which were given but all of which were expected to be known. Perhaps predictably, Challis's paper focused more on applications and physical astronomy, including several questions on subjects such as the wave theory of light and the lunar theory.

Once again, Stokes won the day.[81] Although the prize was worth £25, its real value was in the academic prestige attached to winning. The competition was a much sterner test than the Tripos and although to the outside world a prizeman did not carry the cachet of a senior wrangler, within the confines of the Cambridge mathematical community the honour was recognized as the ultimate achievement.[82]

On 23 January, with all the results announced, Pembroke celebrated Stokes's remarkable success by holding a grand dinner in his honour and making him a Fellow of the College. The world was now his oyster. As one relative wrote to him,[83] he now had only to determine whether he would be 'Prime Minister of England, the Lord Chancellor or Archbishop of Canterbury'.

Appendix: The plan of the Senate-House Examination together with the first and the last question papers

The plan is taken from the Cambridge University Calendar for 1841 (Cambridge: Benjamin Flower), p. 16 while the two Tripos papers are taken from the *Cambridge University Magazine*, vol. II pp. 191–2, 206.

PLAN OF EXAMINATION, January 1841.

Day	Time	Subject	Examiners
Wednesday	9 to 11½	Pure Mathematics	Jun. Moderator and Sen. Examiner.
	1 to 4	Natural Philosophy	Sen. Moderator and Jun. Examiner.
Thursday	9 to 11½	Natural Philosophy	Jun. Moderator and Sen. Examiner.
	1 to 4	Problems	Sen. Moderator.
Friday	9 to 11½	Pure Mathematics	Sen. Moderator and Jun. Examiner.
	1 to 4	Problems	Jun. Moderator.
Saturday	9 to 11½	Problems	Sen. and Jun. Moderators.
	1 to 4	Pure Math. and Nat. Phil.	Sen. and Jun. Examiners.
Monday	9 to 11½	Pure Math. and Nat. Phil.	Jun. Moderator and Jun. Examiner.
	1 to 4	Pure Math. and Nat. Phil.	Sen. Moderator and Sen. Examiner.
Tuesday	9 to 11½	Pure Math. and Nat. Phil.	Sen. Moderator and Jun. Examiner.
	1 to 4	Pure Math. and Nat. Phil.	Jun. Moderator and Sen. Examiner.

SENATE-HOUSE EXAMINATION.

WEDNESDAY, *Jan.* 6, 1841......9 to 11½.

[N.B. *The Differential Calculus is not to be employed.*]

1. THE angles which one straight line makes with another upon the one side of it are either two right angles, or are together equal to two right angles.

2. Describe an isosceles triangle, having each of the angles at the base double of the third angle. Shew that if the points of intersection of the circles, in Euclid's figure, be joined with the vertex of the triangle and with each other, another triangle will be formed similar and equal to the former.

3. State the difference between interest and discount, and find the discount on £397. 6s. 8d. due 9 months hence, at 4 per cent. per ann.

4. Find the number of feet, inches and parts, in the side of a square whose area is 14 feet 11 inches.

5. Define the least common multiple of two quantities, and prove that it measures every other common multiple of them. Find the least common multiple of $3x^2 - 5x + 2$ and $4x^3 - 4x^2 - x + 1$.

6. Solve the following equations:

$$(1) \quad \frac{1}{x-2} - \frac{2}{x+2} = \frac{3}{5}.$$

$$(2) \quad \left.\begin{array}{c} x - y = 12 \\ x^2 + y^2 = 74 \end{array}\right\},$$

and shew that a quadratic equation cannot have more than two roots.

7. Shew that if any one term of the series for $(1 + x)^{\pm n}$, where x is a proper fraction, be numerically less than the one which precedes it, the same is true of all the following terms.

Find the coefficient of x^r in the expansion of $\left(x - \dfrac{1}{x}\right)^{2n-1}$.

8. Prove that $\tan A = \pm \sqrt{\dfrac{1 - \cos 2A}{1 + \cos 2A}}$, and shew which of the two signs ought to be used in particular cases.

9. Explain the use of subsidiary angles in adapting algebraical formulæ to numerical calculation.

Ex. (1) $a \sin x + b \cos x = c$,   (2) $x = \sqrt{a - b} + \sqrt{a + b}$.

10. Find the equation to a straight line passing through a given point in the axis of x, and making an angle of $45°$ with the straight line whose equation is $\dfrac{x}{a} - \dfrac{y}{b} = 1$.

11. In the parabola shew that $SY^2 = AS \cdot SP$, and prove that there is only one point where the focal distance is perpendicular to the tangent.

12. Find the equation to the ellipse, referred to the centre as origin of co-ordinates, and a system of conjugate diameters as axes. Find

the angle between the axes when the equation is reduced to the form

$$x^2 + y^2 = c^2.$$

13. Shew how to transform an equation into one which shall want its second term, Take away the second term from the equation

$$x^3 - 6x^2 + 12x + 19 = 0,$$

and find the three roots of the resulting equation.

14. Prove that in a spherical triangle,
    (1) Any one side is greater than the difference between the two others.
    (2) The sum of any two angles is greater or less than 180° according as the sum of the two opposite sides is greater or less than 180°.

15. Find an expression for the side of a spherical triangle in terms of another side and the angles which are adjacent to that side.

16. How are different systems of logarithms distinguished? Having given the logarithm of a number in one system, shew how to find it in another. Assuming the series for $\log (1 + x)$ prove that

$$\log_e (x + 2h) = 2 \log_e (x + h) - \log_e x$$
$$- \left\{ \frac{h^2}{(x+h)^2} + \tfrac{1}{2} \frac{h^4}{(x+h)^4} + \tfrac{1}{3} \frac{h^6}{(x+h)^6} + \&c. \right\}$$

17. Prove that $\dfrac{\sin \theta}{\theta} = \left(1 - \dfrac{\theta^2}{\pi^2}\right) \left(1 - \dfrac{\theta^2}{2^2\pi^2}\right) \left(1 - \dfrac{\theta^2}{3^2\pi^2}\right)$ &c.

Explain fully the advantages of a set of tables which give the numerical values of $\log_{10} \dfrac{\sin \theta}{\theta}$, when θ is small.

TUESDAY, *Jan.* 12, 1841......1 to 4.

1. EXPLAIN how Newton's method of approximation may be adjusted so as to furnish the real roots of an equation to any required degree of accuracy.

2. Explain how the position of the invariable plane of a system of material particles is determined. Extend the method to the case of a system of bodies of finite magnitude. Upon what suppositions may the position of the invariable plane of the solar system be found, and how far are they correct?

3. A given number of concentric spherical shells, the thickness of each being uniform, are separated by non-conducting media; find the effect of their mutual influence when they are electrized.

4. Explain the nature of developable and twisted surfaces, and shew that the hyperboloid of one sheet is a twisted surface.

5. State the results which have been obtained in the theoretical investigation of the earth's figure independently of any assumed law of the density, and calculate the length of a degree of latitude at any place on the earth's surface in terms of the length of a degree at the equator.

What reason is there for supposing that the variations of pressure in the interior of the earth are proportional to the squares of the densities?

6. Find the variation of $\int_a F(x, y, z, \frac{dy}{dx}, \frac{dz}{dx}, \&c.)$ subject to the condition that $f(x, y, z, \frac{dy}{dx}, \frac{dz}{dx}, \&c.) = 0$, and apply the result to determine the curve of quickest descent in a medium resisting as any function of the velocity.

7. Explain the effect of eccentrical refraction through a lens in producing distortion. A ray of light is incident upon a glass lens in a direction parallel to the axis; find the tangent of the angle which the emergent ray makes with the axis, to a second approximation. Find also the ratio between the radii when the coefficient of the small term, for a given focal length, is a minimum.

8. Assuming the properties of the axes of elasticity, find the velocity of transmission of a wave of light after refraction at the surface of a uniaxal crystal. If the crystal be bounded by planes perpendicular to the axis investigate the difference of retardations of the ordinary and extraordinary rays. State briefly how it is shewn by experiment that plane and circularly polarized light differ from each other.

9. Enunciate the propositions by which it is proved (1) that the eccentricities of the planetary orbits, (2) that their inclinations to the ecliptic, are always small. How are they affected by the circumstance that all the planets revolve in the same direction about the sun? Shew that the variations of the inclination and longitude of the node are given by the equations

$$\frac{d(\tan i_1)}{dt} = \frac{n_1 a_1}{\mu \tan i_1 \sqrt{(1-e_1^2)}} \frac{dR}{d\Omega_1},$$

$$\frac{d\Omega_1}{dt} = -\frac{n_1 a_1}{\mu \tan i_1 \sqrt{(1-e_1^2)}} \frac{dR}{di_1}.$$

CHAPTER 4

Stokes's Optics 1

Waves in Luminiferous Media

OLIVIER DARRIGOL

That George Gabriel Stokes amply contributed to nineteenth-century optics was the common opinion of his contemporaries, and it is the reason why his achievements in this domain are an inevitable component of any memorial volume. The way in which he helped to shape the optics of his time is less obvious. Was he just consolidating the received wave-optics through sharper mathematics and finer experiments? Did he dare significantly depart from Augustin Fresnel's impressive system? Did he extend the object of physical optics? The purpose of this study is to show that Stokes, owing to an unusual mix of mathematical acumen, experimental adroitness, and scientific imagination, worked in these three seemingly incompatible modes, with a relative intensity depending on age and circumstances.

When, in 1837, the eighteen-year-old Stokes entered Pembroke College in Cambridge, the contents and methods of physics were rapidly evolving in the domains of electricity, magnetism, heat and light. A few years after Thomas Young's groundbreaking work on interference phenomena early in the century, the wave theory of light had displaced Isaac Newton's long dominant corpuscular theory. Most decisive in this transition were the precise and powerful laws established by Fresnel between 1815 and 1825 in opposition to the received Laplacian–Newtonian optics. Combining precise experiments, well-chosen dynamical principles and some intuition of the structure of the ether, Fresnel produced an accurate theory of diffraction based on the Huygens–Fresnel principle, a theory of the optics of moving bodies based on the partial dragging of the ether by matter, a new theory of the combined effects of polarization and interference based on the concept of transverse vibration, formulas for the relative intensities of reflected and refracted rays at the interface between two transparent media, and a mathematically sophisticated theory of the propagation of light in anisotropic media. Amazingly, his quantitative laws and formulas do not differ from the ones we would now derive from the electromagnetic theory of light.[1]

Although Fresnel's contemporaries did not benefit from this hindsight, they recognized the power of his theory and the accuracy of its predictions for phenomena that eluded the corpuscular theory. From the mid-1820s onward, the chief British authorities on optics, namely, the

Darrigol, O., *Stokes Optics 1: Waves in Luminiferous Media*. In: *George Gabriel Stokes: Life, Science and Faith*, Mark McCartney, Andrew Whitaker, and Alastair Wood (Eds): Oxford University Press (2019). © Oxford University Press. DOI: 10.1093/oso/9780198822868.001.0004 astronomers John Herschel and George Biddell Airy, adopted and propagated Fresnel's views. Airy applied them to the diffraction of object-glasses and to the explanation of a few intriguing interference phenomena, for instance supernumerary rainbows. He also gave a thorough, highly competent account of Fresnel's theory in the 1831 edition of the *Mathematical Tracts* he wrote for Cambridge students, a source frequently used by Stokes in his optical researches. In Dublin in 1832, William Rowan Hamilton predicted 'conical refraction' at a peculiar incidence of a ray of light on a crystal as a consequence of cusp-shaped singularities on Fresnel's wave surface, a prediction soon confirmed by his colleague Humphrey Lloyd. At the time of Stokes's studies at Cambridge, the Plumian professor James Challis was lecturing on the wave theory of optics. The influential coach William Hopkins ranked Fresnel's theory as the highest achievement of mathematical physics of his time, on a par with Newton's celestial mechanics, and he schooled his students with problems in this new optics.[2]

By the time Stokes began his graduate studies, no mathematically proficient expert on optics could be found in Britain to still defend Newton's theory. This does not mean that British optics was confined to the blind application of Fresnel's theory. There was at least one important player in this field, the immensely productive Scottish experimentalist David Brewster, who still supported the corpuscular concept of light and repeatedly challenged theorists with new observations that seemed to elude the wave theory. On the theoretical side, mathematicians like James MacCullagh and George Green recognized the fragility of Fresnel's concept of the ether and tried to replace it with a dynamical model compatible with Fresnel's best-verified laws. Lastly, there were numerous studies in domains of optics that did not depend on the deeper nature of light: geometrical optics, photometry, and the phenomenology of colour-dependent phenomena including dispersion, emission and absorption spectra, the colours of bodies, and photochemical processes. Some of the most important developments in British nineteenth-century optics belonged to this category, which extended Newton's phenomenological optics.[3]

Stokes's first publications in 1842–3 concerned fluid mechanics, in a highly mathematical vein. He was following a suggestion by his tutor Hopkins, who was using hydrodynamics to train Cambridge students in the methods of the newer French mathematical physics.[4] In these early works, Stokes was not gratuitously displaying his mathematical virtuosity. He meant to prepare his attack on concrete problems such as the damping of pendulum oscillations; and he was prompt to detect possible applications to optics. In 1843, while studying the motion of rigid bodies in a perfect liquid he obtained the formula

$$M = M_x \cos^2 \alpha + M_y \cos^2 \beta + M_z \cos^2 \gamma$$

for the mass to be added to an ellipsoid oscillating in a liquid in the direction $(\cos\alpha, \cos\beta, \cos\gamma)$ referred to the axes of the ellipsoid. This reminded him of Fresnel's formula

$$V^2 = a^2 \cos^2 \alpha + b^2 \cos^2 \beta + c^2 \cos^2 \gamma$$

for the propagation velocity V of a plane transverse wave whose vibration occurs in the direction $(\cos\alpha, \cos\beta, \cos\gamma)$ with respect to the axes of the crystal. As we will see in a moment, this analogy led him to an alternative to Fresnel's theory for the propagation of light in anisotropic media.[5]

This was Stokes's first, unpublished venture in optical theory. In 1845, fluid mechanics and his interest in the damping of pendulum oscillations led him to the theory of viscous fluids, which used to be performed in analogy with the theory of elasticity. In this occasion he reinvented the Navier–Stokes equation and he also discussed media, such as jelly, which behaved like liquids at a large scale of motion and as solids at a small scale of motion. The optical ether, he then submitted, could be a medium of this kind since it had to be rigid to allow for the transverse optical waves and fluid to let the earth and celestial bodies move freely through it. This consideration led him to a new theory of aberration and to one of his numerous polemics with his former professor and bête noire James Challis.

These early trials at the interface between hydrodynamics and optics are the objects of the first section of this chapter. The four following sections cover Stokes's contributions to purer, Fresnel-based, and less speculative optics, in the highly productive years 1847 to 1851. The second section deals with puzzling cases of interference, and the next two with the dynamical theory of diffraction and the 'Stokes parameters' for the most general state of polarization of a beam of light. Stokes's unpublished speculations of 1843, 1848, and 1856 on double refraction and optical rotation are discussed in the final section, together with his lucid report of 1862 on the received theories of double refraction.

In 1851 his research focus had shifted to the frequency-altering alteration of light, which he called fluorescence and discovered by attending to the strange superficial colour of quinine-based solutions reported by John Herschel. In the eyes of Stokes and his contemporaries, fluorescence opened a new era in optics by breaking the old Newtonian dogma of the immutability of simple colours and also by easing the exploration of the ultraviolet spectrum. The first section of Chapter 5 is devoted to this most important contribution and to Stokes's and Thomson's controversial anticipation of solar spectroscopy. The memoirs on fluorescence, published in 1852–3, were Stokes's last great memoirs in optics. In the rest of his life he published shorter articles in which he elucidated various optical wonders, developed new techniques of measurement, and contributed to the improvement of optical instruments. This activity is summarized in the second and third sections of Chapter 5; the fourth section gives an aperçu of Stokes's lecturing in optics from the Lucasian chair and in a more popular venue. Remarks on Stokes's experimental and theoretical styles, on the relative importance of his contribution to optics, and on his attitude toward the contemporary developments are kept for the conclusion of Chapter 5.[6] A bibliography of Stokes's published work on optics and of related publications by other authors appears as an appendix to Chapter 5.

Between Hydrodynamics and Optics

The Jelly-like Ether

On 14 April 1845, Stokes read his famous memoir 'On the theory of the internal friction of fluids in motion, and of the equilibrium and motion of elastic solids', which contained the basic equations for the motion of viscous fluids and of isotropic elastic solids. His main motivation was the quantitative determination of the viscous damping of the oscillations of a pendulum in air, as a contribution to the British metrological project for determining the length of a pendulum that beats the second. Owing to the analogy between transverse pressures (shear stresses) in an elastic solid and in a viscous fluid, it had been common practice, since Claude Louis Navier's seminal work in the early 1820s, to treat solids and real fluids on the same footing as systems of molecules interacting through central forces. In the early 1830s, Augustin Cauchy had used the same model to develop Fresnel's idea of a molecular ether. Thus it is no wonder that Stokes devoted the last section of his bulky memoir to the optical ether.[7]

Fresnel meant his molecular ether to be rigid with respect to the very small displacements implied in the propagation of light, and yet fluid with respect to macroscopic displacements. He judged that displacements over distances larger than the equilibrium distance between successive ether molecules would be relatively free. At the scale of optical vibrations, he believed that the molecular lattice could be such that the pressure required to bring a layer of molecules closer to the next layer would be much higher that the transverse pressure (shear stress) needed to produce a relative shift of these layers, so that longitudinal vibrations would be excluded. This molecular intuition later turned out to be incorrect, and Stokes favoured a continuum approach in which the local deformations of the continuous medium were directly related to the local stresses in the medium, without the Laplacian reduction to intermolecular forces.[8]

For empirical reasons, Stokes admitted states of matter intermediate between the rigid and the fluid state, that is, states enjoying both plasticity and elasticity. In his view, the ether was an extreme case of a medium that was completely plastic or fluid for macroscopic motion and yet highly elastic for the extremely small vibrations implied in light propagation. In order to exclude the never observed longitudinal vibrations, he further assumed the ether to be incompressible. His and George Green's earlier theory of elasticity indeed involved two elastic constants, one determining the response to distortion (without change of volume) and the other determining cubic compression (without distortion). Assuming with Green that the latter constant was infinitely larger than the former, the medium would convey transverse waves only. India rubber, or—better—jelly, were approximations of such a medium.[9]

Three years later, Stokes developed the jelly analogy in the following words:

> Suppose a small quantity of glue dissolved in a little water, so as to form a stiff jelly. This jelly forms in fact an elastic solid: it may be constrained, and it will resist constraint, and return to its original form when the constraining force is removed, by virtue of its elasticity; but if we constrain it too far it will break. Suppose now the quantity of water in which the glue is dissolved to be doubled, trebled, and so on, till at last we have a pint or a quart of glue water. The jelly will thus become thinner and thinner, and the amount of constraining force which it can bear without being dislocated will become less and less. At last it will become so far fluid as to mend itself again as soon as it is dislocated. Yet there seems hardly sufficient reason for supposing that at a certain stage of the dilution the tangential force whereby it resists constraint ceases all of a sudden. In order that the medium should not be dislocated, and therefore should have to be treated as an elastic solid, it is only necessary that the amount of constraint should be very small. The medium would however be what we should call a fluid, as regards the motion of solid bodies through it. The velocity of propagation of normal vibrations in our medium would be nearly the same as that of sound in water; the velocity of propagation of transversal vibrations, depending as it does on the tangential elasticity, would become very small.

In order to get something like the luminiferous ether, Stokes just had to replace the water with a strictly incompressible fluid (thus excluding the longitudinal vibrations), and to adjust the quality and amount of glue to produce to yield the value of the elastic constant corresponding to the velocity of light.[10]

Stellar Aberration

On 14 May 1845, Stokes read a short memoir on the theory of stellar aberration in front of the Cambridge Philosophical Society. The following month, he communicated the same theory at

Fig. 4.1 Stellar aberration according to Bradley. During the travel of a light particle from one extremity of the observation tube AB to the other, the earth carries this tube from the position AB to the position A'B'. Consequently, the tube makes the (small) aberration angle $\alpha = \angle AB'A' \approx (U/c)\sin\theta$ with the true direction of the star if U denotes the velocity of the earth, c the velocity of light, and θ the angle $\angle A'AB'$ between this motion and the direction of the star.

the annual meeting the British Association for the Advancement of Science in Cambridge. James Challis communicated another theory of stellar aberration during the same meeting. The full accounts of these theories later appeared in the July and November issues of the *Philosophical Magazine*.[11]

The astronomer James Bradley had discovered stellar aberration in 1729 in a failed attempt to detect stellar parallax. In his explanation of the new phenomenon, he imagined the direction of the light from the stars to be measured by the inclination of a narrow tube through which the star is observed. Owing to the motion of the earth with respect to the star, the tube must be inclined with respect to the direction of the light in order that a light ray can travel through it without hitting its inner walls (see Fig. 4.1). Besides the idea of the tube, the main ingredient of Bradley's proof is the rectilinear propagation of light, which he justified through the corpuscular theory. In 1746, Leonhard Euler argued that the phenomenon was of a mere kinematic nature, based on the vector composition of the velocity of light with (the opposite of) the velocity of the earth. In that case, aberration was equally well explained in the wave theory of light. Nonetheless, for many years it was commonly believed that stellar aberration confirmed the corpuscular concept of light. When, in 1814, young Fresnel began to reflect on the wave theory of light, he too worried that aberration might contradict it. However, he soon realized that the wave theory gave the same prediction, as long as the ether remained undisturbed by the earth's motion through it. This assumption, which had been only implicit in Euler's study, appeared in 1904 under Thomas Young's pen:[12]

> Upon considering the phenomena of the aberration of the stars, I am disposed to believe, that the luminiferous ether pervades the substance of all material bodies with little or no resistance, as freely perhaps as the wind passes through a grove of trees.

Stokes judged this hypothesis 'very startling' and replaced it with the following (Stokes 1845c, p. 9):

> I shall suppose that the earth and planets carry a portion of the æther along with them so that the æther close to their surfaces is at rest relatively to those surfaces, while its velocity alters as we recede from the surface, till, at no great distance, it is at rest in space.

Under this assumption, Stokes computed how the orientation of a small (approximately plane) portion of wave changed during its travel from the star to the earth. Consider this portion at a given instant t, take the z axis in the direction of its normal motion when the ether does not move, and the x axis in the direction of the projection of the ether motion on the plane of the wave portion (see Fig. 4.2). At time $t + dt$, the orientation of this plane would still be the same if there were no ether motion. Owing to the latter motion, the z coordinate of a point of the ether increases by $v_z(x)dt$, depending on the initial value of the x coordinate of this point and on the velocity component $v_z(x)$ of the ether at this point. Consequently, a wave portion

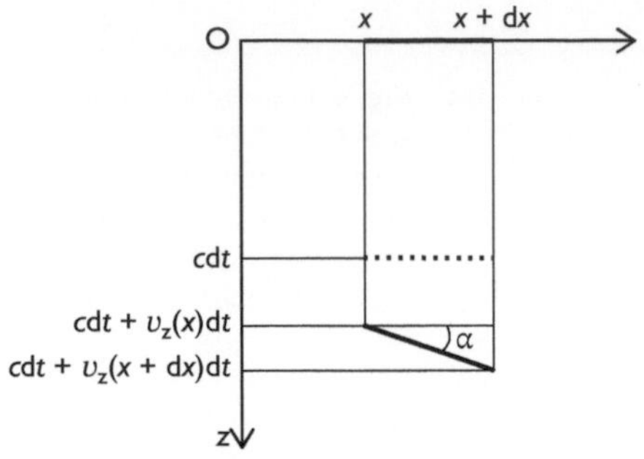

Fig. 4.2 The motion of a portion of wave in a moving ether. The portion (thick line) is on the x axis at time t. If the ether were not moving, the position of the portion at time $t + dt$ would be marked by the dotted thick line. Owing to the motion of the ether in the Oz direction, this position is given by the tilted thick line.

comprised between x and $x + dx$ rotates by the angle $d\alpha = (\partial v_z / \partial x)dt$ (the component of the ether motion in the Ox direction of course does not contribute to the rotation).

The velocity of the ether being small compared to the velocity of light, we may take $dt \approx dz / c$, and the total rotation will be given by

$$\alpha = \int (\partial v_z / \partial x)dz / c.$$

In general, the result of this integration is complicated and it does not agree with the observed aberration. Stokes removes this difficulty by assuming the ether motion to be such that $\mathbf{v} \cdot \mathbf{dr}$ is an exact differential. Then we have $\partial v_z / \partial x = \partial v_x / \partial z$, and the integration yields $\alpha = \bar{v}_x / c$ if $\bar{v}_x$ denotes the velocity of the ether at the surface of the earth in the direction Ox. This velocity is reckoned with respect to the absolute space of fixed stars. By assumption, it is equal to the velocity of the observer in absolute space, since the ether close to the earth is at rest with respect to the earth. For the same reason, the normal to the wave front that reaches the observer is the direction in which the star is seen. The star is therefore seen in a direction inclined by $(U / c)\sin\theta$ toward the motion of the earth, wherein U denotes the velocity of the earth at the point of observation and θ the angle between this velocity and the direction of the star. This is the received law of stellar aberration.[13]

Stokes represented the successive positions ab, cd, etc. of a wave front and the orthogonal trajectory ne on Fig. 4.3. Far from earth, the successive wave fronts are parallel and their common normal gives the true direction of the emitting star. Close to the earth, the orthogonal trajectory of these fronts is curved and its angular deviation gives the aberration. As Stokes later remarked, any temptation to confuse this orthogonal trajectory with a ray of light should be avoided. In essence, Stokes's reasoning relies on wave fronts only, and it crucially depends on two assumptions: 1) the ether motion is irrotational, and 2) the ether is at rest with respect to the earth near the earth. The first assumption yields the desired angular deviation, and the second is necessary for the direction of observation to be perpendicular to the wave front. In 1845, Stokes did not offer any justification for the irrotational character of the ether motion. For the second assumption, he invoked the ether 'being entangled with the earth's atmosphere'.[14]

At first glance, Stokes's new explanation of aberration seemed completely disconnected from Bradley's older explanation. It appeared in the July issue of the *Philosophical Magazine*. James Challis, director of the Cambridge observatory and Plumian professor of astronomy and experimental philosophy, published his own theory in the November issue of the same journal, together with a reflection on its compatibility with the wave theory of light. Challis traced

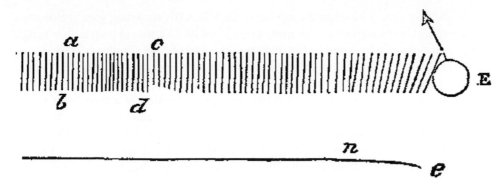

Fig. 4.3 Stokes's drawing of the rotation of a portion of wave plane when approaching the earth. From G. G. Stokes, 'On the Aberration of Light', *Philosophical Magazine* 45 (1845), p. 13.

stellar aberration to the comparison between the direction of the observed star and the direction of an object moving together with the observer, namely, the cross-wire of the astronomer's telescope. After the light from the star has reached the cross-wire, it takes a certain time to reach the eye *e* of the observer. During this time, the cross-wire *w* has moved together with the earth in absolute space. Therefore, the true direction of the cross-wire at the time of observation differs from the direction it had when the light from the star reached it (see Fig. 4.4). The angle between the line *swe′* and the line *e′w′* defines the aberration. In essence, this reasoning is only a rephrasing of Bradley's: Challis's cross-wire and the eye of his observer play the role of the upper and lower extremities of Bradley's tube. In both ways of reasoning, the crucial ingredient is the rectilinear propagation of light. The ether and its possible motion play no role whatsoever. Challis commented:[15]

> According to Mr. Stokes's views, the phenomenon of aberration is entirely owing to the motion which the earth impresses on the æther, and which at the earth's surface he supposes to be equal to the earth's motion. On the contrary, I have to show that the amount of aberration will be the same whatever be the motion of the æther, and if this cannot be shown, the undulatory theory, and not the foregoing explanation, is at fault.

Challis proceeded to show that by a geometric version of Stokes's 'very ingenious and original mathematical reasoning', the propagation of light in a moving ether remained rectilinear (with

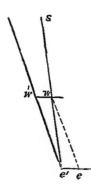

Fig. 4.4 Challis's diagram for his derivation of stellar aberration. From J. Challis, 'A Theoretical Explanation of the Aberration of Light', *Philosophical Magazine* 45 (1845), p. 321 (simplified).

respect to absolute space) whatever the motion of the ether. His reasoning goes in two steps. Firstly, he argues that owing to the ether motion, the absolute direction of propagation of light at a given point of space is altered by the motion of the ether by the amount $\alpha' = -\upsilon_x / c$, υ_x being the projection of the velocity of the ether on the plane perpendicular to the original direction of propagation. This is a trivial consequence of the vector composition of velocities. Secondly, by erroneous geometry Challis finds that a small portion of a wave front rotates by the amount $d\alpha = (\partial \upsilon_x / \partial z)dt$ during propagation in the moving ether in the time dt. Taking the integral of this expression for $dt \approx dz / c$, Challis finds $\alpha = \upsilon_x / c$ for the total rotation of the wave front from a star to a given point of space at which the ether velocity is υ_x. This rotation of the wave front exactly compensates the direct effect of the ether drag (namely: $\alpha + \alpha' = 0$). Therefore, the propagation remains rectilinear whatever the ether motion and Challis's derivation of aberration remains fully justified.[16]

Stokes soon reacted to Challis's claims. Firstly, he criticized the idea that aberration should be traced to a misevaluation of the direction of the cross-wire. In his own theory, for instance, the cross-wire is evidently seen in the true direction. What truly defines the direction of observation of a star is just the direction of the line joining the cross-wire (in the focal plane of the objective) and the centre of the objective of the telescope, as assumed in Bradley's reasoning or later variants. Secondly, Stokes corrected the error in Challis's derivation of the rotation of a wave portion and reasserted the necessity of the condition that the motion of the ether should be irrotational.[17]

In reply, Challis gave a concrete variant of his former derivation of aberration based on a graduated circle, with the provocative comment:

> I think it important to remark, that the foregoing explanation of aberration rests on no hypothesis whatever, being a strict deduction from ascertained facts, without reference to any theory of light. The cause assigned for aberration is, therefore, a *vera causa*, which consequently excludes every explanation of a hypothetical kind, such, for instance, as that which Mr. Stokes proposed in the July number of this Magazine.

Challis next acknowledged his error in calculating the rotation of a portion of wave and conceded to Stokes that the desired result required the irrotational character of the ether motion. For this kind of motion, he repeated his argument that this rotation and the effect of the absolute motion of the ether near the earth on the direction of motion of the wave portion together yielded the usual aberration formula. He concluded with the words:

> There is nothing improbable in the supposition [of irrotational ether motion]: it saves the undulatory theory; but I must protest against its being considered necessary for the explanation of the aberration of light.

In an additional communication, Challis insisted that any theory of aberration based on unnecessary hypotheses failed to deliver a genuine explanation by true causes.[18]

In reply, Stokes remarked that Challis's 'true causes', namely the finite velocity of light and the motion of the earth, implied only that the observed direction of a star differed from the direction of the light from the star when arriving on earth, by an amount given by the usual aberration formula. The latter direction could still differ from that of the light when departing from the star, since the trajectory of the light rays could be curved in the intermediate space.[19]

The remark angered Challis, who retorted that the coincidence between his predicted value of aberration and the measured value was an empirical fact and therefore so too was the constancy of the direction of the light when travelling from the star to us. He added that 'any explanation which rests on a hypothetical motion of the æther, must be *fictitious*' and he haughtily concluded:

> I really think that I have now said quite enough in defence of a very unexceptionable piece of reasoning, and if Mr. Stokes should have anything further to urge, I must decline answering it.

Stokes diplomatically replied that he and Challis were giving different meanings to the word 'explanation'. For Stokes, an explanation implied a deduction from given assumptions, without further ingredient. For Challis, implicit ingredients were tolerable as long as they were of a negative character. In this view the finite velocity of light and the motion of the earth were the true causes of the aberration, and the absence of an additional cause of aberration, namely the curving of the light path from the star, was not be regarded as a cause.[20]

Although Stokes must have been annoyed by Challis's dismissive attitude, he learned something from this exchange with his senior colleague. Combining his own derivation of the rotation of a wave portion, Challis's remark that the absolute motion of a wave portion was altered by the motion of the medium, and the further remark that the direction of the net motion of the wave portion coincided with the direction of a ray of light, he found that a ray of light kept a constant direction as long as the motion of the ether was irrational:

> Hence the direction of the light coming from a star is the same as that of a right line drawn from the star, not merely at such a distance from the earth that the motion of the æther is there insensible, and again close to the surface of the earth, where the æther may be supposed to move with the earth, but *throughout the whole course* of the light; so that a ray of light will proceed in a straight line even when the æther is in motion, provided the motion be such as to render [$\mathbf{v} \cdot \mathbf{dr}$] an exact differential.

As Stokes admitted, he had originally been unaware of 'this curious consequence of [his] theory'. In this communication of July 1846 he failed to credit his unpleasant interlocutor for the insight. He did so many years later, in the first volume of his collected papers:

> Some remarks made by Professor Challis in the course of discussion suggested to me the examination of the path of a ray, which in the case in which [$\mathbf{v} \cdot \mathbf{dr}$] is an exact differential proved to be a straight line, a result which I had not foreseen when I wrote the above paper [of July 1845]…The rectilinearity of the path of a ray in this case, though not expressly mentioned by Professor Challis, is virtually contained in what he wrote.

The rectilinear propagation of light in the moving ether being thus established, the aberration of stars can be applied in Bradley's or a similar manner. As Stokes noted in July 1846, the assumption that the ether does not move with respect to the earth near its surface now becomes unnecessary. The only condition on the ether motion is its irrotational character.[21]

The main purpose of Stokes's memoir of July 1846 was to unravel the physical circumstances that might explain the latter condition. Ignoring the effect of the atmosphere in a first step of the reasoning, he considered the ether motion induced by the translation of the earth through the ether. In conformity with the speculations found in his earlier memoir on fluid friction, he assumed the ether to behave like an incompressible fluid at the macroscopic scale. Ignoring any shear stress in the ether and imagining the motion of the earth to be started from rest in ether at rest, a theorem by Lagrange implies that the induced ether motion should be irrotational. In reality, Stokes tells us, there may be temporary shear stresses in the medium but any resulting motion will be quickly propagated away as light.[22] Based on this remark, Stokes expected the induced ether motion to be irrotational at least when the effect of the atmosphere was ignored. From his pendulum studies he certainly knew that the motion of a perfect liquid induced by a moving sphere implied a shift of the fluid along the walls of the sphere. As for the effect of the atmosphere, in the lack of any established model for the interplay of ether and matter molecules, he offered the conjecture that the molecules of the air acted as a dense cloud of solid corpuscles forcing the ether to be nearly at relative rest in the denser part of the atmosphere.[23] The motion being still irrotational between the air molecules by the former reasoning, Stokes could thus combine irrotational motion through space and lack of relative motion near the surface of the earth. At the same time, he noticed that the latter assumption was unnecessary to the derivation of stellar aberration.[24]

An obvious advantage of this assumption was that it immediately implied that the motion of the earth had no effect on optical experiments performed on earth. In contrast, the Young–Fresnel explanation of aberration implied an ether wind that could in principle alter the results of such experiments. In a letter to François Arago published in 1818, Fresnel had addressed Arago's finding that refraction by a prism did not depend on the earth's motion. Fresnel thereby assumed 1) that the ether remained completely at rest in empty space, 2) that a transparent body moving through the ether carried with it the excess of ether needed to explain the slower velocity of light (the elastic constant of the ether being assumed to be the same in all bodies, and the slower velocity being therefore traced to a higher density), and 3) that waves travelled in a transparent body as in a single medium moving at the average velocity of the normal ether and the excess ether. In a communication of 1846 on Fresnel's theory, Stokes replaced assumptions 2) and 3) with the simpler assumption that the flux of the ether should be conserved at the interface between a vacuum and a transparent medium. In both cases, waves in a transparent body of optical index n moving in the ether with the velocity U travel at the velocity $V = c/n + U(1 - 1/n^2)$ with respect to the transparent body. As Fresnel demonstrated, this partial drag of the waves implies the invariance of the laws of refraction for a prism subjected to the ether wind at the surface of the earth (to first order in U/c). Stokes offered an alternative proof of this result. In addition, he showed that the Fresnel drag explained the lack of fringe shift in an experiment in which Jacques Babinet had brought light to interfere after travelling through two plates of equal thickness but oriented in different directions with respect to the earth's motion. Knowing that his own assumption of a vanishing ether wind directly explained Arago's and Babinet's negative results, Stokes concluded:[25]

> This affords a curious instance of two totally different theories running parallel to each other in the explanation of phænomena. I do not suppose that many would be disposed to maintain Fresnel's theory, when it is shown that it may be dispensed with, inasmuch as we would not be disposed to believe, without good evidence, that the æther moves quite freely through the solid mass of the earth. Still it would have been satisfactory, if it had been possible, to have put the two theories to the test of some decisive experiment.

Stokes never returned to the problem of stellar aberration. During the next twenty years, the optics of moving bodies was rarely discussed. Stokes may have been too old, when, toward the end of the century, this topic became of central interest. For a long time the theories of Stokes and Lorentz both remained viable options, in conformity with the 'curious' equivalence of their predictions. The British typically favoured Stokes's theory, in part owing to the authority of Stokes but also because from the 1860s onward Maxwell's electromagnetic theory of light favoured a full dragging of the ether by matter. The French naturally preferred Fresnel's theory, which received strong support in 1851 when Hippolyte Fizeau confirmed the partial drag of the ether by running water according to Fresnel's formula. The Fizeau experiment, being the only one of this kind until the mid-1880s, did not truly affect the British preference for Stokes's fully dragged ether. In 1864–7, Maxwell performed a variant of Arago's old experiment regarding the effect of the motion of the earth on the refraction of light, with a negative result. He wrongly expected Fresnel's partial drag to imply a positive result and therefore believed to have contradicted both Fresnel and Fizeau. After Stokes corrected him in private, he still did not regard Fizeau's experiment as decisive:

> This experiment seems rather to verify Fresnel's theory of the ether; but the whole question of the state of the luminiferous medium near the earth, and of its connexion with gross matter, is very far as yet from being settled by experiment.

In his 'ether' article of 1878 for the *Encyclopædia Britannica*, Maxwell still judged Stokes's hypothesis for the motion of the ether 'very probable'.[26]

As Wilhelm Veltman and Eleuthère Mascart proved in the early 1870s, the Fresnel drag, Huygens's principle, and the Doppler effect together imply that the motion of the earth should have no effect on optical experiments on earth to first order in the ratio of the velocity of the earth to the velocity of light. In 1881, the American experimentalist Albert Michelson failed to detect a second-order interferometric effect that was to be expected in Fresnel's stationary ether. In 1886 and with the help of Edward Morley, he confirmed the Fresnel drag in a variant of Fizeau's experiment. In this perplexing situation, the young Dutch theorist Hendrik Lorentz published a detailed critical study of the problem of ether motion. Lorentz first argued that Stokes's theory was kinematically impossible, because the flow of an incompressible fluid around a solid sphere cannot be both irrotational and adhering to the sphere. This remark has often been regarded as a fatal objection to Stokes's theory. It truly was not. Firstly, Stokes was aware of the necessary shift of an incompressible inviscid fluid on a moving sphere, and he avoided this shift by assuming a nearly complete drag of the ether by the denser part of the atmosphere. In 1886, the confirmation of Fresnel's partial drag (which is negligible for air) excluded this escape. However, as Lorentz noted, one could accept the shift together with the Fresnel drag. As Stokes had proved, the shift did not affect the amount of stellar aberration as long as the motion remained irrotational; and the Fresnel drag explained the lack of first-order effect of the resulting ether wind. As for Michelson's second-order experiment, it was not yet precise enough to be decisive (after correction of a mistake in Michelson's analysis) and there could be locations on earth in which the ether shift was too small to be detected even in a more precise experiment. At the time of Lorentz's writing, the modified Stokes theory and Fresnel's theory were both compatible with the available experimental results. Lorentz preferred Fresnel's theory for its higher simplicity and because his developing conception of the interplay between matter and ether favoured the stationary ether. The following year, Michelson and Morley confirmed Michelson's negative result. This did not signal the end of the stationary ether. On the contrary, by the end of the century it became clear that 'electron theories' based on a complete separation of the ether from the matter moving through it accounted for the widest range of optical and electromagnetic phenomena. Stokes's idea of ether motion died away not because it was refuted but because theories based on Fresnel's stationary ether turned out to be more powerful.[27]

The Subtleties of Interference

Wave optics still being a young theory in the mid-nineteenth century, it was important to carefully examine its most specific consequences, those deriving from interference. Several of Stokes's contributions to optics in this period fall in this category. He was attentive to any puzzle or incomplete reasoning he could find in British reviews of optical science, and used his theoretical and experimental acumen to fill the gaps.

Brewster's New Polarity of Light

At the annual meetings of the British Association in 1836 and 1837, the old 'Father of Modern Experimental Optics'[28] David Brewster created a sensation by announcing his discovery of a new property of light, which he called 'polarity'. Suppose the light spectrum from a prism to be seen out of focus directly through the eye (so that the image of every line of the spectrum on the eye's retina is blurred). Then a series of dark bands can be formed by covering half of the pupil of the eye with a mica plate of proper thickness on the violet side of the spectrum; but no such band can be formed if the plate is inserted on the red side of the spectrum. In 1840 Airy reproduced this result and showed through a highly ingenious analysis that it could be explained within the wave theory of light as the conjugate effect of interference (between the lights from the covered and uncovered halves of the pupil) and diffraction owing to the finite aperture of the eye. This was the topic of his Bakerian lecture of 18 July 1840. A few months later, Brewster instructed him that the phenomenon also existed when the spectrum was seen in focus, if only thicker mica plates were used. Airy soon gave the theory of this relatively simpler case.[29]

In both cases, the calculation is based on the Huygens–Fresnel principle applied to the truncated spherical wave that converges from the pupil of the eye to the focus of the crystalline lens. Airy obtains the intensity of the light at a given point of the retina in two steps. Firstly, for a given spectral component he gets the corresponding amplitude of the light on the retina by summing the amplitudes of the spherical wavelets emanating from portions of the spherical wave facing the two halves of the pupil, taking into account the phase retardation by the mica plate for the covered half of the pupil. Secondly, he sums the resulting intensity for the various spectral components (which reach the pupil with a variable angle of incidence). The first step leads to an analytical expression depending on the Fresnel integrals in the unfocused case, on a simpler trigonometric formula in the focused case. Airy computed tables for the values of these expressions and drew the corresponding graphs. He then obtained the net intensity distribution by graphically superposing the effects of the various spectral components. As the results were in full agreement with the observations, Airy could claim a major new victory for the wave theory of light.[30]

Stokes became interested in this theory in 1847, when his Oxford colleague and friend Baden Powell informed him about a similar experiment in which bands were formed in the spectrum from a prism filled with transparent liquid by inserting a retarding transparent plate (see Fig. 4.5). Just as in Brewster's experiments, the band can be formed only if the plate is inserted on the violet side of the spectrum. Stokes understood that Airy's theory equally applied to this new experiment. His main service was to simplify Airy's considerations in the focused case. Whereas Airy had used a graphical method to superpose the intensity patterns created by the

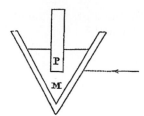

Fig. 4.5 Powell's set-up for producing the Talbot bands. The retarding plate P is immersed in the liquid prism M. From Baden Powell, 'On a New Case of the Interference of Light, *Philosophical Transactions of the Royal Society of London* 138 (1848), p. 214.

various spectral components, Stokes showed that the corresponding integrals could be performed exactly, leading to simple expressions that varied discontinuously as functions of the geometric parameter defining the aperture and the penetration of the retarding plate.[31]

In a simplified, two-dimensional version of Stokes's reasoning, a lens of focal distance f turns the plane-wave spectral components of the light from a prism into circular waves converging in the focal plane of the lens (see Fig. 4.6). A diaphragm next to the lens truncates the circular wave next to the lens; and part of the diaphragm is covered with a retarding plate whose phase shift φ depends on the wavelength λ of the spectral component. First consider the spectral component of normal incidence. Draw the axis Ox from the centre (focus of the lens) of the corresponding circular wave in a direction perpendicular to the axis of the lens. Call x the abscissa of point M of the circular wave and X the abscissa of a point P of the focal plane. In a sufficient approximation, the distance MP is given by $MP \approx f + xX/f$. According to the Huygens–Fresnel principle, the amplitude $A_0(X)$ of the wave at point P is obtained by summing the partial waves emanating from every point of the circular wave within the diaphragm. Call $F(x)$ and $G(x)$ the characteristic functions[32] of the parts of the diaphragm that are respectively covered and uncovered by the retarding plate. The resulting expression of the amplitude is

$$A_0(X) \propto \tilde{\chi}(k), \text{ with } k = 2\pi X / \lambda f \text{ and } \chi(x) = F(x) + e^{i\phi_0} G(x),$$

where the tilde denotes the Fourier transform. For the spectral component of incidence θ the diffraction pattern is shifted by $f\theta$ in the focal plane, so that the corresponding amplitude reads

$$A_\theta(X) \propto \tilde{\chi}(k), \text{ with } k = 2\pi(X - f\theta)/\lambda f \text{ and } \chi(x) = F(x) + e^{i\phi_\theta} G(x).$$

The resulting intensity distribution in the focal plane is

$$I_\theta(X) \propto \tilde{\chi}(k)\,\tilde{\chi}^*(k) = \tilde{F}\tilde{F}^* + \tilde{G}\tilde{G}^* + 2\,\mathrm{Re}[\tilde{F}^*\tilde{G}\,e^{i\varphi_\theta}].$$

At this point, Stokes assumes the spectrum to be much larger than the width of the diffraction amplitudes $\tilde{F}$ and $\tilde{G}$ around the image point $X = f\theta$, and he replaces the phase φ_θ with

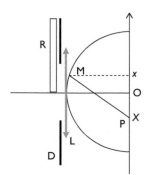

Fig. 4.6 Diagram for Stokes's computation of the Talbot–Brewster bands. The lens L produces the spherical wave centred on the focus O. This wave is truncated by the diaphragm D and it is partially phase-shifted by the retarding plate R.

its linear approximation $\varphi(X/f)+\alpha k$ within the diffraction patch. For the superposition of the various spectral components, these approximations lead to the net intensity

$$I(X) \propto \int_{-\infty}^{+\infty}(\tilde{F}\tilde{F}* + \tilde{G}\tilde{G}*)\mathrm{d}k + 2\,\mathrm{Re}\left[e^{i\varphi(X/f)}\int_{-\infty}^{+\infty}\tilde{F}*\tilde{G}e^{i\alpha k}\mathrm{d}k\right].$$

By well-known properties of the Fourier transform, this is equivalent to

$$I(X) \propto \int_{-\infty}^{+\infty}(F^2+G^2)\mathrm{d}x + 2\cos[\varphi(X/f)]\int_{-\infty}^{+\infty}F(x)G(x+\alpha)\mathrm{d}x.$$

Since the retardation phase φ varies with the wavelength associated with the value of X, this expression implies an oscillation of the intensity in the focal plane, with an amplitude depending on the integral $\int_{-\infty}^{+\infty}F(x)G(x+\alpha)\mathrm{d}x$ that measures the overlap between the F domain and the G domain after the latter has been shifted by $-\alpha$. Take the simplest case in which the intervals of the F and G domains are the intervals $[-a, 0]$ and $[0, a]$ respectively. If $0 < \alpha < 2a$, the F domain and the shifted G domain overlap (with a complete overlap at $\alpha = a$). If the two domains are permuted and if the coefficient α has the same positive value, the overlap integral always vanishes and there are no interference bands. This explains Brewster's polarity.[33]

Newton's Rings Revisited

Around the same time, Stokes addressed an older challenge to the wave theory of light regarding total reflection at the interface between two transparent media, when the index n of the first medium with respect to the second medium is larger than one (for instance, when light is reflected within a piece of glass immersed in air). In 1823, Fresnel had remarked that the Huygens–Fresnel construction correctly predicted the absence of reflection at incidences beyond the critical incidence $\sin^{-1}(1/n)$ for which the sine relation $n\sin i = \sin r$ no longer admits a real solution. However, he did not expect the construction to give correct results in the immediate vicinity of the interface and he expected the incident wave to slightly penetrate the second medium in cases of total reflection. As experimental evidence for this penetration, he reported an observation he had made after pressing two prisms that jointly formed a parallelepiped of glass, with a defect of contact between the two prisms owing to a slight curvature of one of the touching surfaces (see Fig. 4.7). The prisms having been cut at an angle slightly superior

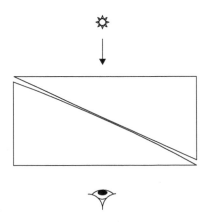

Fig. 4.7 Fresnel's double prism.

to the critical angle, Fresnel observed a distant source of light through them and saw a patch of light around the point of contact. If the vibrations did not at all penetrate the air in the interval between the two prisms, Fresnel reasoned, then there should be total reflection everywhere expect at the point of contact and no light should be transmitted through the double prism. The observed patch was to be traced to a slight penetration of the waves, which enabled them to enter the second prism near the point of contact. From the fact that the size of the patch was comparable to the size of the first Newton ring in the conditions of normal observation, Fresnel concluded that the penetration length of the waves was of the order or the wavelength.[34]

Stokes was aware of Fresnel's argument through Lloyd's approving account in his British Association report of 1835 on physical optics. In 1848 Stokes managed to calculate the intensity of reflected (or transmitted) light through Matthew O'Brien's earlier extrapolation of Fresnel's laws for the intensity of reflected and refracted light beyond critical incidence. Whatever the exact dynamics of the media and the exact boundary conditions, O'Brien and Stokes reasoned, Fresnel's laws result from the equality of the refracted vibration to the sum of the incident and reflected vibrations at the surface of the two media together with the equality of some spatial derivatives of the same quantities. The three vibrations being of the form $\mathbf{a}e^{i(\mathbf{k}\cdot\mathbf{r}-\omega t)}$, the equality of the spatial frequencies in the plane of contact between the two media implies $k_x = k_x'$ for the wave numbers of the incident and refracted waves along the intersection Ox of the incidence plane and the contact plane. Taking Oz perpendicular to the plane of incidence and taking into account the relations $k = nk'$ and $k_x = k\sin i$, this implies

$$k_z' = \sqrt{k'^2 - k_x^2} = k'\sqrt{1 - n^2 \sin^2 i}.$$

For $i > \sin^{-1}(1/n)$, k_z' has an imaginary part $i\alpha$ which results in an exponential decrease of the wave intensity at the rate α. The boundary conditions and the resulting expressions of the reflected and refracted amplitude should remain the same except that the variables related to k_z' now have an imaginary part. This is how O'Brien and Stokes justified their extrapolation of Fresnel's formulas for the amplitude of the reflected and refracted waves. Stokes then mimicked the usual derivation of Newton's rings, in which the air gap at a given distance of the point of contact is replaced with a parallel air plate of the same thickness and the final reflected wave is obtained by summing a series of partial waves, the first of which results from a single reflection on the first interface; the second from refraction at the first interface followed by reflection at the second interface and refraction at the first interface; the third from refraction at the first interface followed by reflection at the second interface, reflection at the first interface, reflection at the second interface, and refraction at the first interface; and so forth (see Fig. 4.8). The resulting phenomena are complex, since they depend on the polarization of the incident light and since the transition from Newton's rings to a single dark spot involves a delicate mixture of periodic and exponential behaviour. Stokes verified his predictions for the central dark spot beyond critical incidence through quantitative experiments.[35]

The Principle of Reversion

As Stokes knew from Airy's *Mathematical Tracts*, in the subcritical case of Newton's rings the central dark spot is *completely* dark. The first proof of this result, given by Poisson in 1823,

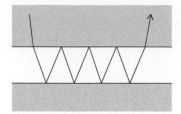

Fig. 4.8 Multiple reflection of a light ray in an air gap.

depended on Fresnel's expression for the reflection and refraction coefficients at the interface between these two media and on the aforementioned possibility of multiple reflections in the air gap. With Airy and Stokes, call b the reflection coefficient at the glass/air interface, c the refraction coefficient at the same interface, e the reflection coefficient at the air/glass interface, and f the refraction coefficient at the same interface. The total amplitude of the reflected light is then given by

$$a = b + cef + ce^3 f + ce^5 f + \ldots = b + \frac{cef}{1-e^2}.$$

Fresnel's laws imply

$$b = -e \ \left(\text{Arago's law}\right) \text{ and } cf = 1 - e^2,$$

from which $a = 0$ results.[36]

In early 1849, Stokes proved that the two former relations simply result from the 'principle of reversion' according to which any solution of the equations of the dynamical medium admits a time-reversed equation. Consider the process in which a wave of amplitude 1 yields the reflected wave of amplitude b and the refracted wave of amplitude c at the interface glass/air. In the time-reversed process, the wave of amplitude b yields a reflected wave of amplitude b^2 and a refracted wave of amplitude bc at the same interface; and the wave of amplitude c yields a reflected wave of amplitude ce and a refracted wave of amplitude cf at the air/glass interface (see Fig. 4.9). By the principle of reversion, the sum of the similar waves of amplitudes b^2 and cf must be the time-reverse of the original wave of amplitude 1; and the sum of the waves of amplitudes ce and bc must vanish. Therefore, we must have

$$b^2 + cf = 1 \text{ and } ce + cb = 0,$$

which implies the desired relations. This argument was not only an ingenious way of short-circuiting complex calculations. It was, in Stokes's times, a pioneering use of discrete symmetry considerations as a constraint on the form of physical laws.[37]

Fig. 4.9 Diagram for Stokes's application of time-reversal to partial reflection. The first diagram represents the direct process; the second and third represent two components of the reverse process. By time-reversal, the superposition of these two components should reproduce the amplitudes of the first process.

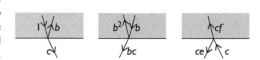

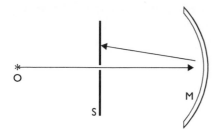

Fig. 4.10 Newton's set-up for observing the colours of thick plates. The light from the source O, passing through the hole in the screen S, is diffusely reflected on the spherical glass mirror M. The coloured rings appear on the (white) side of the screen facing the mirror.

Thick Plates

Newton's rings are the first example of a phenomenon discovered by Newton, interpreted in his theory of fits of easy reflection, and later understood as a simple case of interference. The other example is the 'colours of thick plates', that is, the coloured rings obtained by the diffuse reflection of a narrow beam of light on a thick and curved glass mirror (see Fig. 4.10). In both cases Thomas Young gave an elementary explanation based on two-ray interference. In the thick-plate case, the first ray enters the glass of the mirror perpendicularly, is reflected by the silvered surface, and is scattered away from the axis by impurities on the glass surface upon its return; the second ray is scattered when hitting the glass surface, then reflected by the silvered surface, and refracted when crossing the glass surface again. Stokes's purpose, when he addressed this old problem in 1851, was to give the complete, quantitative theory of this phenomenon for various shapes of the silvered mirror. Stokes clearly indicated the nature of this study:[38]

> Although the present paper is a little long, the reader must not suppose that the theory of the rings and bands is anything but simple. The length arises partly from the detail in which the subject has been considered, partly from the generality of some of the investigations, partly from the description of experiments which accompanies the theoretical investigations.

Altogether, Stokes's memoirs on interference-based phenomena belonged to what Thomas Kuhn would have called paradigmatic physics: they explored in detail the consequences of a well-established theory. The basic ways of reasoning were not new and Stokes mostly borrowed them from Airy, O'Brien, Young and Herschel. What makes them remarkable is Stokes's mathematical astuteness, his attention to details, and his experimental virtuosity in verifying the finest theoretical predictions. Stokes pushed the mathematical analysis much further than his forerunners; he gave exquisitely detailed descriptions of the physical consequences; and he explained or verified many peculiarities of the colourful interference patterns that never fail to fascinate the students of this sort of phenomenon.

The Dynamical Theory of Diffraction

On 12 November 1849, Stokes read a major memoir entitled 'On the dynamical theory of diffraction' in front of the Cambridge Philosophical Society. As was mentioned, his earlier explanation of Brewster's 'new polarity' of light involved effects of diffraction. Like every-one else before him, Stokes computed these effects by means of the Huygens–Fresnel prin-ciple. According to Fresnel's intuitive reasoning, the amplitude of the light from an opening on a screen is obtained by summing the amplitudes of circular wavelets emanating from every point of the opening, with proper amplitude and phase. Fresnel's considerations were largely intuitive. He did not write any wave equation and did not try to solve the associated problem of wave propagation in any rigorous mathematical manner. Stokes's ambition was to solve the wave equation under given boundary conditions at least in a reasonable approximation. In a word, he wanted to give a 'dynamical theory' of diffraction.[39]

From the Wave Equation to the Diffraction Formula

Stokes's strategy was based on the principle of superposition and on an integral formula by Poisson giving the vibration at a given point at time t as a function of the vibration and its time derivative at time zero at any point of space. Suppose the screen to be flat and the incoming wave to be a plane wave whose wave planes are parallel to the screen. According the superposition principle, the vibration in the diffraction zone can be obtained by sum-ming the effect of successive slices of the incoming wave. When a given slice reaches the screen, Stokes reasons, this slice is truncated so that its later effect in the diffracting zone (beyond the screen) depends only on its magnitude and velocity within the opening. Stokes's diffraction formula is just the expression of the cumulative effect of these truncated wave slices. In his memoir, Stokes directly dealt with the problem of transverse waves propagating in a two-constant elastic solid. In the simpler case of longitudinal waves in an elastic fluid, his formulas reduce to the following.

The wave equation for the vibration $u(\mathbf{r},t)$ is

$$\ddot{u} - c^2 \Delta u = 0.$$

The Poisson formula yields the disturbance u at a given point (say $\mathbf{r} = \mathbf{0}$) at the instant t as the following combination of the averages of the initial values of u and $\dot{u}$ over the sphere of radius ct centred on this point:[40]

$$4\pi u(\mathbf{0},t) = t \int_{r=ct} d\Omega \dot{u}(\mathbf{r},0) + \frac{d}{dt}\left[t \int_{r=ct} d\Omega u(\mathbf{r},0) \right],$$

where $d\Omega$ denotes the element of solid angle in the direction of $\mathbf{r}$. This formula may be rewrit-ten as

$$4\pi u(\mathbf{0},t) = t \int_{r=ct} d\Omega \dot{u}(\mathbf{r},0) + \int_{r=ct} d\Omega u(\mathbf{r},0) + ct \int_{r=ct} d\Omega \frac{\partial u}{\partial r}(\mathbf{r},0).$$

Stokes slices the plane wave $U(\mathbf{r},t) = f(ct - x)$ into thin layers of thickness $c\tau$ and constructs the diffracted wave by summing the contributions of each of these layers at the times $n\tau$ at which they meet the diffracting plane $x = -d$. For one of these layers, he considers the portion dV that lies above the unscreened surface element dS of this plane. For large r, the second term in the previous equation is negligible, and the contribution of the element dV to the disturbance at $\mathbf{r} = \mathbf{0}$ is

$$4\pi \, \mathrm{d}u(\mathbf{0},t) = \sum_n \mathrm{d}\Omega_n \left[\frac{r}{c}\dot{U}(\mathbf{r},t-r/c) + r\frac{\partial U}{\partial r}(\mathbf{r},t-r/c) \right],$$

where $\mathrm{d}\Omega_n$ is the solid angle determined by the intersection of the sphere $r_n = nc\tau$ with the volume $\mathrm{d}V$. The identities

$$\mathrm{d}V = c\tau \, \mathrm{d}S = \sum_n c\tau \, r_n^2 \mathrm{d}\Omega_n, \quad \dot{U} = -c\frac{\partial U}{\partial x} = -cU', \quad \frac{\partial U}{\partial r} = -\cos\theta\frac{\partial U}{\partial x},$$

where θ is the angle that the vector $\mathbf{r}$ makes with the x-axis, then lead to the formula

$$u(\mathbf{0},t) = -\int \frac{\mathrm{d}S}{4\pi r}(1+\cos\theta)U'(\mathbf{r},t-r/c)$$

for the total disturbance expressed as an integral over the plane $x = -d$.[41]

For monochromatic waves, this formula agrees both with Fresnel's earlier intuitive considerations (except for the angular factor) and with Gustav Kirchhoff's later theory of diffraction based on an extension of Green's theorem. With transverse vibrations in view, Stokes solved the same problem for a vector vibration $\mathbf{u}(\mathbf{r},t)$ obeying the equation

$$\ddot{\mathbf{u}} - b^2\Delta\mathbf{u} - (a^2 - b^2)\nabla(\nabla\cdot\mathbf{u}) = 0$$

of a bi-constant elastic solid. In order to reduce this problem to the simpler problem of scalar waves, he inaugurated the decomposition of a vector field into its longitudinal and transverse parts (for which $\nabla\times\mathbf{u}$ and $\nabla\cdot\mathbf{u}$ respectively vanish). In the end he took the Green limit $a \to \infty$ for which the longitudinal waves disappear.

An Attempt to Decide the Direction of Optical Vibrations

For the (transverse) secondary vibration emitted by the surface element $\mathrm{d}S$, Stokes found that the vibration occurred in the plane defined by the vibration of the incoming wave and the vector $\mathbf{r}$ joining the point of observation to the surface element, and that its amplitude contained an additional factor $\sin\phi$, where ϕ is the angle that the vector $\mathbf{r}$ makes with the incoming vibration. As Stokes emphasized, the resulting distribution of the polarization of the diffracted light as a function of the polarization of the incoming light depends on whether the optical vibration occurs in the plane of polarization or in the perpendicular direction. By carefully studying the large-angle diffraction from a fine grating borrowed from his mineralogist friend William Hallowes Miller, Stokes decided in favour of the second alternative.[42]

In order to appreciate the importance of this announcement, it should be remembered that there were two classes of theories of the ether as an elastic medium.[43] In the first class, pioneered by Fresnel and developed by Cauchy, Green, and many others, the transverse vibration occurs in the direction perpendicular to the conventional plane of polarization of light (defined as the incidence plane for light polarized by reflection); in the second class developed by Franz Neumann, James MacCullagh and Kirchhoff, the vibration occurs in the polarization plane. The two kinds of theories (roughly) lead to the same empirical laws of reflection and refraction at the interface between two transparent media, but with different boundary conditions. According to Stokes, diffraction experiments could decide between the two options. The experiments were difficult and—since Stokes assumed the diffraction to take place before the wave reached the grooves of the grating—their interpretation was complicated by refraction in the glass of the grating. Stokes also noted that Joseph Fraunhofer, in his early experiments with gratings, had observed asymmetries of the diffracted light incompatible with Fresnel's simple theory. Stokes believed these asymmetries to be compatible with his theory of diffraction once the asymmetries of the diffracting grooves were taken into account. In general, he believed that the theoretical difference between the polarizations predicted under Fresnel's and under Neumann's assumptions were too large to be inverted by the finer circumstances of the diffraction.[44]

In 1852, Stokes found another reason to adopt Fresnel's choice in the polarization of light scattered by a suspension of small particles. Empirically this polarization was known to be in the scattering plane (the plane containing the incident ray and the scattered ray). Stokes imagined the scattering motes to be particles floating in the ether (here regarded as a liquid) and modifying its vibration through their higher inertia. Granted that the motes were much smaller than a wavelength, Stokes reasoned, the modification would be parallel to the original vibration. If the vibration occurred in the plane of polarization, the vibration of the scattered ray would then be in the scattering plane and it could therefore not be induced by the secondary vibration of the ether (for a scattering angle of 90°, the vibration of the scattered ray would be perpendicular to the secondary vibration). Although this simple reasoning depended on an unproven model for the action of the motes, it must have increased Stokes's confidence in his diffraction-based argument.[45]

In 1856, the Stuttgart physicist Carl Holtzmann shook this confidence with grating diffraction experiments that confirmed Neumann's assumption of a vibration in the plane of polarization. Stokes did not question Holtzmann's results. He simply noted that Holtzmann had used a different kind of grating, in which lines of soot replaced the grooves, and that diffraction at large angle could well depend on the nature of the material of the screening matter. He now made clear that his theory of diffraction rested 'on the hypothesis that in diffraction at large angle, as we know to be the case in diffraction at a small one, the office of the opaque body is *merely to stop* a portion of the incident light'. He even indicated that he had always regarded this hypothesis as 'rather precarious', a dubious claim despite his earlier mention of Fraunhofer's contradictory observations.[46]

In 1860, the Danish physicist Ludvig Lorenz confirmed Stokes's theoretical prediction in a theory of his own, and performed his own polarization experiments with smoke gratings similar to Holtzmann's. He found his results to agree with Fresnel's choice and he suggested that Holtzmann's contrary results resulted from an experimental error Lorenz had himself committed in a first series of experiments. When Stokes's memoir of 1849 reappeared in 1883 in the second volume of Stokes's collected papers, Stokes cited Lorenz's results approvingly, and no longer hinted at any difficulty in applying his theory to large-angle diffraction. Three years later the French physicist Louis Georges Gouy studied the refraction of light by the edge of a razor plate at large angles and found the polarization of the diffracted light to strongly depend on the nature of the material of the edge and on its sharpness. This confirmed and extended Fraunhofer's and Fizeau's earlier observations that large-angle diffraction could have unsuspected polarizing effects, and this definitely contradicted the predictions of Stokes's and Kirchhoff's theories.[47]

In his Sorbonne lectures of 1887–8 on the mathematical theories of light, Henri Poincaré noted this contradiction and showed that these theories depended on an approximation that no longer applied to large-angle diffraction. He and Arnold Sommerfeld later gave exact theories of diffraction by a metallic half plane that could account for Gouy's results. In his lectures, Poincaré professed a complete equivalence of the various dynamical theories of light, whether the direction of vibration was in the plane of polarization or in the perpendicular direction. The equation of motion being the same up to simple changes of variables, the boundary conditions could always be adjusted to produce this equivalence. In the electromagnetic theory of light, which Poincaré taught in his next set of lectures but to which Stokes never paid much attention, the electric field vector **E** must be chosen perpendicular to the conventional plane of polarization, and the question whether this vector or the perpendicular magnetic vector **B** represents a mechanical displacement becomes subsidiary. The issue gradually vanished when physicists ceased to look for mechanical models of the ether.[48]

In retrospect, Stokes's considerable theoretical and experimental efforts to confirm Fresnel's choice of the direction of optical vibrations were in vain. His theory of diffraction has been largely forgotten, although its predictions completely agree with Kirchhoff's theory of 1882. It nonetheless was the first successful attempt to justify Fresnel's intuition of the diffraction process by more rigorous dynamical reasoning. Stokes's way of reasoning by slicing-up the incoming wave was ingenious, and can today be made entirely rigorous by distribution theory.[49] The only uncontrolled element of the reasoning is the assumption that the slices of the wave are simply truncated by the screen, without significant alteration of their amplitude within the screen's opening. Kirchhoff's theory depended on a similar assumption. Its historical success has to do with its reliance on the Green–Helmholtz theorem, which yields the desired formula in a more direct relation with the boundary conditions.

On the Most General State of Polarization of a Light Beam

In the *Mathematical Tracts* Stokes admired and frequently cited, Airy had interpreted natural light as a mixture of elliptically polarized wave trains with random orientation of the axis of the ellipse. Puzzled by this concept, Stokes performed a simple calculation of the optical behaviour of the mixture and found it to be partially polarized, against Airy's intuition to the contrary. He was thus led to study the more general question of the properties of an arbitrary incoherent mixture of polarized waves, which forms the subject of his memoir 'On the composition and resolution of streams of different light from different sources'. As he writes:[50]

> Certain difficulties seem to have arisen respecting the connexion between common and elliptically polarized light which it needed only a more detailed study of the laws of combination of polarized light to overcome ; and accordingly the subject may be deemed not wholly devoid of importance.

In 1822, Fresnel had characterized the most general state of polarization of a transverse plane sine wave as one in which the displacement of a particle of the ether in a wave plane is elliptical. In symbols, the vibration has the general form $\mathrm{Re}[\mathbf{E}e^{i(\omega t - kz)}]$ in which the complex vector **E** can be written as

$$\mathbf{E} = a_x e^{i\varphi_x}\,\hat{\mathbf{x}} + a_y e^{i\varphi_y}\,\hat{\mathbf{y}}.$$

The choice of the perpendicular unit vectors $\hat{\mathbf{x}}$ and $\hat{\mathbf{y}}$ in the xy plane is arbitrary, and the amplitudes and phases depend on it. Rectilinear polarization and circular polarization are special subcases of elliptical polarization. As Fresnel demonstrated, any elliptical polarization can be produced by passing rectilinearly polarized light through a quartz plate of proper cut and thickness (the projections of the original vibration on two perpendicular axes in the plane of the plate being subjected to a different retardation). Conversely, the state of polarization of a wave can be tested by passing it through a quartz plate and a polarizer. From a formal point of view, this combination may be characterized by a unit complex vector $\mathbf{u}$ representing the state of polarization it would itself produce. In a simple generalization of Malus's law, the intensity $|\mathbf{E}|^2$ of the light of original amplitude $\mathbf{E}$ is reduced to $|\langle \mathbf{E} | \mathbf{u} \rangle|^2$ after passing the combination. Stokes did not use the complex Cauchy–Jones vector $\mathbf{E}$ and he of course did not know about Hermitian products. He characterized a polarized wave by a pair of trigonometric expressions giving its amplitude in two orthogonal directions, and he had a complicated trigonometric expression in place of $|\langle \mathbf{E} | \mathbf{u} \rangle|^2$. In the following, I go on exploiting the Hermitian structure in order to simplify the derivation of Stokes's results.

According to Stokes, an arbitrary beam of monochromatic light may be regarded as an incoherent superposition of waves polarized in the above sense, with the amplitudes $\mathbf{E}_1, \mathbf{E}_2, \ldots \mathbf{E}_r, \ldots$. The resultant intensity after passing the testing combination of polarizer and quartz plate is given by

$$I_{\mathbf{u}} = \sum_r |\langle \mathbf{E}_r | \mathbf{u} \rangle|^2 = \sum_r |\langle \mathbf{u} | \rho_r | \mathbf{u} \rangle|^2 = |\langle \mathbf{u} | \rho | \mathbf{u} \rangle|^2, \text{ with } \rho_r = |\mathbf{E}_r \rangle \langle \mathbf{E}_r | \text{ and } \rho = \sum_r \rho_r.$$

Therefore, the most general state of polarization of the beam is represented by a Hermitic operator ρ implying four independent constants. Evidently, the mixing of two polarized waves is sufficient to produce the most general state. In addition, this state can always be represented as a mixture of a completely unpolarized beam with a completely polarized beam. Indeed, in a basis $(\mathbf{u}, \mathbf{v})$ for which the operator ρ is diagonal, we have

$$\rho = \begin{pmatrix} I & 0 \\ 0 & J \end{pmatrix} = J \begin{pmatrix} 1 & 0 \\ 0 & 1 \end{pmatrix} + (I - J) \begin{pmatrix} 1 & 0 \\ 0 & 0 \end{pmatrix}.$$

Taking $I > J$, the first term of this decomposition represents a completely unpolarized beam of intensity J and the second term represent a beam of intensity $I - J$ polarized along $\mathbf{u}$. In the original basis $(\hat{\mathbf{x}}, \hat{\mathbf{y}})$, the coefficients of the ρ matrix are completely determined by four measurements: the total intensity A of the beam, the difference B of the intensities produced by polarizers in the directions $\hat{\mathbf{x}}$ and $\hat{\mathbf{y}}$, the difference C of the intensities produced by polarizers in the directions $(\hat{\mathbf{x}} + \hat{\mathbf{y}})/\sqrt{2}$ and $(\hat{\mathbf{x}} - \hat{\mathbf{y}})/\sqrt{2}$, and the difference of the intensities produced by the circular polarizers associated with the complex vectors $(\hat{\mathbf{x}} + i\hat{\mathbf{y}})/\sqrt{2}$ and $(\hat{\mathbf{x}} - i\hat{\mathbf{y}})/\sqrt{2}$. These definitions yield

$$A = \rho_{xx} + \rho_{yy}, \ B = \rho_{xx} - \rho_{yy}, \ C = \rho_{xy} + \rho_{yx}, \ D = i(\rho_{xy} + \rho_{yx}),$$

which imply

$$\rho_{xx} = \frac{A + B}{2}, \ \rho_{yy} = \frac{A - B}{2}, \ \rho_{xy} = \frac{C - iD}{2}, \ \rho_{yx} = \frac{C + iD}{2}.$$

The coefficients (A, B, C, D), which Stokes uses to characterize a partially polarized beam, are now called the Stokes parameters and they are commonly used to characterize incoherent polarized light, for instance in astronomy.[51]

There is an evident analogy between Stokes's incoherent mixtures of plane-polarized waves and mixtures of pure states in quantum mechanics, so much so that physicists nowadays use Stokes parameters to represent the spin polarization of electron beams. Stokes's 'capital theorem' that an incoherent mixture of any number of polarized waves can always be reduced to a mixture of two states of polarization foreshadows the quantum-mechanical theorem according to which a mixture of an arbitrary number of pure states can always be reduced to a mixture of two states when the state space has two dimensions only. What Stokes modestly described as 'an uninviting subject of investigation' that may nonetheless be 'deemed not fully devoid of importance' truly pointed to an important structure of modern physics.[52]

Double Refraction and Optical Rotation

The British Association Report of 1862

In 1862 Stokes accepted an invitation to report on physical optics for the British Association. Much time had elapsed and too much had happened since Lloyd's report on the same topic in 1835. For this reason, Stokes decided to confine himself to one subtopic only. Somewhat surprisingly, he avoided diffraction and fluorescence, the two subtopics to which he had most contributed, and he picked double refraction. In the early years of the wave theory of light, Fresnel's most spectacular success had been his mathematical theory of the propagation of light in crystals, which, together with the Huygens–Fresnel principle, accounted for the double refraction of light at the surface of any crystal, with amazing precision. Fresnel's theory was based on general principles, on his faith in Huygens's earlier results for uniaxial crystals, and on some intuition of the elastic response that could exist in a molecular ether. As we know in retrospect, the laws and constructions he obtained are exactly the ones resulting from the electromagnetic theory of light. Before such derivation became possible, a few mathematical physicists tried to retrieve Fresnel's laws by means of a dynamical theory of the ether regarded as an elastic medium. The purpose of Stokes's memoir was to critically assess the main attempts of this kind.[53]

Stokes begins with Cauchy's theory of 1830, in which the transparent medium is identified with a set of point-like particles interacting through central forces. In an anisotropic medium of this kind there are six elastic constants if the medium has three mutually orthogonal planes of symmetry (as is empirically true for any optical crystal). By means of three ad hoc supplementary relations between these constants, Cauchy could approximately derive Fresnel's laws of propagation for approximately transverse vibrations. The reasoning required the vibration to lie in the plane of polarization, against Fresnel's assumption. In 1836, Cauchy produced a second theory in which the medium had three additional elastic constants due to an initial state of stress, and the direction of vibration agreed with Fresnel's choice. Stokes rejected both theories for failing to provide exactly transverse waves, for appealing to purely ad hoc constraints on the elastic constants, and for allowing longitudinal waves of velocity and intensity comparable to those of the observed transverse waves.[54]

In his early theory of aberration, Stokes had already noted that the ether had to be incompressible in order to exclude longitudinal vibrations. Cauchy's molecular theory of elasticity did not allow incompressibility, since it implied an elastic response of the same order of magnitude for the mutual sliding and the compression of successive layers of molecules. In the isotropic case, this theory yielded a single independent elastic constant and therefore excluded media like gum, jelly and the ether, which resist compression much more than they resist distortion. Stokes repeated this criticism he had made many years earlier in his memoir on fluid friction, and he again praised Green for introducing the necessary degree of generality in the theory of elasticity.

Green had addressed wave propagation in an anisotropic elastic medium in an optical memoir of 1839. As Stokes explained, Green used a Lagrangian approach in which the equations of motion derived from expressions for the kinetic and potential energies of the continuous medium. He assumed the most general form of the elastic energy density in a stressed medium. As there are six independent stress components, the quadratic form representing the elastic energy has twenty-one coefficients instead of fifteen in the Cauchy case. This additional freedom allowed Green to choose the constants so that strictly transverse vibrations were possible and Fresnel's laws exactly held for these vibrations. In one version of his theory, there is no initial stress in the medium and the direction of vibration is in the plane of polarization; in another, there is initial stress (implying six additional constants of elasticity) and Fresnel's choice can be adopted for the direction of vibration. In both cases, the elastic constants can be chosen so that the longitudinal waves have an infinite velocity.[55]

Stokes judged Green's theory to be an enormous improvement over Cauchy's, although he did not regard it as quite definitive. In its second version, it shared with Cauchy's the difficulty of requiring ad hoc conditions on the elastic constants. Moreover, Green's theory failed to account for Fresnel's laws regarding the intensity of reflected and refracted light, although Green used the correct boundary conditions at the interface between two different elastic solids (the equality of the displacements of the medium and the balance of pressures).[56] His predecessors had been able to retrieve Fresnel's exact laws only at the price of ad hoc boundary conditions. Owing to his narrow focus on propagation, Stokes did not address these difficulties regarding the intensity laws. Had he done so, he would have had to tone down his praise of Green's theory.

There are two ways out of this difficulty. The first, imagined by Cauchy in 1839 and reactivated by William Thomson in 1888, is still based on the theory of elastic solids but relies on a new way of eliminating the unwanted longitudinal waves. Whereas Fresnel, Thomson and Stokes assumed an incompressible ether in which the longitudinal waves had infinite velocity, in his third theory Cauchy adjusted the two elastic constants of an isotropic solid so that the this velocity would vanish.[57] He found that in this case the true boundary condition yielded Fresnel's laws of the intensities and reflected and refracted light at the interface of two isotropic media. Stokes did not even mention this theory in his report, because he knew from Green that the compressibility of the ether was negative in this case, implying instability. When Thomson rehabilitated this theory in 1888 under the name of 'labile ether', his first task was to show that an infinite medium of negative compressibility was not necessarily unstable. Retrospectively, the labile ether theory can be regarded as a mechanical interpretation of Maxwell's equations in which the vector **D** (electric displacement) is the mechanical vibration. This explains why it properly accounts for Fresnel's laws.[58]

Another way to retrieve Fresnel's laws was to depart from the received theory of elastic solids and to choose the form of the elastic energy density so that it would lead to the desired laws. This is what the Irish physicist and mathematician James MacCullagh did in 1839, the year of Cauchy's third theory and of Green's aforementioned theories. His expression of the elastic energy, $(1/2)K(\nabla \times \mathbf{u})^2$ in the isotropic case, leads to the equation of motion

$$\rho\, \ddot{\mathbf{u}} = -\nabla \times (K\nabla \times \mathbf{u})$$

through the Lagrangian procedure. The corresponding boundary conditions are the continuity of **u** and the continuity of the parallel component of $K\nabla \times \mathbf{u}$ at the interface between two different media. These equations and conditions can also be regarded as a mechanical interpretation of Maxwell's equations in which the vector $K\nabla \times \mathbf{u}$ plays the role of the electric field $\mathbf{E}$ (and $\rho\dot{\mathbf{u}}$ the role of the magnetic field $\mathbf{B}$). MacCullagh also considered anisotropic media for which the constant K is replaced by a symmetric operator, and he was the first to obtain the correct laws of intensity for the reflected and refracted light at the surface of a crystal.[59]

In his review, Stokes gave a full account of MacCullagh's theory but denounced its incompatibility with the orthodox theory of elasticity. MacCullagh's theory indeed leads to an antisymmetric stress system in the medium, whereas according to an old argument by Cauchy any mechanical stress system has to be symmetric. From a physical point of view, this means that the elastic response of the ether depends on the absolute rotation of its elements, a mechanical nonsense in Stokes's opinion. Many years elapsed before Joseph Larmor explained that such resistance to absolute rotation could be generated by mounting tiny flywheels in the ether, as Thomson had already done for other purposes.[60]

At the time of Stokes's report, Maxwell's 'On physical lines of force' was being published in the *Philosophical Magazine*. Although this memoir contained a first form of Maxwell's equations and a first attempt to derive optics from electromagnetism, the full derivation of Fresnel's laws on the basic of these equations waited the contributions of Hendrik Antoon Lorentz in 1875 (through Helmholtz's version of the theory) and George Francis FitzGerald in 1880 (through MacCullagh's ether).[61] Surely Stokes cannot be blamed for ignoring possible connections with electromagnetic theory in his report. With more justification, he has sometimes been blamed for the long neglect of MacCullagh's theory. In Stokes's conservative view, the ether had to be a limiting case of substances already known to us, such as jelly. In MacCullagh's modernist view, qualitative departure from the mechanical behaviour of ordinary substances was tolerable—and even to be expected—for a medium as subtle as the ether. Stokes recognized this difference as well as the merits of MacCullagh's approach:[62]

> Indeed MacCullagh himself expressly disclaimed having given a mechanical theory of double refraction. His methods have been characterized as a sort of mathematical induction, and led him to the discovery of the mathematical laws of certain highly important optical phenomena. The discovery of such laws can hardly fail to be a great assistance towards the future establishment of a complete mechanical theory.

To summarize, in his report Stokes asserted his preference for a continuum theory of the ether based on analogy with ordinary elastic bodies. He rejected Cauchy's molecular approach because the imagined molecular structure did not yield a sufficiently general kind of elasticity and because, more broadly, he preferred macroscopic methods based on Cauchy's concepts of stress and strain in a continuous medium and on the associated Lagrangian. This explains his enthusiasm for Green's theory, although it could not account for all of Fresnel's laws. He did not mention his diffraction-based argument in favour of theories, such as Fresnel's and (one of) Green's, in which the optical vibration is perpendicular to the plane of polarization. He had other reasons to disfavour the complementary class of theories: they relied on arbitrary boundary conditions (in Neumann's and Kirchhoff's cases) or they assumed a kind of elasticity incompatible with the general laws of mechanics (in MacCullagh's case). He dealt only with theories in which the combination of ether and matter was regarded as a single medium with specific constants of inertia and elasticity. He judged this supposition 'at first sight unnatural' and he briefly mentioned a few attempts by Lloyd, Cauchy, Challis and Robert Moon to take into account the interaction of the ether with the molecules of matter. He clearly did not think much of these attempts. At the same time, he did not believe that the received one-medium theories were the end of the story: 'In concluding this part of the subject, I may perhaps be permitted to express my own belief that the true dynamical theory of double refraction has yet to be found.'[63]

An Unpublished Theory of Double Refraction

Stokes was not himself a total stranger to the double-medium molecular theories. Around 1843 he had speculated that in a model of rigid molecules immersed in a liquid ether (as also used in his theory of aberration of 1845), the ether could acquire an effective anisotropic inertia if the molecules had an ellipsoidal shape. As was earlier mentioned, he reasoned by analogy with the effective inertia of immersed bodies owing to their dragging a portion of the fluid, which he derived in a contemporary memoir on fluid dynamics. In the resulting theory of double refraction, the equation of the optical vibrations is

$$[\rho]\ddot{\mathbf{u}} = K\Delta\mathbf{u} \text{ (with } \nabla\cdot\mathbf{u} = 0),$$

wherein K is the elastic constant and $[\rho]$ is the symmetric positive operator representing the anisotropic inertia. Consider a monochromatic plane wave of the form $a\hat{\mathbf{x}}e^{i(\omega t - \mathbf{k}\cdot\mathbf{r})}$, where $\hat{\mathbf{x}}$ is the unit vector giving the direction of the vibration. The wave equation implies

$$\hat{\mathbf{x}}\cdot[\rho]\hat{\mathbf{x}}\omega^2 = Kk^2, \text{ and } V^{-2} = k^2/\omega^2 = K^{-1}\hat{\mathbf{x}}\cdot[\rho]\hat{\mathbf{x}}$$

for the plane-wave velocity V as a function of the direction of the vibration. This is to be compared with the consequence

$$V^2 = \rho^{-1}\hat{\mathbf{x}}[K]\hat{\mathbf{x}}$$

of Fresnel's theory in which the inertia is isotropic and the elastic response anisotropic.[64]

Combining the new expression of the phase velocity with the Huygens–Fresnel construction, Stokes arrived at an alternative theory of light propagation in anisotropic media. As he later explained,

> I refrained…from putting forward that theory…because, on calculating the difference of refraction of the extraordinary ray on this theory and according to Huyghens's construction, at about 45° from the axis, where the difference would be greatest, I found it barely small enough, as seemed to me, to have escaped detection.

In his report on double refraction, Stokes briefly alluded to 'a certain physical theory' that led to $V^{-2} = K^{-1}\hat{\mathbf{x}}\cdot[\rho]\hat{\mathbf{x}}$ and called for accurate measurements of extraordinary refraction in order to decide between the still open theoretical alternatives. He also described a new method measurement, which he himself applied in 1866 to Iceland spar. His results, published in 1872, confirmed Fresnel's theory to the fourth decimal and definitely excluded his own theory (whose predictions departed from Huygens's by 1 per cent for an incidence of 45°).[65]

Optical Rotation

Double refraction being a consequence of the anisotropy of the medium on the propagation of light, it was and still is customary to discuss it in parallel with the effect of other asymmetries on the propagation of light. These include the rotation of the plane of polarization of light when traversing 'optically active' media such as sugar solutions or when passing through magnetized glasses (the Faraday effect, discovered in 1845). Stokes did not address these challenges to optical theory in his report on double refraction or in any of his published writings. Yet he had privately discussed them with William Thomson. In 1848 he 'worked inductively' from the experimental data on these effects to arrive at the following equations of motion for the optical medium:

$$\ddot{\mathbf{u}} = c^2 \Delta \mathbf{u} + \alpha \Delta (\nabla \times \mathbf{u})$$

in the case of ordinary optical rotation, and

$$\ddot{\mathbf{u}} = c^2 \Delta \mathbf{u} + \gamma (\mathbf{H} \cdot \nabla)(\nabla \times \dot{\mathbf{u}})$$

for the Faraday effect in a magnetic field of intensity **H**. The first equation is a straightforward generalization of equations MacCullagh had obtained in 1836 for optical rotation. The second is compatible with the generic form of the equations proposed by Airy in 1846, and it is strictly identical with the equation derived by Maxwell in his own theory of the Faraday effect, given in 1862 in 'On physical lines of force'. Stokes privately communicated his equations to Thomson but never published them.[66]

When, in early 1856, Thomson asked Stokes whether some crystals spontaneously produced the Faraday effect, Stokes replied that they could not do so 'without violating the principle of reversion'. Remember that in 1849 he had used this principle to explain the perfect blackness of the central spot of Newton's rings. He meant that in a mechanical system for which all forces depend on the spatial configuration only, to any possible process there corresponds another possible process obtained by time reversal. Therefore, the rotation of the plane of polarization of light during its propagation should be exactly reversed during the reverse propagation, as is indeed the case for optically active substances. In contrast, the Faraday effect produces the same rotation in both directions of propagation. This is so, Stokes tells us, because in a magnetized medium it is no longer true that all forces depend on the spatial configuration only.[67]

A few months later in the *Philosophical Magazine*, Thomson argued that the proximate cause of the Faraday effect was a rotation occurring at small scale in the medium, with the same symmetry as the electric currents responsible for the magnetic field. His analysis was based on the well-known equivalence of the Faraday effect with a different propagation velocity for right-hand and left-hand circularly polarized light travelling in the direction of the magnetic field. In Thomson's mind, there could be only two possible mechanical causes for this difference: either a spatial heterogeneity, or a local circulation of the medium around the magnetic lines of force. The peculiar symmetry of the Faraday effect excluded the first option. Whereas Thomson believed his argument to be 'unanswerable', Stokes remained unconvinced that the Faraday effect required motions going on independently of the optical vibration, because his own attempt to derive a mechanical model of this kind had led to a wrong law for the dependency of the optical rotation on the wavelength (in contrast with his model-independent wave equation of 1848).[68]

Maxwell did not share Stokes's reticence. On the contrary, he took Thomson's argument as the proof that magnetic fields always implied a kind of rotation in the electromagnetic medium, even in a vacuum. As is well known, this idea was the basis of the mechanical model that led him to the full set of Maxwell's equations. Naturally, he used this model for a theory of the Faraday effect and thus rediscovered the magnetically modified wave equation that Stokes had communicated to Thomson a few years earlier. This was, for Stokes, a story of missed opportunities: he found and did not publish a correct equation of propagation for the optical vibration in a magnetized medium; he invented the symmetry argument that excluded any explanation of the Faraday effect based on a static medium and yet let Thomson publish an inferior and plausibly not unrelated proof of the necessity of rotations in the medium; he failed to find a mechanical model for his equation of propagation, and Maxwell did it for him on the basis of Thomson's rotations.

CHAPTER 5

Stokes's Optics 2

Other Phenomena in Light

OLIVIER DARRIGOL

Fluorescence and Spectra

The Discovery of Fluorescence

In the early nineteenth century, numerous experiments had been performed to test Newton's old theory of the colour of bodies, according to which the colour is caused by small constitutive particles or molecules acting as thin transparent plates. In the 1830s, David Brewster and John Herschel refuted Newton's theory and proved that the colours of bodies resulted from selective absorption of one part of the solar spectrum. They investigated various coloured substances, crystals, and solutions, some of which exhibited what we now identify as fluorescence. In 1845, Herschel read two short notes in front of the Royal Society to announce a phenomenon 'unique in physical optics'. On a dilute solution of sulfate of quinine (and tartric acid) exposed to natural light, he observed (see Fig. 5.1):

> Though perfectly transparent and colourless when held between the eye and the light, or a white object, [the solution] yet exhibits in certain aspects, and under certain incidences of the light, an extremely vivid and beautiful celestial blue colour, which, from the circumstances of its occurrence, would seem to originate in those strata which the light first penetrates in entering the liquid, and which, if not strictly superficial, at least exert their peculiar power of analysing the incident rays and dispersing those which compose the tint in question, only through a very small depth within the medium.

In conformity with the Newtonian doctrine of the immutability of simple colours, Herschel expected this phenomenon to be due to intense selective and diffusive absorption of blue light in a thin superficial layer of the fluid. This interpretation would of course imply a deficiency of blue light in the transmitted light. As Herschel did not observe any tint of the transmitted light, he tried to increase the expected deficiency by iterating the superficial absorption. To his surprise, the transmitted light was completely unable to induce the superficial blue in a second solution of quinine sulfate. Herschel then turned his attention

Darrigol, O., *Stokes's Optics 2 Other Phenomena in Light*. In: *George Gabriel Stokes: Life, Science and Faith*, Mark McCartney, Andrew Whitaker, and Alastair Wood (Eds): Oxford University Press (2019). © Oxford University Press.
DOI: 10.1093/oso/9780198822868.001.0005

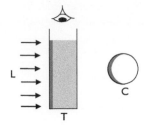

Fig. 5.1 Herschel's epipolic dispersion. The light L from a window in a dark room impacts the tube T that contains a quinine solution. The crescent C of intense blue light is seen from above, next to the illuminated part of the walls of the tube.

to the dispersed blue light. Passing it through a prism, he found it to be heterogeneous. He also had this blue light enter another quinine solution: to his surprise, this light freely traversed the solution and no dispersion occurred. The phenomenon being so new and so strange in his eyes, he gave it the name of 'epipolic dispersion', from the Greek ἐπιπολή for surface. He called the transmitted light 'epipolized' since it enjoyed the new qualitative property of no longer inducing the superficial blue.[1]

Before Herschel's observation, there had been many reports of odd effects of induced colour that we would now interpret as cases of fluorescence. Stokes was aware of older experiments in which Brewster had passed a narrow beam of light through a bright green solution of chlorophyll and seen orange light emerging laterally from the beam. He also knew that Brewster had observed and studied similar effects occurring in fluorite (then called fluor-spar). Brewster interpreted these observations as cases of differential light scattering by impurities. Herschel's observations posed a challenge of higher order, at least for one like Stokes who believed that light, being a transverse wave in the ether, could not have any qualitative property other than frequency and polarization.[2]

After vainly struggling to explain Herschel's observations under Newton's principle of the immutability of simple colours, Stokes realized that these observations were easily explained by assuming that the colour of spectrally simple (monochromatic) light was altered when diffused from the active layer of the quinine solution. In particular, this explained why the diffused light travelled freely through the quinine solution whereas the light that caused it was quickly absorbed. It also explained why the transmitted light could no longer induce superficial diffusion. Stokes then decided to test his assumption through simple arrangements of filters, prisms, diaphragms, lenses and Nicols around a quinine solution. In August 1851, he wrote to Thomson:

> I have not been idle in the scientific line; for I have been following out a rather remarkable physical discovery "Tho' I says it that oughtn't" which I made about week after Easter. I have communicated it to Fischer [Frederick William Lewis Fischer, then professor of natural and experimental philosophy at St Andrews][3] and will do so to you when you come; but unless the sun shines I cannot show you the experiments.

Stokes intensively pursued this matter for about a year, and read from the resulting hundred-page memoir on 27 May 1852 in front of the Royal Society. It won him the Society's Rumford medal in the same year.[4]

In a first series of experiments, Stokes simply placed a smoke-coloured glass, which he knew to absorb the most refrangible rays (violet and ultraviolet), successively before the quinine solution and before the eye (see Fig. 5.2). He found that the secondary blue light disappeared in the first case whereas it was unaltered in the second case. In a second, more decisive kind of

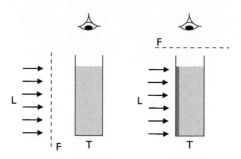

Fig. 5.2 Stokes's first set-up for proving the change of refrangibility. When the filter F, which blocks violet and ultraviolet light, is placed between the light source and the tube, Herschel's blue crescent is no longer seen. It persists when the filter is placed between the eye and the tube.

experiment, he used a chain of prisms to separate the various components of the incoming light. In one of these trials, he simply had the separated light impact the surface of the solution and observed that the light from the visible part of the spectrum excluding the violet penetrated the solution without any abnormal absorption or scattering; in contrast the violet and ultraviolet part of the spectrum induced the intense superficial blue, with a few line-shaped gaps corresponding to the Fraunhofer lines of this part of the solar spectrum. Stokes multiplied experiments in which monochromatic light induced the superficial blue, which he then analysed through a second prism. Even in this pure case, the scattered blue light had a spectrum of finite breadth, with refrangibilities below the refrangibility of the incoming light. This is what came to be known as *Stokes's law*. Its first statement reads: 'The refrangibility of light is *always lowered* by internal dispersion.' Stokes found that the phenomenon occurred in many organic substances and many minerals, although the modified light was often mixed with ordinary scattered light.[5]

In continuity with Herschel's and Brewster's terminology, Stokes called the new phenomenon 'true internal dispersion' and distinguished it from the 'false internal dispersion' or scattering caused by impurities or motes. Note that Stokes defined the new phenomenon as 'the change of refrangibility of light' and not as the change of frequency or wavelength as we would now; not that he had any doubt about the truth of the wave theory of light, but he wanted his discovery to be expressed as a basic, theory-independent fact. This fact had enormous significance for it shook a central dogma of the optics of colours since Newton: the immutability of simple colours during the interaction of light with matter. Even though fluorescence-related observations had long been available, Stokes was first to characterize them as a change of refrangibility and to empirically establish this change. It is therefore appropriate to call him the discoverer of fluorescence.[6]

Stokes introduced the word 'fluorescence' (in place of 'true internal dispersion') in a footnote after suggesting 'diffusive reflexion' for cases of 'true internal dispersion' in which the dispersion is superficial (because the concentration of the solution is too strong or the incoming light is too weak):

> I confess I do not like this term ['diffusive reflexion']. I am almost inclined to coin a word, and call the appearance *fluorescence*, from fluor-spar, as the analogous term *opalescence* is derived from the name of a mineral.

Stokes briefly noted the partial analogy with phosphorescence, the main difference being in the delay of the phosphorescent light after exposure to sunlight. Since it was already known that the composition of the light emitted by a phosphor differed from the composition of the light that excited the phosphorescence, it could be argued that fluorescence was not much of a novelty after all. Stokes deflected this potential belittling of his achievement by noting that in the case of phosphorescence, the phosphorus had naturally been considered as an independent source of light with its own characteristic colour since the emission persisted long after the exposition of the phosphorus to daylight.[7]

Stokes promptly realized the potential of his discovery. It offered a new way to detect ultraviolet light even in the remotest parts of the spectrum, more direct and more practical than the photographic emulsions that had been so far used for this purpose (hence the name 'chemical rays' then used for the ultraviolet light).[8] It could thus be used to determine the ultraviolet absorption spectrum of various substances, the ultraviolet component of the solar spectrum (see Fig. 5.3), and the relative abundance of ultraviolet light in various sources including flames and the electric arc. In addition, information on the frequency of the absorbed light and on the frequency of the accompanying dispersed light could be used for the purpose of chemical analysis: 'It is quite possible that internal dispersion might turn out for importance as a chemical test.' With much zeal and apparent pleasure, Stokes investigated the fluorescence of all the substances on which Brewster had observed internal dispersion, and he inspected many opaque bodies too. He found the power to excite fluorescence to be extremely common in organic materials: 'Thus, wood of various kinds, cork, horn, bone, ivory, white shells, leather, quills, white feathers, white bristles, the skin of the hand, the nails, are all more or less sensitive.'[9]

In a sequel to his memoir, Stokes described a very simple and cheap way of detecting the fluorescence of a substance, by means of two complementary high-pass and low-pass filters. Indirect daylight is let through a diaphragm to illuminate an object. The high-pass filter is placed between the diaphragm and the object; the low pass filter between the object and the eye. In the absence of fluorescence, no light reaches the eye since the second filter blocks the light admitted by the first. When the object is fluorescent, it appears luminous. The luminosity is optimal if the cut-off frequency of the filters coincides with the frequency threshold of the fluorescence. The second filter may then be replaced by a prism in order to determine the spectrum of the fluorescence light. Stokes described four approximate realizations of pairs of complementary filters through coloured solutions and tinted glasses. Again, he hoped that the method 'might be directly exploited by chemists in discriminating between different substances'.[10]

Stokes repeatedly concretized this hope in the following years. In 1859, he proved that horse-chestnut bark, which Herschel and himself had used to prepare highly fluorescent extracts, contained a mixture of two different active chemicals in various proportions. Similarly, he used

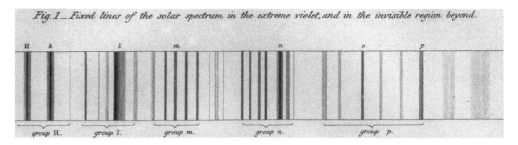

Fig. 5.3 Stokes's mapping of the ultraviolet part of the solar spectrum. From G. G. Stokes, 'On the Change of Refrangibility of Light', *Philosophical Transactions of the Royal Society of London* 142 (1852), plate.

Fig. 5.4 Stokes's 'long spectrum' of a few metals. R denotes the extreme red. Only the lines in the invisible part of the spectrum are represented. From G. G. Stokes, 'On the Long Spectrum of Electric Light', *Philosophical Transactions of the Royal Society of London* 152 (1862), p. 606.

optical fluorescence to distinguish between alizarine and purpurine in 1860, between biliverdin and chlorophyll, and between the reduced and oxidized forms of haemoglobin in 1864. He then advertised the merits of fluorescence-based and absorption-based optical techniques for chemical analysis at the Royal Institution and in front of the Royal Chemical Society.[11]

Fluorescence also served to explore absorption and emission spectra in the ultraviolet region. While preparing a lecture for the Royal Institution in 1853, Stokes exploited the powerful electrical apparatus of the Institution to produce 'long spectra' in which the source was the light from an electric arc, whose spectrum turned out to extend very far beyond the ultraviolet. Having earlier determined that quartz was much more transparent to ultraviolet light than regular glass, Stokes formed his spectra with quartz prisms and lenses. In the following years, he discovered the line spectra produced by the metals used as electrodes in the electric arc, and he also studied the absorption spectra of various organic spectra in the ultraviolet regions. After becoming aware of similar results that his friend Miller had obtained in Oxford by photographic means, he published the result in 1862 including the plate of Figure 5.4 for the spectra of three metals.[12]

A Molecular Mechanism

In his memoir of 1852, Stokes kept theory to a minimum, to the point of preferring 'refrangibility' to 'wavelength' or 'frequency' in his labelling of simple colours. At the end of his memoir, he nonetheless speculated about a mechanism for the frequency of the waves to be diminished in their interaction with matter. He imagined that the molecules of the fluorescent matter were non-linear oscillators, brought to oscillate at their own frequency by the incoming wave. In his opinion, the strongly non-linear character of the intramolecular oscillations could be used to explain both the spread of their frequency and their being slower than the frequency of the incoming light. The argument involved a fanciful mechanism which Stokes offered only as a tentative guess. A few years later, he would give up this mechanism and instead develop an analogy with the solar heating of a dark object followed by infrared emission at a lower frequency. In this case, the emission is obviously due to local heating spread through the object by conduction, with decreasing frequency of the emitted radiation as the temperature diminishes. Stokes similarly imagined that fluorescent light depended on the transfer of the excitation of a given molecule to neighbouring molecules. He modelled this process by the periodic excitation of one of the masses of a discretely loaded elastic string. In such a string, excitations of frequency inferior to the cut-off frequency of the string are quickly propagated away. In contrast, excitations of frequency above the cut-off reach a sufficient intensity and slowly diffuse to the other masses along the string, thus implying more and more frequencies below the cut-off.[13]

One thing Stokes judged beyond doubt was that the fluorescence was caused by 'molecular disturbances' themselves induced by the incoming light. For the sake of simplicity, he assumed a similar mechanism to be responsible for the absorption of light, except that in the latter case the energy absorbed from the incoming light had to be redirected to a different channel (possibly undetected fluorescence). In some cases of absorption (by chlorophyll for instance), Stokes assumed that the molecules did not differ much from harmonic oscillators, which implied strong absorption of light at the frequency of the oscillators. Talbot and Herschel had earlier connected selective absorption to resonance, but in a contrary manner: they assumed that resonance facilitated the transmission of light.[14]

Solar Spectroscopy

In February 1854, William Thomson asked Stokes a few questions, including 'Have you made any more revolutions in science?' (a probable allusion to the discovery of fluorescence) and:

> I want to ask you about artif[l] lights and the solar dark lines. Is there any other substance than soda that is related to *D*? Are bright lines corresponding to it to be seen where soda is not present? Have any terrestrial relations to any other of the solar dark lines been discovered (or to the dark lines of any of the stellar spectra)?

As Thomson and Stokes knew, in 1817 the Bavarian telescope maker Joseph Fraunhofer had observed fixed dark lines in the spectrum of the light from the sun and from the stars with a flint prism and a theodolite. He had also observed a bright narrow doublet of yellow lines in the spectrum of the light from burning lamps, and found it to be exactly at the same position as a doublet of dark lines in the spectrum of sunlight, which he called D. This repeatedly confirmed coincidence remained a mystery at the time of Thomson's query to Stokes. It was also known that the addition of sodium to a flame enhanced the D line, but the line was commonly seen in many sources of light.[15]

In his reply to Thomson's letter, Stokes mentioned that the flame of pure alcohol did not contain the D line, and opined that its frequent occurrence in other sources was probably due to traces of sodium. He also mentioned Brewster's observation of emission lines (in a flame from saltpetre) that fitted a few other Fraunhofer lines, and he referred Thomson to Abbot Moigno's *Répertoire d'optique* for Miller's observations of dark bands and lines in the absorption spectrum of coloured vapours. Moigno introduced his account of Miller's experiments with the words:

> The search for the cause of the dark lines of the solar spectrum is still fraught with difficulties and obscurities. The stakes of a solution of this frustrating question are so high, concerning the nature of light, that we should welcome any relevant facts, even if they do not lead to a complete solution.

Moigno thus alluded to an analogy between the Fraunhofer lines and the absorption lines of vapours.[16]

Reacting to Stokes's letter, Thomson wrote:

> I think it is really a splendid field of investigation, that of the relations betw. the bright lines of artif[l] light and the dark lines of the solar spect[m]. Don't you think anyone who takes it up might

find a substance for almost each one of the principal dark lines, by examining the effects of all salts on the flame of burning alcohol, or on other artif¹ lights?...Will you not take up the whole subject of spectra, of solar & artif¹ lights, since you have already done so much on it. I am quite impatient to get another undoubted substance besides vapour of soda in the sun's atmosphere. What you tell me looks very like as if there is potash too [since there is potassium in saltpetre].

In his next letter, Stokes recalled an (unpublished) experiment in which Miller had confirmed with utmost precision the coincidence between the D line in the solar spectrum and the D line from a lamp, with the comment:

> It seemed to me that a plausible physical reason might be assigned for it by supposing that a certain vibration capable of existing among the ultimate molecules of certain ponderable bodies, and having a certain periodic time belonging to it, might either be excited when the body was in a state of combustion, and thereby give rise to a bright line, or be excited by luminous vibrations of the same period, and thereby give rise to a dark line by absorption.—But we must not go on too fast. This explanation I have not seen, so far as I remember, in any book, nor do I know a single experiment to justify it. I am not aware that any absorption bands seen in the spectrum of light transmitted across any vapour that has been examined have been identified with D.

Just like Moigno, Stokes was here suggesting that the solar dark lines had to do with absorption by vapours in the sun's atmosphere. In addition, he offered a mechanism relating emission and absorption, in the spirit of his earlier explanation of fluorescence by induced vibrations. He did so with due caution, knowing that Miller had not found the D doublet in his absorption spectra. A few months later, he wrote a short note to Thomson to tell him about an experiment of 1849 in which Léon Foucault had seen the dark D doublet in the spectrum of light originally devoid of this doublet and passed through an electric arc whose own light included a bright D doublet. This observation confirmed his intuition of a general correlation between emission and absorption. Yet Stokes did not follow Thomson's suggestion to further investigate the matter.[17]

Four years later, Gustav Kirchhoff and his chemist colleague Robert Bunsen at Heidelberg studied the spectrum of pure salts in a Bunsen burner with unprecedented care and precision and were thus able to produce line spectra that turned out to be characteristic of the metal in the salt. In order to test whether these lines fitted some of the Fraunhofer lines, Kirchhoff and Bunsen passed sunlight through the salt-flame and examined its spectrum. When the sunlight was relatively weak, they observed bright lines fitting a few Fraunhofer lines; for stronger sunlight, they were very surprised to see that the salt-flame enhanced the dark lines. High temperature being the cause of the emission of light both for the sun and for the salt-flame, Kirchhoff sought light in a theoretical study of the thermal equilibrium of two facing radiators. For the radiative equilibrium to be possible, he found that the ratio of the absorptive power and the emissive power for a substance at a given temperature for radiation of a given frequency had to be a universal function of temperature and frequency. This law explained the observed correlation between emission lines and absorption lines. It became the theoretical basis for Bunsen and Kirchhoff's grand project of a chemical analysis of solar and stellar atmospheres.[18]

From the above, it is clear that Thomson and Stokes privately anticipated crucial components of the Kirchhoff–Bunsen project: they expected a correlation between absorption lines in the solar atmosphere and the emission lines in artificial light and they foresaw the use of this correlation for chemical analysis from a distance. Stokes also imagined a mechanism justifying this correlation, and he found it to be (partially) confirmed by Foucault's observation of the spectrum of light passed through an electric arc. When Stokes learned about Kirchhoff's discovery of the inversion of lines in salt-flames, he published a translation of Foucault's and Kirchhoff's relevant texts as well as the following variant of his absorption–emission mechanism:[19]

> We know that a stretched string which on being struck gives out a certain note (suppose its fundamental note) is capable of being thrown into the same state of vibration by aerial vibrations corresponding to the same note. Suppose now a portion of space to contain a great number of such stretched strings, forming thus the analogue of a 'medium.' It is evident that such a medium on being agitated would give out the note above mentioned, while on the other hand, if that note were sounded in air at a distance, the incident vibrations would throw the strings into vibration, and consequently would themselves be gradually extinguished, since otherwise there would be a creation of *vis viva*. The optical application of this illustration is too obvious to need comment.

Some of Stokes's friends used his private communications to Thomson and the fact that Thomson had long been teaching the correspondence between dark and bright lines in Glasgow to argue Stokes's priority in the discovery of spectral analysis as applied to celestial bodies.[20] Thomson himself wrote in 1862 about 'Stokes's principles of Solar and Stellar Chemistry' he had been teaching 'for the last eight or nine years' and referred to Kirchhoff and Bunsen's 'independent discovery of Stokes's theory'. Stokes himself rejected this honour. To be true, his considerations were formulated in a tentative manner only; they remained unpublished; and they did not serve to found a new experimental project. In a letter to Roscoe of February 1862, Stokes emphasized the lack of publication:

> My share in the history of solar chemistry, I look upon it, is simply nil ; for I never published anything on the subject, and if a man's conversations with his friends are to enter into the history of a subject there is pretty nearly an end of attaching any mention or discovery to any individual.

In addition, Stokes later explained that prior to Kirchhoff's announcement he had not expected the lines-inversion phenomenon in flames because in his view the absorption of light at the frequency of the D line required the presence of free sodium, to be expected at the extremely high temperature of the sun but not in the much lower temperature of flames (in contrast, he believed the emission of the D line to be possible even for bound sodium atoms). Therefore, he credited Kirchhoff for the discovery of inversion in salt-flames and for the general thermodynamic argument that made absorptive power a necessary consequence of emissive power in a heated substance.[21]

A last problem to which Stokes's resonance-based interpretation of selective absorption might have been relevant is anomalous dispersion. In some exceptional cases, the optical index of a transparent bodies increases with the wavelength. The first systematic study of this anomaly occurred in 1871, when the German experimentalist August Kundt found the anomalous dispersion of dies to be related to 'surface colour' in Herschel's sense. In the light of Stokes's memoir of 1852, this suggested that anomalous dispersion, like surface colour, depended on intense selective absorption. Kundt indeed found that an absorption line occurred at the wavelength separating two contiguous regions of normal and anomalous dispersion. This finding confirmed Thomas Young's old idea that dispersion in general had to do with the coupling between the vibrations of the ether and the vibrations of material oscillators embedded in the ether, and Wolfgang Sellmeier's newer idea that the frequency of these oscillators determined both the frequency of the absorption line and the separation between regions of normal and anomalous dispersion. Sellmeier and Hermann Helmholtz authored the first theories of dispersion based on molecular oscillators. Even though Stokes had pioneered the concept of optically induced molecular resonance, he remained unconcerned by these developments. When Thomson, in early 1884, asked him whether he knew about Helmholtz's memoir of 1875 on this topic, his reply was: 'I have just dipped into a paper or two, I forget by whom, as to anomalous dispersion. I took for granted the explanation offered would be my making mu [the optical index] imaginary.'[22]

To sum up, Stokes's reflections on the mutual relations of absorption and emission and on a tentative mechanical explanation of these relations had potential bearing on the spectroscopic mystery of the coincidence of stellar absorption lines and terrestrial emission lines; they could also have informed the interpretation of anomalous dispersion. Yet, Stokes satisfied himself with private remarks and refused to venture too far from the context of his original work on fluorescence.

Optical Wonders

All through his scientific life, Stokes was alert to colourful optical curiosities reported by some of his contemporaries. In the case already discussed of Herschel's epipolic dispersion, this led him to a fundamental discovery. In other cases, he did his best to explain the singular phenomenon by received theory, or at least to clarify the circumstances.

In 1844 the Austrian mineralogist Werner Haidinger discovered the *Lichtpolarisationsbüschel*—soon to be called Haidinger's brushes—while examining the light refracted by some of his crystals. Whenever polarized light from a uniformly lit object (a portion of the sky for instance) enters the eye, a trained observer can see a tiny cross-shaped double brush, oriented according to the direction of polarization and coloured in blue and yellow. In 1850, Stokes explained these colours after investigating the brushes produced by monochromatic polarized light. In 1855, he rejected Jules Jamin's explanation through successive refractions in the eye and propounded that the phenomenon depended on the way the nervous fibres in the retina were stimulated by light. This may be regarded as a vague anticipation of Helmholtz's later explanation through the dichroism of some tubes in the yellow spot of the retina.[23]

Dichroism is the property of some materials, such as tourmaline, to absorb light in one direction of polarization and transmit the light polarized in the perpendicular direction. In 1852 Stokes received from the surgeon and chemist William Bird Herapath crystals of a special salt of quinine (iodoquinine sulfate) that enjoyed double refraction, double absorption, and double metallic reflection. He studied this intriguing mix of anisotropic properties and suggested that the remarkably strong dichroism of this salt might be used some day to replace tourmaline in the fabrication of polarizers, if only the quinine salt crystals could be prepared in sufficient quality and quantity. This is what Edwin Land achieved around 1930 with his invention of Polaroid sheet polarizers. Stokes returned to herapathite in 1854 in a discussion with Haidinger about the relation between the tints of reflected and absorbed light in dichroic crystals. He then rejected Haidinger's idea of determining the direction of the optical vibration (in the plane of polarization) through dichroism (and informed Haidinger about his own diffraction-based determination).[24]

In 1853, Stokes explained the anomalous rings observed by William Crookes in photographs of the interferential rings produced by a quartz plate and analyser, by superposing the effect produced by the various spectral components of the incident light on the photographic emulsion. In 1855, he studied the coloured optical checkerboard produced by projecting on a screen the image of a regularly perforated blind. Whereas Haidinger had attributed this curious phenomenon to interference, Stokes proved it to be only an effect of the superposition of the defocused image of the perforation for each colour of the spectrum. In 1876, he decided to publish the anomalous angular variations of the Newton rings produced by the contact of a slightly incurved glass prism with a metal plate, a phenomenon he had discovered long ago but of which he still had no interpretation. In 1882, he explained the dark line often seen near the limit of a dark object on photographs by the conflicting electrochemical action of out-of-focus rays (the objective of the camera being designed so as to focus the rays of the centre of the chemically most active part of the spectrum).[25]

Lastly, in 1885 Stokes described what Rayleigh called 'one of the most curious phenomena in the whole range of Optics', namely, the iridescent colours of light reflected by crystals of chlorate of potassium. In 1854, Herapath had sent the strange crystals to Stokes but they were too small and too thin to allow for a systematic study. In the 1880s, Stokes finally obtained larger crystals and he could trace the iridescence to thin twin-strata within the crystal. He still could not explain why the reflected light was unpolarized and why its spectrum was almost entirely contained in a single narrow band. Rayleigh later suggested that a periodic series of twin strata might be responsible for the phenomenon.[26]

Although this is not exactly optics, it may be worth mentioning that the old Stokes, like many of his contemporaries, manifested a vivid interest in Wilhelm Röntgen's discovery of X-rays in 1895. He propounded his own interpretation of these rays as a succession of pulses in the ether, based on the British corpuscular interpretation of cathode rays and on the known fact that X-rays resulted from the impact of the cathode rays on the glass wall of the tube (or on the anticathode in later sources). Stokes's pulses were similar to light except for the lack of periodicity, which explained why no one had been able to refract or diffract them. In the first few months following Röntgen's discovery, there were many different views on the nature of cathode rays and X-rays. Stokes's interpretation of X-rays or electromagnetic variants of them became dominant after the discovery of the electron around 1897. For most physicists of the late-nineteenth century, X-rays were not a completely new entity, they were just an extreme form of light either pulse-like or of very high frequency.[27]

In his reactions to optical curiosities, Stokes displayed his ability to dissect complex appearances in cleverly designed experiments as well as his ingenuity in finding theoretical explanations. Excepting the case of epipolic dispersion, the curiosities did not lead him to any important discovery or to any major new development. In the later part of his career they were an enjoyable distraction from his administrative duties, and they kept alive his reputation as the master of British optical science.

Measurement and Instruments

Another characteristic of Stokes's works in optics was his frequent involvement in the improvement of instruments and techniques of measurement. In 1849, he showed how the astigmatism of the human eye could be measured by a system of two cylindrical lenses. One of the lenses is a plane-concave cylinder, the other a plane-convex cylinder, and their mutual orientation can be varied to compensate for any amount of pure astigmatism (implying opposite values of the convergence in two perpendicular directions of the object plane). In the same year, he corrected Fraunhofer's determination of the B line of the solar spectrum for the purpose of better measurement of dispersive powers. In 1850, he designed a new method to measure the phase-shift in metal reflection. In 1851, he showed how to measure elliptic polarization through the combination of a quarter-wave quartz plate and a Nichol prism (a method anticipated by MacCullagh). In 1855, he gave a new rule for the construction of achromatic lenses: the focal length (as a function of the wavelength) should be a minimum at the centre of the brightest part of the spectrum. The photographic variant of this rule, based on the centre of the chemically most active part of the spectrum (which varies according to the amount of ultraviolet light emitted by the object), was the object of a later publication in 1873. In 1862, he published the theory of the transmission of light through a pile of inclined glass plates, a device frequently used as a polarizer. He repeatedly discussed ways of measuring the optical properties of crystals, in 1846 in a simplified version of the common method for determining the optical constants, then in 1862 in a new method for testing the laws of anisotropic propagation with high precision. As has been mentioned, he later used this method to decide between Fresnel's theory and his own theory based on anisotropic inertia. In 1877, he published a theoretical discussion of still another method, invented by the geologist Henry Clifton Sorby and based on interposing a crystal plate between a cross-wire and the objective of a microscope.[28]

In 1862, Stokes received sample glass prisms from old William Vernon Harcourt, one of the founders of the British Association, who was then developing new optical glasses. Although Stokes's original purpose was to study the fluorescence of the new glasses, he conceived improved determinations of their dispersive power and worked with Harcourt on strictly achromatic combinations of specially prepared glasses. When Harcourt died in 1871, Stokes had this to say on the collaboration:

> I may certainly say for myself, and I think it will not be deemed at all derogatory to the memory of my esteemed friend and fellow-labourer if I say of him, that I do not think that either of us working singly could have obtained the results we arrived at by working together.

In 1874, Stokes helped the Irish optical instrument maker Howard Grubb design a perfectly achromatic objective by means of Harcourt's glasses. The following year he conceived a titanic-glass alternative to Harcourt's phosphoric glasses (which were too soft) and had it made by the Birmingham glassmaker John Hopkinson. The telescope Grubb built with the Harcourt–Stokes objective was much freer of chromatic aberration than any other telescope, but it could only be a demonstration instrument because the striations of the glass affected its performance. It was not any of Stokes's British friends but the Carl Zeiss company in Jena that was to bring to fruition Stokes and Grubb's dream of a perfectly achromatic telescope in the 1880s. More anecdotally, in 1878 Stokes offered a new method for determining the dispersing power of bad (striated) glass by achromatizing a prism of this glass with another prism. He also corrected an error in a certain Prof. R. Keith's determination of the effective aperture of the objective of a microscope when immersed in a liquid.[29]

Despite the number of publications of this sort, Stokes is not remembered as the inventor of groundbreaking new instruments or experimental techniques. Yet there is no doubt that his multiple interventions in this field and his generosity in advising colleagues and students greatly contributed to the high level of British experimental optics. In addition, his highly precise measurements of double refraction protected Huygens's and Fresnel's laws from future theoretical challenges. In Lord Kelvin's opinion this was Stokes's greatest service to optics:[30]

Perhaps more than all, by his accurate *measuring* work, from which he drew an exceedingly rigorous verification of the accuracy of Huygens' geometrical construction for the double refraction of Iceland spar, Sir George Stokes has done much to make the Undulatory Theory of Light sure and strong as it is—a codification of laws divined by Huygens and Fresnel.

Lectures

Stokes lectured on optics for many years from the Lucasian chair he inherited in 1849 from the mathematician Joshua King. His teaching was clear, eloquent, systematic, and pleasantly illustrated by simple but always successful experiments entirely of his own.[31] These qualities of his lectures and the importance of his contributions to optics being generally recognized, in the 1860s he was expected to write the major treatise in this field, as Thomson and Tait were doing for general dynamics, as Maxwell would do for electromagnetism, and Rayleigh for the theory of sound. This never happened; and the lecture notes of his Cambridge students were never published, even though Kelvin expressed this wish in an obituary.[32] Stokes's only systematic review of the present state of optics was his British Association report of 1862, which covered only one part of the field.[33]

In the 1880s, Stokes accepted the Burnett foundation's invitation to lecture on optics in Aberdeen. In the previous century, the rich and devout merchant John Burnett had bequeathed part of his fortune to recompense periodic essays proving the existence and goodness of the deity. In 1881, his trustees reinterpreted his will as the financing of series of lectures on various domains of knowledge, with religious morals. Stokes was their first choice. He divided his lectures in three parts, the first on the nature of light, the second on its applications, and the third on its benefits. The audience being a popular one, he avoided mathematical developments. The lectures were nevertheless conceptually precise and historically well informed, and they gave a good sense of Stokes's interests and approaches in optics.[34]

In the first part of these lectures, Stokes recalled the historical conflict between Newton's corpuscular theory and the wave theory, and he gave all the evidence in favour of light being a transverse vibration, including interference, diffraction, polarization, double refraction, and chromatic polarization. In the second part, he discussed the chemical and astronomical applications to which he had himself much contributed: selective absorption, phosphorescence, fluorescence, solar and stellar spectroscopy, the Doppler effect and other optics-based inferences about the motion of celestial bodies. He briefly mentioned the various dynamical theories of the ether: Fresnel's, Green's, MacCullagh's, Gabriel Lamé's and, 'last but not least,' Maxwell's electromagnetic theory, citing their convergence as proof of the robustness of the phenomenological laws of optics. Stokes also mentioned the phenomena of chemically or magnetically induced optical rotation, to the study of which he had not publicly contributed and which his friends Thomson and Maxwell had used to speculate on molecular and ethereal mechanisms. Stokes broadly accepted that dispersion, double refraction, optical rotation and magneto-optical phenomena depended on the way the molecules of matter interacted with the ether. But he refrained from speculating on the precise mechanism of this interplay:[35]

> What may be the precise nature of the molecular vibrations, what may be the mode of connexion by which the vibrations of the ether agitate the molecules, or the molecules in their turn are able to agitate the ether, what may be the cause of the diminished velocity of propagation

in refracting media, what may be the mechanical cause of the difference of the velocity of propagation of right and left-handed circularly polarized light in media like sirup of sugar, which is manifested by a rotation of the plane of polarization of plane-polarized light, still more what may be the nature of the action of magnetism in respect of the propagation of light through bodies—all these are questions concerning the true answers to which we can affirm nothing, though plausible conjectures may in many cases be framed.

In the third part of his lectures, on the beneficial aspects of light, Stokes took 'light' to include any radiation from the sun (including the infrared and ultraviolet components). He told his audience how the resulting thermal effects explained the winds, weather, and climate that made human life possible (this was before global warming). He explained how light indirectly provided the energy needed by living organisms, through the synthesis of nutriments or through photosynthesis. He discussed light as the means of vision, including the structure of the retina, photochemical reactions in the retina, the three-component theory of the perception of colours, and accommodation mechanisms. Lastly, he satisfied Burnett's wish by giving examples of astonishingly well-adapted designs in nature, for instance the human eye, and taking them as proofs of the existence of a designing mind (he believed he could reconcile this view with Darwinian evolution). This was not mere complacency. Stokes was deeply religious and he fully developed his design argument in his Gifford lectures on natural theology in the 1890s.[36]

Conclusions

Stokes's work in optics may be divided in four different categories: 1) speculative theory relying on molecular or hydrodynamic mechanisms; 2) paradigmatic wave optics based on Fresnel's laws and concepts, with a mathematical bonus; 3) elucidation and exploitation of odd colour phenomena; 4) precision measurement and observation.

In the first category we find the early theory of stellar aberration and a few unpublished considerations on the origin of double refraction, on the Faraday effect, and on the interpretation of the Fraunhofer lines. Despite the early controversy with Challis, Stokes's explanation of stellar aberration was long regarded as a serious alternative to Fresnel's. His speculations about the Faraday effect and about the Fraunhofer lines are cases of missed opportunity, for they probably inspired William Thomson's influential analysis of the Faraday effect and they certainly determined his early teaching of the basic principle of solar spectroscopy.

In the second category, we find the theory of Brewster's new polarity of light, extensions of the received theories for Newton's rings and for the colours of thick plates, the Stokes parameters for the state of polarization of a light beam, and the dynamical theory of diffraction. These studies all imply sophisticated mathematical analysis; an emphasis on general principles of superposition, interference and symmetry; and a keen sense of representational economy. Stokes's parameters of polarization and his general diffraction formula (later re-derived by Kirchhoff by different means) still belong to standard optics.

The third category contains Stokes's discovery of fluorescence while elucidating a mysterious phenomenon of superficial colour in a quinine solution, as well as his discussion of many strange colour effects occurring during the reflection and absorption of light by various substances and crystals. This type of work does not imply any mathematics: it is very empirical, and it borders on natural history. It relies on the single principle that spectrally pure light is entirely characterized by its refrangibility, intensity and polarization. For the most part, it does not depend on the wave theory of light. In the case of fluorescence, it has important applications to ultraviolet spectroscopy and chemical analysis.

The fourth category, covering fine techniques of measurement and observation, is more significant than we would retrospectively think, not only because of the extent of Stokes's largely invisible collaboration with makers of optical glass and instruments, but also because his own precision measurements helped consolidate Fresnel's laws. Although we now know that these laws are exact consequences of the electromagnetic theory of light and that the latter theory is extremely well verified, in Stokes's youth it was still possible to doubt the strict validity of these laws and there were a few alternative theories in which they were only approximations. Stokes consolidated Fresnel's edifice by proving the accuracy of its laws of double refraction at a precision of 10^{-4}.

Stokes's style is characterized by an unusual combination of mathematical sophistication and rare experimental ability. On the mathematical side, he mastered and developed the techniques needed in solving the partial differential equations encountered in fluid mechanics and elasticity theory. From time to time he published purely mathematical memoirs in which he solved problems raised in an optical context, for instance an evaluation of Airy's integral for the distribution of light near a caustic (which explains the supernumerary rainbows) or the determination of the equation of wave surfaces for which Kirchhoff's law extends to radiation emitted and absorbed in an anisotropic medium. Kelvin praised the 'mathematical supersubtlety' of his friend in this domain while noting: 'With Stokes, mathematics was the servant and assistant, not the master. His guiding star in science was natural philosophy.'[37]

Stokes was a keen observer of nature and a natural experimentalist. His optical experiments were simple, economical and cleverly conceived. He made them with rudimentary equipment and without the help of an assistant, even when he prepared his lectures. Typically, he used sunlight as a source, let it enter his room through a hole on a blind, and then placed diaphragms, reflecting or transparent objects, prisms, lenses and Nicols on the light path. The method was unpractical because of the rarity of sunlight under the English climate and because of the constant motion of the sun. Stokes nevertheless preferred sunlight because of the relative stability and breadth of the solar spectrum and because the Fraunhofer lines offered natural markings in the spectrum. On rare occasions, he borrowed more sophisticated apparatus from friends or performed experiments in better equipped laboratories. The overall tendency was best captured by one of his students, J. J. Thomson:[38]

> In his lectures on light we marvel at the experimental skill with which the most difficult experiments in optics are successively performed with the simplest apparatus; indeed, it has been said that if you give Stokes the sun and three-quarters of an hour there is not an experiment in optics which he cannot perform.

Stokes's ideal of a theory was a coherent mathematical construct that accounted for empirical laws and results with a minimum of hypotheses. Even though he occasionally ventured to speculate on hidden mechanisms, he tried, as Newton had done before him, to distinguish different levels of analysis according to their proximity with experience. Closest to experience, there are laws implying only the basic notions of ray, reflection, refraction and simple (spectral) colour. Stokes was comfortable with the next level in which the wave-like nature of light is assumed, the existence of the optical ether is taken for granted and an equation is written for the propagation of the vibrations. He did not doubt that still another level was needed in which the mechanical structure of the ether and its mode of interaction with the molecules of matter would be specified. But working at this level was too adventurous for his taste. To him it was more important to consolidate the phenomenological levels than to develop the more speculative ones. A striking, conscious example of this attitude is found in his derivation of the perfect blackness of the central spot of Newton's rings from the sole 'principle of reversion' without any further assumption on the mechanical structure of the implied media. He commented:

> In sifting the evidence for the truth of any set of hypotheses, it becomes of great importance to consider whether the phenomena explained, or some of them, are explicable on more simple and general hypotheses, or whether they appear absolutely to require the more particular restrictions adopted.

Some hypotheses, such as the molecular structure of the ether in Fresnel's and Cauchy's theories, could not survive the sifting process. Others, such as the molecular structure of matter, could survive and then be deemed real. Even in the latter case, Stokes preferred to avoid molecular reasoning in the discussion of phenomena that did not require it.[39]

Stokes's interests in optics were very broad: in the first twenty years of his scientific life, he contributed to nearly every aspect of this science. In contrast, he did not contribute to the two most spectacularly innovative domains of the physics of this time: electrodynamics and thermodynamics (although he sometimes taught the latter). He considered himself as an expert in the older sciences of hydrodynamics, optics and gravitation, and judged it unwise to get involved in the tempestuous evolution of other domains. This is quite evident in his correspondence with Thomson, who on the contrary contributed to every domain of physics and enjoyed sailing under stormy weather. This is also apparent in Stokes's lack of involvement in the electromagnetic theory of light. He praised it but never used it or taught it. In general, he studied optics for itself and not in its relation with other domains of physics, save for fluid mechanics and elasticity theory on which it traditionally depended. The few exceptions to this reticence, for instance his reflections on the Faraday effect or on thermal radiation, remained private. At the same time, he was eager to apply physical optics beyond physics: to chemistry, astronomy, solar physics, and to the physiology of vision.

As Lord Rayleigh once noted, Stokes's scientific production in optics and elsewhere sharply declined soon after his discovery of fluorescence. Rayleigh blames the administrative duties that began to fall on Stokes's shoulders when he was elected Secretary of the Royal Society in 1854. It could also be that his marriage in 1857 altered his lifestyle in a manner unfavourable to sustained scientific inquiries. A third possibility, emphasized by Jed Buchwald, is that Stokes was prejudiced against the microphysical approaches that changed optics in the last third of the century. In this period, the main challenges to the received optics were anomalous dispersion, magneto-optical effects and the optics of moving bodies. Hermann Helmholtz, Hendrik Lorentz, Joseph Larmor and others met these challenges through a new understanding of the relation between ether and matter, involving the interaction of electrically charged particles (ions and electrons) with waves propagated in a strictly stationary ether. Stokes remained completely indifferent to these developments, even though his earliest works in optics concerned moving bodies. This could be a consequence of the 'over-cautious' temperament that, according to Rayleigh, prevented Stokes from pursuing hypothetical visions. Or it could simply be the lack of curiosity that often comes with age.[40]

This apparent decline did not hamper Stokes's reputation as the greatest British contributor to optical science in the nineteenth century. In seven glorious years, between 1845 and 1852, he had abundantly contributed to physical optics, with lasting results including an ingenious theory of stellar aberration, the first dynamical theory of diffraction, a theory of the most general state of polarization that anticipated the modern distinction between pure and mixed states, and the discovery of the radically new phenomenon of fluorescence. In the following years he remained the chief British authority in all sorts of optical questions, he consolidated Fresnel's laws and he guided the younger generation though his teaching and his numerous private communications. As Kelvin put it in his address for his friend's Copley medal: 'In optics and the undulatory theory of light, Stokes has been the teacher and guide of his contemporaries.'[41]

After Stokes passed away, Rayleigh judged that optics had been his 'greatest subject'. Or was it fluid mechanics? Rayleigh's view may have been reflecting the diminishing importance of fluid mechanics: it no longer enjoyed the paradigmatic role it had in earlier British physics. At any rate, it was Stokes's works in optics that raised him to the highest rank of British science: his memoir on diffraction 'marked him for the Lucasian chair', as Rayleigh put it, and his memoir on fluorescence won him the Rumford medal and the secretaryship of the Royal Society. The Cambridge classical scholar Richard Claverhouse Jebb knew that optics had illuminated most of Stokes's scientific life when, on the occasion of his jubilee in 1899, he wrote the verses:[42]

> Clear mind, strong heart, true servant of the light,
> True to that light within the soul, whose ray,
> Pure and serene, hath brightened on thy way,
> Honour and praise now crown thee on the height
> Of tranquil years.

BIBLIOGRAPHICAL APPENDIX: OPTICS

The following abbreviations are used: *AP*, *Annalen der Physik*; *BAR*, British Association for the Advancement of Science, *annual report*; *CR*, Académie des sciences, *Comptes rendus hebdomadaires des séances*; *PM*, *Philosophical magazine*; *PRS*, Royal Society of London, *Proceedings*; *PT*, Royal Society of London, *Philosophical transactions*; *SMSCi*, G. G. Stokes, *Memoirs and scientific correspondence*, ed. J. Larmor, 2 vols. (Cambridge, 1907), vol. i; *SMPPi*, G. G. Stokes, *Mathematical and physical papers*, 5 vols. (Cambridge, 1880–1905), vol. i; *TCPS*, Cambridge Philosophical Society, *Transactions*.

For the memoirs published in *TCPS*, I have used the date of reading, because the corresponding *TCPS* volume typically appeared several years later and because (as far as can be judged from Stokes's correspondence) the memoirs were usually printed and circulated soon after the date of reading.

STOKES'S WRITINGS ON OPTICS

1843 [read on 20 May 1843]. On some cases of fluid motion. *TCPS* 8 (1849). Also in *SMPP*1, 17–68.

1845a [read on 14 April 1845] On the theories of the internal friction of fluids in motion, and of the equilibrium and motion of elastic fluids. *TCPS* 8 (1849). Also in *SMPP*1, 75–129.

1845b On the aberration of light. *BAR* (1845), 9.

1845c On the aberration of light. *PM*, 27: 9–15.

1846a Remarks on Professor Challis's theoretical explanation of the aberration of light. *PM*, 28: 15–17.

1846b On Fresnel's theory of the aberration of light. *PM*, 28: 76–81.

1846c On the aberration of light. *PM*, 28: 335–36.

1846d On a formula for determining the optical constants of doubly refracting crystals. *Cambridge and Dublin mathematical journal*, 1. Also in *SMPP*1, 148–52.

1846e On the constitution of the luminiferous aether, viewed with reference to the phenomenon of the aberration of light. *PM*, 29: 6–10.

1848a On the constitution of the luminiferous aether. *PM*, 22. Also in *SMPP*2, 8–13.

1848b On the theory of certain bands seen in the spectrum. *PT*, 138. Also in *SMPP*2, 14–35.

1848c [read 11 Dec 1848] On the formation of the central spot of Newton's rings beyond the critical angle. *TCPS*, 8 (1849). Also in *SMPP*2, 56–81.

1849a On the perfect blackness of the central spot in Newton's rings, and on the verification of Fresnel's formulae for the intensities of reflected and refracted rays. *Cambridge and Dublin mathematical journal*, 4. Also in *SMPP*2, 89–103.

1849b On a mode of measuring the astigmatism of a defective eye. *BAR*. Also in *SMPP*2, 172–5.

1849c On the determination of the wave length corresponding with any point of the spectrum. *BAR*. Also in *SMPP*2, 176–7.

1849d [read 26 Nov 1849] On the dynamical theory of diffraction. *TCPS*, 9 (1856). Also in *SMPP*2, 243–328.

1849e Supplement to a paper 'On the theory of certain bands seen in the spectrum.' *PM*, 34: 309–11.

1850a On the mode of disappearance of Newton's rings in passing the angle of total internal reflexion. *BAR*. Also in *SMPP*2, 358–9.

1850b On metallic reflexion. *BAR*. Also in *SMPP*2, 360.

1850c On a fictitious displacement of fringes of interference. *BAR*. Also in *SMPP*2, 361.

1850d On Haidinger's brushes. *BAR*. Also in *SMPP*2, 362–4.

1850e On the numerical calculation of a class of definite integrals and infinite series. *TCPS*, 9. Also in *SMPP*2, 329–57.

1851a [read 19 May 1851]. On the colours of thick plates. *TCPS*, 9 (1856). Also in *SMPP*3, 155–196.

1851b On a new elliptic analyser. *BAR*. Also in *SMPP*3, 197–202.

1852a On the total intensity of interfering light. *Transactions of the Royal Society of Edinburgh*, 20. Also in *SMPP3*, 228–32.

1852b [read 16 Feb and 15 Mar 1852] On the composition and resolution of streams of different light from different sources. *TCPS*, 9 (1856). Also in *SMPP3*, 233–58.

1852c [read 27 May 1852] On the change of the refrangibility of light. *PT*, 142. Also in *SMPP3*, 266–413.

1852d On the optical properties of a recently discovered salt of quinine. *BAR*. Also in *SMPP4*, 18–21.

1853a [read 16 Jun 1853] On the change of refrangibility of light.--II. *TCPS*, 9 (1856). Also in *SMPP4*, 1–17.

1853b On the change of refrangibility of light, and the exhibition thereby of chemical rays. Royal Institution of Great Britain, *Proceedings*, 1. Also in *SMPP4*, 22–9.

1853c On the cause of the occurrence of abnormal figures in photographic impressions of polarized rings. *PM*, 6. Also in *SMPP4*, 30–7.

1853d On the metallic reflexion exhibited by certain non-metallic substances. *PM*, 6. Also in *SMPP4*, 38–49.

1854 [read 27 May1852]. Abstract of a paper 'On the change of the refrangibility of light.' *PRS*, 6. Also in *SMPP3*, 259–66.

1855a [from Stokes to Haidinger, 9 Feb 1854] Die Richtung der Schwingungen des Lichtäthers im polarisirten Lichte. Mittheilung aus einem Schreiben des Hrn. Prof. Stokes, nebst Bemerkungen von W. Haidinger. *AP*, 96. Also in *SMPP4*, 50–4.

1855b [from Stokes to Haidinger, 9 Feb 1854] Mittheilung aus einem Schreiben des Hrn. Prof. Stokes, über das optische Schachbrettmuster. *AP*, 96. Also in *SMPP4*, 55–60.

1855c [from Stokes to Haidinger, 9 Feb 1854] Einige neuere Ansichten über die Natur der Polarisationsbüschel. *AP*, 96. Also in *SMPP4*, 60.

1855d On the achromatism of a double object-glass. *BAR*, Also in *SMPP4*, 63–4.

1856 Remarks on Professor Challis's paper, entitled 'A theory of the composition of colours etc.' *PM*, 12. Also in SMPP4, 65–69 [corrects Challis's confusions regarding the interpretation of various experiments on colors by Brewster, Herschel, Stokes, and Maxwell].

1857 On the polarization of diffracted light. *PM*, 13. Also in *SMPP4*, 74–6.

1859a On the existence of a second crystallizable fluorescent substance (paviin) in the barge of the horsechestnut. *Quarterly Journal of the Chemical Society of London* 11. Also in *SMPP4*, 113–16.

1859b On the bearing of the phenomena of diffraction on the direction of polarized light, with remarks on the paper of Professor F. Eisenlohr. *PM*, 18. Also in *SMPP4*, 117–18.

1860 Optical characters of purpurine and alizarine [etc.]. In Edward Schunck, 'On the colouring matters of madder', *Quarterly Journal of the Chemical Society of London* 12: 198–221, on 219–21. Also in *SMPP4*, 122–6.

1861 Note on internal reflexion. *PRS*, 11. Also in *SMPP4*, 137–44.

1862a On the intensity of the light reflected from or transmitted through a pile of plates. *PRS*, 11. Also in *SMPP4*, 145–56.

1862b Report on double refraction. *BAR*. Also in *SMPP4*, 157–202.

1862c On the long spectrum of electric light. *PT*, 152. Also in *SMPP4*, 203–35.

1864a On the supposed identity of billiverdin with chlorophyll, with remarks on the constitution of chlorophyll. *PRS*, 13. In *SMPP4*, 236–7.

1864b On the discrimination of organic bodies by their optical properties. Royal Institution of London, *Proceedings*, 4. Also in *SMPP4*, 238–48.

1864c On the application of the optical properties of bodies to the detection and discrimination of organic substances. *Journal of the Chemical Society*, 2. Also in *SMPP4*, 249–63.

1864d On the reduction and oxidation of the colouring matter of the blood. *PRS*, 13. Also in *SMPP4*, 264–76.

1869 On a certain reaction of quinine. *Journal of the Chemical Society*. Also in *SMPP4*, 327–33.

1871 Notice on the researches of the late Rev. W. Vernon Harcourt on the conditions of transparency in glass and the connexion between the chemical constitution and optical properties of different glasses. *BAR*. Also in *SMPP4*, 339–43.

1872 On the law of extraordinary refraction in Iceland spar. *PRS*, 20. Also in *SMPP4*, 336.

1873a Sur l'emploi du prisme dans la vérification de la loi de la double réfraction. *CR*, 77. Also in *SMPP4*, 337–8.

1873b On the principles of the chemical correction of object-glasses. *Photographic journal*. Also in *SMPP4*, 344–54.

1874a Note to Thomas Grubb, 'On the improvement of the spectroscope.' *PRS*, 22. Also in *SMPP4*, 355.

1874b On the construction of a perfectly achromatic telescope. *BAR*. Also in *SMPP4*, 356–7.

1875 On the optical properties of a titano-silicic glass. *BAR*. Also in *SMPP4*, 358–60.

1876a On the early history of spectrum analysis [Stokes to Whitmell, 23 Dec 1875]. *Nature*, 12. Also in *SMPP4*, 133–6.

1876b On a phenomenon of metallic reflexion. *BAR*. Also in *SMPP4*, 361–4.

1877 On the foci of lines seen through a crystalline plate. *PRS*, 26. Also in *SMPP5*, 6–23.

1878a On the question of a theoretical limit to the apertures of microscopic objectives. *Journal of the Royal Microscopical Society*, 1. Also in *SMPP5*, 36–9.

1878b On an easy and at the same time accurate method of determining the ratio of the dispersions of glasses intended for objectives. *PRS*, 27. Also in *SMPP5*, 40–51.

1880–1905 *Mathematical and physical papers*. 5 vols. Cambridge: Cambridge University Press.

1882 On the cause of the light border frequently noticed in photographs just outside the outline of a dark body seen against the sky: with some introductory remarks on phosphorescence. *PRS*, 34. Also in *SMPP5*, 117–24.

1885 On a remarkable phenomenon of crystalline reflection. *PRS*, 38. Also in *SMPP5*, 164–79.

1887 *Burnett lectures on light in three courses, delivered at Aberdeen in November, 1883, December, 1884, and November, 1885*. London: Macmillan.

1891 On an optical proof of the existence of suspended matter in flames. Royal Society of Edinburgh, *Proceedings*. Also in *SMPP5*, 232–4.

1892 On the reactions occurring in flames. *Proceedings of the Chemical Society*. Also in *SMPP5*, 235–7.

1896 On the nature of the Röntgen rays. Cambridge Philosophical Society, *Proceedings*, 9. Also in *SMPP5*, 254–5.

1897 On the nature of the Röntgen rays. Manchester Literary and Philosophical Society *Memoirs and proceedings*, 41. Also in *SMPP5*, 256–77.

1903 [Note to] William Crookes, 'The ultra-violet spectrum of radium.' *PRS*, 62. Also in *SMPP5*, 293–5.

1907 *Memoirs and scientific correspondence*. 2 vols., ed. Joseph Larmor. Cambridge: Cambridge University Press.

• •

OTHER AUTHORS

Airy, George Biddell

1831 *Mathematical tracts on the lunar and planetary theories, the figure of the earth, precession and nutation, the calculus of variations, and the undulatory theory of optics. Designed for the use of students in the university*. 2nd edn. Cambridge: Deighton.

1840 The Bakerian lecture: On the theoretical explanation of an apparent new polarity in light. *PT*, 130: 225–44.

1841 Supplement to a paper 'On the theoretical explanation of an apparent new polarity in light.' *PT*, 131: 1–10.

1846 On the equations applying to light under the action of magnetism. *PM,* 28: 469–77.

Becquerel, Edmond

1867 *La lumière, ses causes et ses effets.* 2 vols. Paris: Firmin Didot.

Powell, Baden

1846 Remarks on some points of the reasoning in the recent discussions on the theory of the aberration of light. *PM*, 29: 425–40.

1848 On a new case of the interference of light. *PT*, 138: 213–26.

Bradley, James

1729 A letter from the Reverend Mr James Bradley Savilian Professor of astronomy at Oxford, and F. R. S. to Dr. Edmund Halley Astronom. Reg. &c giving an account of a new discovered motion of the fix'd stars. *PT*, 35 (1727–8) [read and pub. in 1729]: 637–61.

Brewster, David

1834 On the colours of natural bodies. *PT*, 12: 538–45.

1837 On a new property of light. *BAR* (1837), 12–13.

1838 On a new kind of polarity in homogeneous light. *BAR* (1838), 12–13.

1838 On a new phenomenon of colour in certain specimens of fluor spar. *BAR* (1838), 10–12.

1845 On a new polarity of light, with an examination of Mr Airy's explanation of it on the undulatory theory. *BAR* (1845), 7–8.

Buchwald, Jed

1985 *From Maxwell to microphysics: Aspects of electromagnetic theory in the last quarter of the nineteenth century.* Chicago: University of Chicago Press.

1988 The Michelson experiment in the light of electromagnetic theory before 1900. In Stanley Goldberg and Roger Stuewer (eds), *The Michelson era in American science 1870–1930*, 55–70. New York: American Institute of Physics.

1989 *The rise of the wave theory of light: Optical theory and experiment in the early nineteenth century.* Chicago: University of Chicago Press.

1992 Why Stokes never wrote a treatise in optics. In Peter Harman and Alan Shapiro (eds), *An investigation of difficult things: Essays on Newton and the history of exact sciences*, 451–76. Cambridge: Cambridge University Press,

Cantor, Geoffrey

1983 *Optics after Newton: Theories of light in Britain and Ireland, 1704–1840.* Manchester: Manchester University Press.

Cauchy, Augustin

1831 Mémoire sur la théorie de la lumière. Académie des sciences, *Mémoires*, 10: 293–316.

1836 Notes de M. Cauchy sur l'optique, adressées à M. Libri. Première note. *CR*, 2: 341–9.

1839a Mémoire sur la polarisation rectiligne et la double réfraction. Académie des sciences, *Mémoires*, 18: 153–216.

1839b Mémoire sur la polarisation des rayons réfléchis ou réfractés par la surface de séparation de deux corps isophanes et transparents. *CR*, 9: 676–91.

Challis, James

1845a On the aberration of light. *BAR* (1845), 9.

1845b A theoretical explanation of the aberration of light. *PM*, 27: 321–7.

1846a On the aberration of light, in reply to Mr Stokes. *PM*, 28: 90–3.

1846b On the principles to be applied in explaining the aberration of light. *PM*, 28: 176–7.

1846c On the aberration of light, in reply to Mr Stokes. *PM*, 28: 393–4.

1848 On the course of a ray of light from a celestial body to the earth's surface, according to the hypothesis of undulations. *PM*, 32:168–70.

Chen, Xiang

2000 *Instrumental traditions and theories of light: The uses of instruments in the optical revolution.* Dordrecht: Kluwer.

Chappert, André

2004 *L'édification au XIXe siècle d'une science des phénomènes lumineux.* Paris: Vrin.

Darrigol, Olivier

2000 *Electrodynamics from Ampère to Einstein.* Oxford: Oxford University Press.

2005 *Worlds of flow: A history of hydrodynamics from the Bernoullis to Prandtl.* Oxford: Oxford University Press.

2012 *A history of optics from Greek antiquity to the nineteenth century.* Oxford: Oxford University Press.

2015 Poincaré's light. In Bertrand Duplantier and Vincent Rivasseau (eds), *Henri Poincaré, 1912–2012*, 1–50. Basel: Birkhäuser.

Eisenstaedt, Jean

2005 *Avant Einstein: Relativité, lumière, gravitation.* Paris: Le Seuil.

Euler, Leonhard

1746 Mémoire sur l'effet de la propagation successive de la lumiere dans l'apparition tant des planètes que des comètes. Académie Royale des sciences et des belles-lettres de Berlin, *Mémoires*, 2: 141–81.

Feffer, Stuart

1994 Microscopes to munitions: Ernst Abbe, Carl Zeiss, and the transformation of technical optics, 1850–1914. PhD dissertation. University of California at Berkeley.

Fizeau, Hippolyte

1851 Sur les hypothèses relatives à l'éther lumineux, et sur une expérience qui paraît démontrer que le mouvement des corps change la vitesse avec laquelle la lumière se propage dans leur intérieur. *CR*, 33: 349–55.

Frank, James

1983 The conservation of energy, theories of absorption and resonating molecules, 1851–1854: G. G. Stokes, A. J. Ångström and W. Thomson. *Notes and records of the Royal Society*, 38: 79–107.

Fraunhofer, Joseph

1817 Bestimmung des Brechungs- und Strahlungsvertreuungs-Vermögens verschiedener Glasarten, in Bezug auf die Vervollkommnung achromatischer Fernrohre. *Annalen der Physik*, 56: 264–313.

Fresnel, Augustin

1818 Sur l'influence du mouvement terrestre dans quelques phénomènes optiques. *Annales de chimie et de physique*, 9: 57–66, 286.

1823 Quelques observations sur les principales objections de Newton contre le système des vibrations lumineuses et sur les difficultés que présente son hypothèse des accès. *Bibliothèque universelle des sciences, belles-lettres, et arts*, 22 (1823), 73–91.

Gouy, Louis Georges

1886 Recherches expérimentales sur la diffraction. *Annales de chimie et de physique*, 52: 145–92.

Green, George

1838 On the laws of reflexion and refraction of light at the common surface of two non-crystallized media. *TCPS*, 7 (1842), 1–24.

1839 On the propagation of light in crystallized media. *TCPS*, 7 (1842), 121–40.

Haidinger, Werner

1844 Über das direkte Erkennen des polarisirten Lichtes und der Lage der Polarizationsebene, *AP*, 63: 29–39.

Harvey, Edmund Newton

1957 *A history of luminescence from the earliest times until 1900*. Philadelphia: American Philosophical Society.

Helmholtz, Hermann

1867 *Handbuch der physiologischen Optik*. Leipzig: Voss.

Herschel, John

1845a Ἀμόρφωτα. No, I.—On a case of superficial colour presented by a homogeneous liquid internally colourless. *PT*, 135: 143–5.

1845b Ἀμόρφωτα. No. II.—On the epipolic dispersion of light, being a supplement to a paper entitled, 'On a case of superficial colour presented by a homogeneous liquid internally colourless.' *PT*, 135: 147–53.

Holtzmann, Carl Alexander

1856 Das polarisirte Licht schwingt in der Polarisationsebene. *AP*, 99: 446–51.

Huggins, William

1868 Further observations on the spectra of some of the stars and nebulae, with an attempt to determine therefrom whether these bodies are moving towards or from the earth, also observations on the spectra of the sun and of comet II. *PT*, 158: 529–64.

Jackson, Myles

2000 *Spectrum of belief: Joseph von Fraunhofer and the craft of precision optics*. Cambridge, MA: MIT Press.

Janssen, Michel, and John Stachel

2004 The optics and electrodynamics of moving bodies. Preprint 265, Max Planck Institut für Wissenschaftsgeschichte. Berlin.

Jebb, Richard Claverhouse

1899 To Sir George Gabriel Stokes. *Cambridge review*, 20: 370.

Jungnickel, Christa, and Russell McCormmach

1986 *Intellectual mastery of nature: Theoretical physics from Ohm to Einstein*, vol. 1: *The torch of mathematics, 1800–1870*, vol. 2: *The now mighty theoretical physics, 1870–1925*. Chicago: University of Chicago Press.

Knudsen, Ole

1976 The Faraday effect and physical theory. *Archive for history of exact sciences*, 15: 235–81.

Land, Edwin Herbert

1951 Some aspects on the development of sheet polarizers. *Journal of the Optical Society of America*, 41: 957–63.

Lloyd, Humphrey

1835 Report on the progress and present state of physical optics. *BAR* (Dublin, 1835), 295–414.

Lorentz, Hendrik Antoon

1886 Over den invloed dien de beweging der aarde op de lichtverschijnselen uitofent. Koninklijke Akademie van Wetenschappen, Amsterdam, *Verslagen*. Transl. as 'De l'influence du mouvement de la terre sur les phénomènes lumineux' in *Archives Néerlandaises des sciences exactes et naturelles* (1887) and *Collected papers*, 4: 153–214.

Lorenz, Ludvig Valentin

1860 Bestimmung der Schwingungsrichtung des Lichtäthers durch die Polarisation des gebeugten Lichtes. *AP*, 111: 315–28.

MacCullagh, James
1836 On the laws of double refraction in quartz. Royal Irish Academy, *Transactions*, 17(1837), 461–69.
1839 An essay towards the dynamical theory of crystalline reflexion and refraction. Royal Irish Academy, *Transactions*, 21 (1848), 17–50.

Malley, Marjorie
1991 A heated controversy on cold light. *Archive for history of exact sciences*, 42: 173–86.

Maxwell, James Clerk
1861–62 On physical lines of force. *PM*, 21 (1861), 161–75, 281–91, 338–48; 23 (1862), 12–24, 85–95.
[1867] On the influence of the motion of the heavenly bodies on the index of refraction of light [Letter to Huggins of 12 June 1867]. In Huggins 1868: 532–5
1878 Ether. *Encyclopaedia Britannica*, 9th edn, vol. 8 (1878), pp. 568–72.

McGucken, William
1969 *Nineteen-century spectroscopy: Development of the understanding of spectra 1802–1897*. Baltimore, MD: Johns Hopkins Press.

McMaster, William
1954 Polarization and the Stokes parameters. *American journal of physics*, 22: 351–62.

Moigno, François Napoléon Marie (abbé)
1850 *Répertoire d'optique moderne ou analyse complète des travaux modernes relatifs aux phénomènes de la lumière.* 4 vols. Paris: Franck.

O'Brien, Matthew
1842 On the reflexion and refraction of light at the surface of an uncrystallized body. *TCPS*, 8 (1849), 7–30.

Poisson, Siméon Denis
1819 Mémoire sur l'intégration de quelques équations linéaires aux différences partielles, et particulière-ment de l'équation générale du mouvement des fluides élastiques, Académie des sciences, *Mémoires*, 3 (1919). 121–76.
1823 Sur le phénomène des anneaux colorés. *Annales de chimie et de physique*, 22: 337–47.

Rayleigh, Lord (John William Strutt)
1888 On the reflexion of light at a twin plane of a crystal. *PM*, 26: 256–65.
1903 Sir George Gabriel Stokes, Bart. 1819–1903. *PRS*. Also in *SMPP5*, ix–xxv.
1904 [Allocution at the Stokes memorial ceremony of 7 July 1904]. *SMSC1*, 318–20.

Schaffner, Kenneth
1972 *Nineteenth-century aether theories*. Oxford: Oxford University Press.

Shapiro, Alan
1993 *Fits, passions, and paroxysms: Physics, method, and chemistry and Newton's theories of colored bodies and fits of easy reflection*. Cambridge: Cambridge University Press.

Tait, Peter Guthrie
1876 *Lectures on some recent advances in physical science*. London: Macmillan.

Talbot, William Henry Fox
1837 An experiment on the interference of light. *PM*, 10: 364.

Thomson, Joseph John
1899 The Stokes jubilee. *Cambridge review*, 20: 370–1.
Thomson, William (Lord Kelvin)
1856 [received 10 May] Dynamical illustrations of the magnetic and helicoidal rotary effects of transpar-ent bodies on polarized light. *PRS*, 8: 150–8.
1862 Physical considerations regarding the possible age of the sun's heat. *PM*, 23: 158–60.
1888 On the reflexion and refraction of light. *PM*, 26: 414–25, 500–1.
1893 Copley medal. Sir G. Gabriel Stokes, Bart., F.R.S. *PRS*, 54: 389–91. Cited in *SMSC1*, 268–70.

1902 Obituary. *Nature.* Cited in *SMSC1*, 307–9.

Whittaker, Edmund

1910 *A history of the theories of aether and electricity: From the age of Descartes to the close of the nineteenth century.* London: Longmans, Green, & Co.

Wilson, David B.

1972 George Gabriel Stokes on stellar aberration and the luminiferous ether. *British journal for the history of science,* 6: 57–72.

1987 *Kelvin and Stokes: A comparative study in Victorian physics.* Bristol: Adam Hilger.

1990 (ed.) *The correspondence between Sir George Gabriel Stokes and Sir William Thomson, Baron Kelvin of Largs.* 2 vols. Cambridge: Cambridge University Press.

2003. Arbiters of Victorian science: George Gabriel Stokes and Joshua King. In Kevin Knox, and Richard Noakes (eds), *From Newton to Hawking: A history of Cambridge University's Lucasian Professors of Mathematics,* 295–342. Cambridge: Cambridge University Press.

Young, Thomas

1804 Experiments and calculations relative to physical optics. *PT,* 94 (1804). Also in *Miscellaneous works,* 3 vols. (London: John Murray, 1855), vol. 1, pp.179–91.

CHAPTER 6

Stokes's Fundamental Contributions to Fluid Dynamics

PETER LYNCH

Introduction

George Gabriel Stokes was one of the giants of hydrodynamics in the nineteenth century. He made fundamental mathematical contributions to fluid dynamics that had profound practical consequences. The basic equations formulated by him, the Navier–Stokes equations, are capable of describing fluid flows over a vast range of magnitudes. They play a central role in numerical weather prediction, in the simulation of blood flow in the body and in countless other important applications. In this chapter we put the primary focus on the two most important areas of Stokes's work on fluid dynamics, the derivation of the Navier–Stokes equations and the theory of finite amplitude oscillatory water waves.

Stokes became an undergraduate at Cambridge in 1837. He was coached by the 'Senior Wrangler-maker', William Hopkins, and, in 1841, Stokes was Senior Wrangler and first Smith's Prizeman. It was following a suggestion of Hopkins that Stokes took up the study of hydrodynamics, which was at that time a neglected area of study in Cambridge. Stokes was to make profound contributions to hydrodynamics, his most important being the rigorous establishment of the mathematical equations for fluid motions, and the theoretical explanation of a wide range of phenomena relating to wave motions in water.

Stokes's Collected Papers

Stokes's collected *Mathematical and Physical Papers* (*MPP*)[1] were published in five volumes over an extended period from 1880 to 1905 (Fig. 6.1). They contain articles originally published in journals, additional notes prepared by Stokes and miscellaneous material such as

Lynch, P., *Stokes's Fundamental Contributions to Fluid Dynamics*. In: *George Gabriel Stokes: Life, Science and Faith*, Mark McCartney, Andrew Whitaker, and Alastair Wood (Eds): Oxford University Press (2019). © Oxford University Press. DOI: 10.1093/oso/9780198822868.001.0006

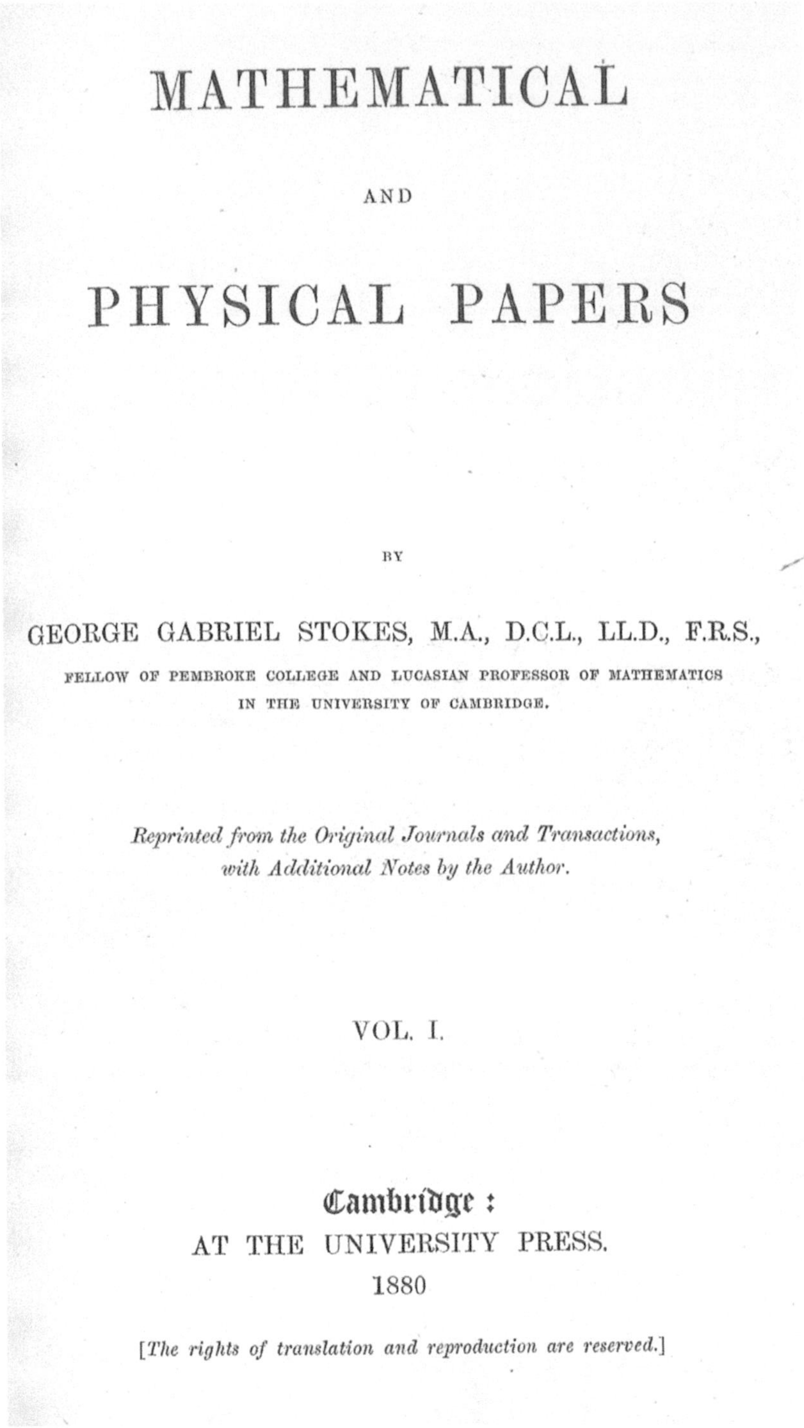

MATHEMATICAL

AND

PHYSICAL PAPERS

BY

GEORGE GABRIEL STOKES, M.A., D.C.L., LL.D., F.R.S.,

FELLOW OF PEMBROKE COLLEGE AND LUCASIAN PROFESSOR OF MATHEMATICS
IN THE UNIVERSITY OF CAMBRIDGE.

Reprinted from the Original Journals and Transactions,
with Additional Notes by the Author.

VOL. I.

Cambridge:
AT THE UNIVERSITY PRESS.
1880

[*The rights of translation and reproduction are reserved.*]

Fig. 6.1 The title page of Volume 1 of Stokes's collected *Mathematical and Physical Papers*.

examination papers. Stokes published about 140 scientific papers. Of these, some twenty-three were on hydrodynamics. The papers are in chronological order of original publication. The first three volumes, published respectively in 1880, 1883 and 1901, were prepared by Stokes himself. The last two were edited by his successor in the Lucasian Chair, Joseph Larmor, and published in 1904 and 1905 after Stokes had died. The final volume includes an interesting obituary of Stokes by Lord Rayleigh.

The majority of papers in Volume I of *MPP* are on fluid motion. The volume contains two of Stokes's most profound papers, one on the fundamental equations of motion now known as the Navier-Stokes equations, and one on oscillatory wave motion in fluids. The first paper in the collection, 'On the steady motion of incompressible fluids', is starkly mathematical in style. The paper is concerned mainly with fluid motion in two dimensions. Little is presented by way of motivation or physical background. In this work, Stokes introduced the notion of fluid flow stability. He pointed out that the existence of a solution does not imply that it can be sustained, as there may be many other motions compatible with the given boundary conditions. Stokes wrote 'There may even be no steady state of motion possible, in which case the fluid would continue perpetually eddying'. He was beginning to grapple with the recondite problem of turbulence. Ever since, stability of fluid flow has been a fundamental hydrodynamical concept.

The second paper, 'On some cases of fluid motion', opens with an expository section of four pages before the author launches into mathematical details. Stokes writes that 'Common observation seems to show that, when a solid moves rapidly through a fluid … it leaves behind a succession of eddies in the fluid'.[2] He then presents a comprehensive account of some fourteen problems in fluid flow. In this paper, Stokes begins to consider friction in fluid flow, discussing no-slip boundary conditions and their consequences.

The issue of internal friction is central in Stokes's monumental paper, 'On the theories of the internal friction of fluids in motion, and of the equilibrium and motion of elastic solids'. This paper[3] extends over fifty-five pages. Stokes notes that it is commonly assumed that the mutual action of two adjacent fluid elements is normal to the surface separating them. The resulting equations yield solutions agreeing with observations in a range of applications. However, there is an entire class of motions for which this theory, which makes no allowance for the tangential action between elements, is wholly inadequate. This effect arises from the 'sliding of one portion of a fluid along another, or of a fluid along the surface of a solid'. Stokes notes that the tangential force plays the same role in fluid motion that friction does with solids. Stokes gives the example of water flowing down a straight inclined chute. The then-current theory of fluid flow would indicate a uniform acceleration of the water, something that is completely at odds with experience.

In his masterful paper in 1847, 'On the theory of oscillatory waves',[4] Stokes investigates the dynamics of surface waves in the case where the height of the waves is not assumed to be infinitesimally small. In a supplement to this paper, also included in Volume I of *MPP*, Stokes showed that, for the highest possible wave capable of propagation without change of form, the surfaces at the crest enclose an angle of 120°.

Volume II of *MPP* contains three sets of *Notes on Hydrodynamics*. These were part of a series of notes for students prepared by William Thomson (Fig. 6.2) and Stokes. The three sets by Stokes are entitled 'On the dynamical equations', 'Demonstration of a fundamental theorem' and 'On Waves'. In the first of these, on the dynamical equations, Stokes gives a homely illustration of fluid viscosity: 'The subsidence of the motion in a cup of tea which has been stirred may be mentioned as a familiar instance of friction'. In the third set, on waves, he revisited his paper on oscillatory waves.

Fig. 6.2 William Thomson (1824–1907) (later Lord Kelvin). Thomson and Stokes corresponded with each other for over fifty years, and it is in a letter from William Thomson to Stokes in 1850 that what we call Stokes's theorem first appears. Stokes went on to ask the proof of the theorem as a question in his Smith's Prize paper of 1854. From Agnes Gardner King, *Kelvin the Man* (London: Hodder & Stoughton, 1925), frontispiece.

MPP Volume III opens with an extensive study of the effects of air friction on the motion of a pendulum. This paper,[5] published in 1850, is 141 pages in length. Surprisingly, it was considered by Stokes to be one of his greatest contributions to science. The remaining papers in Volume III are on the physics of light. Volume IV of *MPP* contains little of relevance to fluid dynamics. In Volume V we find 'On the highest wave of uniform propagation',[6] published in 1883, which considers the waves of maximum steepness that can propagate without change of form. The volume also contains a second supplement to Stokes's great 1847 paper on oscillatory waves. Finally, the questions for the Mathematical Tripos and Smith's Prize for the period from 1846 to 1882 are included in Volume V. In the Smith's Prize paper of February 1854, Question 8 asked for a proof of what is now known as Stokes's Theorem, a standard result in vector calculus.

It is abundantly clear from *MPP* that Stokes's greatest work was done before he had reached his thirty-fifth birthday. Stokes's work on fluid dynamics was done during two distinct periods, from 1842 to 1850 and, after a thirty-year gap, between 1880 and 1898. In his obituary, Lord Rayleigh remarks that 'if the activity in original research of the first fifteen years had been maintained for twenty years longer, much additional harvest might have been gathered in'.

The Navier–Stokes Equations

The Navier–Stokes equations are the universal mathematical basis for fluid dynamics problems. Navier's original derivation in 1822 was not immediately accepted, and gave rise to some heated discussions and debate. An excellent review can be found in Darrigol.[7] There were several attempts, following Navier's publication in 1822, to develop a rigorous derivation of equations for viscous fluid flow. Most notable were those of the French scientists Poisson, Cauchy and Saint-Venant. George Green also made substantial contributions to the problem. Although Stokes was not the first to derive the equations in their final form, his derivation was founded on more general and physically realistic assumptions. Stokes also found several particular solutions to the viscous equations. Euler had obtained his fluid equations in 1755:

$$\rho\left(\frac{\partial \mathbf{V}}{\partial t} + \mathbf{V} \cdot \nabla \mathbf{V}\right) = -\nabla p + \mathbf{F}$$

Unfortunately, these equations produced absurd results in a wide range of practical situations where fluid resistance was important. This was recognized by d'Alembert and by Euler himself. D'Alembert expressed his concerns thus: 'I do not see how one can satisfactorily explain, by theory, the resistance of fluids.' He remarked that the theory leads to 'a singular paradox which I leave to future geometers for elucidation'.[7] Thus, at the beginning of the nineteenth century, fluid dynamics was incapable of explaining a wide range of important fluid flow phenomena. Hydraulic engineers had an armoury of empirical techniques, but these were not firmly based on fundamental physical principles.

Navier's equation, first written down in 1822, was freshly discovered at least five times, by Navier, Cauchy, Poisson, Saint-Venant and Stokes. Cauchy and Poisson paid no attention to Navier's work. Saint-Venant and Stokes acknowledged it but regarded it as lacking in precision and rigour. We can write the Navier–Stokes equations in modern notation as

$$\frac{\partial \mathbf{V}}{\partial t} + \mathbf{V} \cdot \nabla \mathbf{V} + \frac{1}{\rho} \nabla p - \nu \nabla^2 \mathbf{V} + \mathbf{F} = 0 \tag{1}$$

with the assumption of nondivergent flow, $\nabla \cdot \mathbf{V} = 0$. These are equations (13) in Stokes's paper of 1845, and he comments that they are 'applicable to the determination of the motion of water in pipes and canals, to the calculation of the effect of friction on the motions of tides and waves, and such questions'.

Stokes notes in his introduction that, having derived his equations, he discovered that Poisson had arrived at the same equations, and that the same equations had also been obtained in the case of an incompressible fluid by Navier. However, both Poisson and Navier had used methods markedly different from Stokes.

In addition to the equations applying to the interior of the fluid, Stokes also considered the conditions that must be satisfied at solid boundaries. There was widespread controversy about the appropriate boundary conditions, without which problems could not be formulated, let alone solved. At the time Stokes presented his memoir on the fluid equations (1845), he already believed that the most natural assumption for the relative velocity at a rigid boundary was that it must vanish.

Controversy

James Challis, the Plumian Professor at Cambridge, had the great misfortune to have observed Neptune on two occasions a month before Urbain Le Verrier's predictions were confirmed, but to have failed to identify it as a planet. He blamed pressure of other work for this oversight. During his undergraduate years, Stokes attended some of the lectures of Challis on fluid dynamics. He differed strongly with Challis in several important ways, and their disagreements led to the publication of several acrimonious exchanges. Challis published some fourteen papers on hydrodynamics, characterized by Craik[8] as 'mostly worthless'. Challis argued that the assumption of irrotational flow implied rectilinear motion. Of course we can easily show that there is no essential link between curvature and rotation of the flow. Linear flow with lateral shear has non-vanishing vorticity; moreover, a circular vortex with azimuthal velocity varying inversely with radial distance from the centre, as in the external region of a Rankine vortex, is irrotational. Challis also maintained that Euler's equations for incompressible fluid flow were incomplete, a view strongly contested by Stokes. Ultimately, Stokes tired of the ongoing conflict. In a letter to William Thomson in 1851, Stokes wrote about an 'awful heterodoxy' of Challis in the *Philosophical Magazine*.[9] He concluded 'I am half inclined to take up arms, but I fear the controversy would be endless'.

Stokes's Applications of the Equations

The solution of the full Navier–Stokes equations is quite beyond any analytical attack. However, when drastic approximations are made, systems amenable to analysis may result.

Stokes's First and Second Problems

For steady flows with parallel streamlines, the nonlinear terms vanish and a full solution is normally easily obtained. The associated initial value problems, where the motion is started impulsively, are also amenable to solution as the advection terms drop out again. The flow due to the impulsive motion of a flat plate parallel to itself is known as Stokes's First Problem.[10] The fluid motion known as Stokes's Second Problem is the flow around an infinite flat plate that moves sinusoidally in its own plane. There is a natural time-scale here, imposed by the period of the forcing, and there is no similarity solution for this problem. Stokes found the solution that obtains after the initial transient response has decayed.

The Pendulum

Stokes's motivation for studying fluid resistance came from his interest in the use of pendulums for geodesic measurements. Friedrich Bessel had published an influential memoir taking into account the effects of atmospheric drag on the motion of the pendulum. This triggered a series of practical experiments, but a full theoretical understanding was lacking.

The pendulum has provided an invaluable scientific apparatus and has played a vital role in horological science and in geodesy. The precise measurement of time has been of crucial importance in the scientific world and also in many practical situations. A most notable example was the determination of longitude, essential for the purposes of navigation.

Theoretical results often assume that the apparatus is *in vacuo*, so that the effects of air must be considered when comparing experimental results with theoretical values. Stokes's extensive paper addresses this question. In his study 'On the effect of the internal friction of fluids on the motion of pendulums',[5] Stokes assumed that the viscosity of air is proportional to the density. It was only later that Maxwell showed that the viscosity is insensitive to density over a wide range.

Creeping Flow around a Sphere

In his study of the effect of air resistance on the motion of a pendulum, Stokes was led to examine the resistance on a sphere of radius a moving at speed U through a viscous fluid. He gave a solution for the creeping flow around a sphere. He considered axisymmetric laminar flow and, assuming high viscosity, neglected the inertial terms in the equation of motion, deriving an equation

$$\left[\frac{\partial^2}{\partial r^2} + \frac{\sin\theta}{r^2}\frac{\partial}{\partial\theta}\left(\frac{1}{\sin\theta}\frac{\partial}{\partial\theta}\right)\right]^2 \psi = 0.$$

He solved this with appropriate boundary conditions to find the velocity and pressure fields. The pressure maximum is at the forward stagnation point and the minimum is at the rear stagnation point. Stokes then found an expression for the drag force,

$$D = 6\pi\mu a U,$$

showing that the resistance is proportional to the velocity. One third of the drag is due to pressure and two thirds to skin friction. The result that, for low Reynolds number flow, the drag force varies linearly with speed is frequently referred to as Stokes's law of resistance. This result was later crucial for Millikan in designing his oil-drop experiment to measure the charge on an electron.

For a two-dimensional obstacle such as a cylinder, the Stokes balance

$$\frac{1}{\rho}\nabla p = \nu\nabla^2\mathbf{u}$$

has no solution satisfying the boundary conditions at infinity. This result is often called Stokes's Paradox. This led Stokes to conclude that a steady slow flow around a cylinder cannot exist in nature. The explanation is that the small parameter $1/\mathrm{Re}$ multiplies the highest-order term in the governing equation and the perturbation problem is singular. An improvement of Stokes's solution, using a linear approximation of the inertial term at large distances, was provided by Oseen in 1910. His result may be written

$$D = 6\pi\mu a U[1 + 3Ua/8\nu]$$

Stokes used his law of resistance to explain why tiny droplets of moisture in a cloud can remain suspended over a long timescale. Stokes remarks that 'The pendulum thus, in addition to its other uses, affords us some interesting information on the department of meteorology'. This is in relation to the application of his analytical results to droplets of moisture falling through the atmosphere. For the small droplets forming a cloud, the terminal velocity is so small that 'the apparent suspension of the clouds does not seem to present any difficulty'.

Modern Applications of Navier–Stokes Equations

Climate change and its consequences are amongst the most pressing problems facing humanity today. There are enormous uncertainties concerning the future climate, and the best way we have for reducing these is by means of predictions based on computer simulations. The computer models for simulating weather and climate are known as Earth System Models. They are of great complexity, embracing a wide range of physical phenomena, with components for the atmosphere, the oceans, the land surface and sub-surface and the cryosphere. There are strong interactions between all these sub-systems. At the heart of every Earth System Model lies a dynamical core. The 'kernel of the core' comprises the Navier–Stokes equations. The same models are used regularly for short and medium range weather forecasts. Over recent decades, there has been a dramatic improvement in the accuracy and scope of computer forecasts, with enormous benefits for human society.[11] Thus, the fundamental work of Stokes underlies one of the greatest scientific advances of the twentieth century.

Of course, the Navier–Stokes equations have far wider applicability. They are used by aeronautical engineers to optimize aircraft design, by ship-builders to improve safety and minimize energy loss, by hydraulic engineers and by biologists studying blood flow in the body. Scientists use the Navier–Stokes equations in fundamental studies of turbulence, and the properties of the solutions of these equations are among the great unsolved problems of mathematics.

Stokes's Theorem

Students' first encounter with Stokes's name is usually through a fundamental theorem in vector calculus. Stokes's theorem relates the surface integral of a vector field over an open surface to the line integral of the field around the boundary. In modern notation, it may be written

$$\int_A \nabla \times \mathbf{V} \cdot \mathbf{n} dA = \oint_{\partial A} \mathbf{V} \cdot \mathbf{ds}$$

Taking $\mathbf{V}$ to be the flow velocity, this result states that the areal integral of vorticity over a surface is equal to the circulation around the boundary. It also expresses the fact that the circulation around the closed boundary curve $C = \partial A$ is equal to the flux of vorticity across the surface. Thus, it implies that for irrotational flow the circulation vanishes. In more old-fashioned notation, Stokes would have written that, for irrotational flow, $udx + vdy + wdz$ is an exact differential.

Stokes's theorem is a generalization of the fundamental theorem of calculus, which states that the integral of a function f over a closed interval $[a,b]$ can be evaluated as the difference between the values of the antiderivative of f at the ends of the interval. In turn, Stokes's theorem itself has been generalized to become an important principle in differential geometry: the integral of a differential form over the boundary of an orientable manifold is equal to the integral of its exterior derivative over the manifold; symbolically,

$$\int_{\partial\Omega} \omega = \int_{\Omega} d\omega.$$

The theorem has an interesting history. The basic result was contained in a letter from William Thomson to Stokes in 1850. Stokes set the theorem as a question on the Smith's Prize exam for 1854, which led to his name becoming attached to the result. In a footnote in Volume V of *MPP*, Larmor mentions earlier researchers who had integrated the curl of a vector field over a surface. Neither Stokes nor Thomson published a proof of the theorem. The first proof appeared in an 1861 publication of Hermann Hankel. The general result, in modern form, was formulated by Élie Cartan.

The Theory of Oscillatory Waves

The linear theory of water waves was developed in the eighteenth and early nineteenth century by French mathematicians, most notably Laplace, Lagrange, Poisson and Cauchy. Nonlinear waves were studied in Germany by Franz Joseph von Gerstner, who found the first exact solution for finite amplitude waves in deep water. In the 1830s and 1840s, several British physicists helped to advance the theory of waves. These included James Challis, George Green, John Scott

Russell, Philip Kelland, Samuel Earnshaw and George Biddel Airy. The origins of water wave theory are reviewed comprehensively by Craik[8] and Darrigol.[12]

In 1837, the year Stokes went up to Cambridge, the British Association for the Advancement of Science set up a Committee on Waves, to carry out observations and conduct experiments. John Scott Russell and Sir John Robinson were the directors of the committee. Within a few years, they had produced several reports. A topic that has attracted great attention was the 'solitary wave' observed on a canal by Scott Russell. This defied theoretical explanation until 1876, when Lord Rayleigh derived a solution by retaining both dispersion and nonlinearity.

The monumental *Report on Waves* published by Russell and Robinson in 1841 proved invaluable to scientists grappling with the theory of waves. Prominent amongst these was Stokes. The intriguing observations and experimental results of Scott Russell provided an impetus to Stokes to study wave dynamics, and in 1847, just ten years after the establishment of the BAAS committee, Stokes had completed his monumental work 'On the theory of oscillatory waves'. This is one of the great classical papers of hydrodynamics.

In a dispersive medium, different wave components travel at different speeds, moving in and out of phase with each other. Therefore, a small amplitude disturbance of unchanging form must be sinusoidal: if there are two or more wave components, they will will travel at different speeds, changing the shape of the wave form. This led Stokes to believe that Russell's solitary wave was a mathematical impossibility.

In a nondispersive medium, where the phase speed is independent of wavenumber, nonlinear interactions can lead to unbounded growth of amplitude if there is no counteracting effect. There is an opportunity for nonlinear steepening to be attenuated by dispersion, and for the dispersive effects to be balanced by nonlinear steepening. As a result, finite amplitude waves of constant form become possible.

Stokes was influenced by the earlier work on waves of George Biddell Airy,[13] and also by the researches of George Green. Airy's survey article 'Tides and Waves' appeared just as Stokes was setting out on his hydrodynamical researches. This survey contains the now-standard linear theory of water waves, including the dispersion relation which may be written in modern form as

$$c^2 = \frac{g}{k} \tanh kh$$

where k is the wavenumber and h the mean depth. A similar result had been obtained much earlier by Laplace.

Stokes Waves

In his 1847 paper,[4] Stokes studied wave motions in the case where the amplitude was sufficiently large that the nonlinear interactions could not be neglected. This was the first comprehensive analysis of waves of finite amplitude. He showed that periodic waves of finite amplitude are possible in deep water. Stokes considered weakly nonlinear periodic waves in water of intermediate or large depth. He devised a perturbation approach and derived solutions to third order in a small quantity, the product of amplitude and wavenumber or wave *steepness*, $\varepsilon = ka$. Thus, the amplitude was assumed to be small relative to the length of the waves.

Stokes's solution for the free surface elevation is

$$y = a\left[\cos k(x-ct) + \frac{1}{2}\varepsilon \, \cos 2k(x-ct) + \frac{3}{8}\varepsilon^3 \, \cos 3k(x-ct) + O(\varepsilon^3)\right]$$

All the Fourier components propagate at the same speed c, given by

$$c = \left(1 + \frac{1}{2}\varepsilon^2\right)\sqrt{\frac{g}{k}}$$

so that the wave profile is unchanging in time. It is noteworthy that the phase speed c depends upon the amplitude a. As there are components of different scales, the wave form is not longer a pure sinusoid. The ridges are steeper and narrower than the troughs. This wave profile might have been noticed by any keen observer of waves, but it took the genius of Stokes to provide a theoretical explanation.

One limitation of Stokes's weakly nonlinear analysis of 1847 was its inadequacy in describing the solitary waves observed by Scott Russell. Indeed, this gave rise to ongoing controversy leading to doubt being cast upon Scott Russell's results. Neither Airy nor Stokes was convinced about the importance that Russell ascribed to his 'Great Primary Wave'. It is regrettable that, as a consequence of the growing authority of Stokes, recognition of the value of Russell's work took many decades. It was only much later that the work of Joseph Boussinesq and Lord Rayleigh, which took account of both dispersion and nonlinearity, provided a solid analysis of solitary waves.

Some fifty years after Stokes's finite amplitude wave solution was found, it was shown by Korteweg and de Vries[14] in 1895 that the Stokes wave is a large-depth approximation to the cnoidal wave solutions of the equation formulated by them. Russell's solitary waves correspond to the infinite period limit of these solutions.[15] Many investigations following Stokes have shown that periodic wave trains of unchanging profile, such as the one discovered by him, are found in a wide range of physical systems and indeed are typical in nonlinear dispersive systems.

The analysis of nonlinear waves is complicated because boundary conditions must be specified at the free surface, the position of which is unknown until the problem is solved. Stokes circumvented this obstacle by using a perturbation approach—the Stokes expansion—that enabled him to express the boundary condition in terms of quantities at the known mean surface elevation. To avoid spurious 'secular variations', Stokes also expanded the dispersion relationship as a perturbation series. This approach, now known as the Lindstedt–Poincaré method,[16] is widely applied.

Waves of Maximum Height

In 1866 Stokes was appointed to the newly created Meteorological Council. This led him to consider several practical problems involving ocean waves. He investigated methods of accurately observing and measuring wave heights and periods from ships. He also studied ways to determine the location of distant storms using measurements, recorded in ships' logs, of the swells generated by them.

Stokes, who grew up on Ireland's western shore, was a skilled swimmer and a keen observer of nature. During his many holidays in Ireland, he undertook his own observational studies of waves and swell. In his mathematical study of surface waves he recalled his youth: 'In watching many years ago a grand surf which came rolling in on a sandy beach near the Giant's Causeway, without any storm at the place itself, I recollect being struck with the blunt wedge-like form of the waves where they first lost their flowing outline, and began to show a little broken water at the very summit. It is only I imagine on an oceanic coast, and even there on somewhat rare occasions, that the form of the waves of this kind, of nearly the maximum height, can be studied to full advantage.'

It was during the second period of study of sea waves, in the 1880s, that Stokes examined the question of the highest possible periodic wave. He concluded that the wave of maximum height had a sharp crest, with the water surfaces ahead of and behind the peak meeting at an obtuse angle. This also accorded with observations that he had made during his holidays in Ireland.

Stokes showed that the maximum wave steepness is $H/\lambda \approx 0.1412$ or $\sqrt{2}-1$. He returned to this question in research described in a supplement in MPP[17] and showed that the angle at the crest of these waves of maximum steepness is 120°. Stokes's solutions had, and continue to have, application to practical problems in coastal and off-shore engineering. For larger waves or shallower water, cnoidal theory, where the solutions are expressed in terms of Jacobi elliptic functions, may yield more accurate results.

The reference frame chosen by Stokes has the Oy-axis pointing downwards. This has led to some confusion. On page 211 of Volume I of MPP, Stokes's formula for the surface height appears as

$$y = a \cos mx - \frac{1}{2}ma^2 \cos 2mx + \frac{3}{8}m^2a^3 \cos 3mx \qquad (2)$$

(he writes m for the wavenumber k). Stokes remarks that the term of third order is almost insensible. The form of the equation is identical to that in the original paper in the *Transactions*.[4] The profile drawn by Stokes is for an amplitude $a = 7\lambda/80$, where λ is the wavelength. This corresponds to wave steepness $\varepsilon \approx 0.55$. Figure 6.3, which is consistent with Stokes's illustration, shows plots of the second and third order expressions. The thick curve is the second-order approximation, the thin curve includes the third-order term. The figure confirms the fact that the effect of the third-order term is quite small or, as Stokes put it, 'almost insensible'.

Expressions consistent with (2) are reproduced in Darrigol[18] and in Craik.[19] The correct form of the equation is found in many texts, but often with the Oy-axis pointing upwards. This choice is made by, for example, Lamb[20] and Whitham.[21] However, Lamb also gives the fourth-order approximation with the negative sign.[22] One may wonder whether this error has found its way into computer codes. It may be noted that Stokes likened the wave profile to a prolate cycloid. The wave that he drew closely resembles an inverted curtate cycloid.

Stokes Wave for a = 7λ/80. Black: Second Order. White : Third Order.

Fig. 6.3 Form of the Stokes wave, with sharpened crests and flattened troughs. Thick curve: second-order approximation. Thin curve: third order approximation.

Stokes Drift

For nonlinear waves, there is an ambiguity in the partitioning of the solution into wave and mean-flow parts. Stokes identified two ways of defining wave speed or *celerity*. In the first approach, the wave is considered in a frame moving with the mean horizontal velocity. In the second approach, the mean horizontal mass transport in the reference frame is zero.

If we consider small amplitudes, linear wave theory indicates that fluid particles move vertically up and down as a wave travels horizontally. However, observations show that an object floating on the sea surface in the absence of wind moves slowly in the direction of the waves. This is a finite-amplitude effect, now known as *Stokes drift*.

The trajectory of the floating object is not a closed curve but has the form of an epicycloid. The mean velocity at a fixed point is zero, but the mean Lagrangian velocity of a fluid parcel is non-vanishing: the parcel's forward movement at the top of the trajectory is greater than its backward movement at the bottom. Although this is a second-order effect, it is often significant and has important practical consequences. For deep water gravity waves with amplitude a, frequency ω and wavenumber k, the mean Lagrangian speed is

$$\bar{U}_{\mathrm{L}} = a^2 \omega k \exp 2kz_0$$

where the initial coordinates (x_0, z_0) serve to label the particle. This is also called the mass transport velocity.

Group Velocity

Keen wave observers will have noticed how difficult it is to follow the movement of an individual wave crest in a deep pond. Waves occur in groups or bunches and wave crests seem to appear from nowhere at the rear of the group, move through it and vanish somewhere ahead. The first report on this phenomenon may have been by Scott Russell around 1844.[12] Russell's observations generated little interest at the time. Independently, William Rowan Hamilton considered a similar phenomenon in the context of optics.[23] Stokes had been studying swells in calm conditions and argued that the wave period could be used to determine the location of the storm that gave rise to the swell. He was aware that longer waves travel faster than shorter ones and that, as a consequence, the observed period of waves from a distant storm decreased with time. William Froude also noted the distinction between the speed of individual crests and that of the wave group. Froude pointed out that the relevant speed for this estimate should be that of the group, not the phase speed of the waves.

In 1876, Stokes wrote to Airy about what he believed to be an original result: the overall speed of the wave group in deep water is only half the speed of the individual waves. This is easily shown, taking the phase speed in deep water to be $c = \omega / k = \sqrt{g / k}$. The group velocity is then $c_g = d\omega / dk = \frac{1}{2}\sqrt{g / k} = \frac{1}{2}c$. Stokes posed this problem as a question for the Smith's Prize that same year. The theory of group velocity was further advanced by Lord Rayleigh, who was also inspired by experiments of Osborne Reynolds. Rayleigh demonstrated the important relationship between group velocity and energy propagation.

Group velocity is of immense importance in weather forecasting. The large wave-like disturbances in the atmosphere at middle latitudes, known as Rossby waves, travel at an approximate speed of $c = U - \beta / k^2$, where U is the mean zonal flow and β is a constant. The group velocity is easily shown to be $c_g = U + \beta / k^2$, which is *greater than the phase speed*. Wave minima or troughs are commonly linked to stormy weather. Through the action of group velocity, a new storm can appear 'spontaneously' downstream of an existing chain of storms. The propagation of energy is more rapid than the movement of the individual storms.

Conclusion

Stokes's study of oscillatory waves, which initiated the nonlinear theory of dispersive waves, was far in advance of contemporary developments. He showed that periodic wave trains are possible in nonlinear systems and that their speed of propagation varies with the amplitude. This had deep influence on subsequent research. Stokes's work on waves, in addition to his other achievements, led to his appointment in 1849 as Lucasian Professor, a position he held for more than fifty years. In 1854, he became Secretary of the Royal Society and, from that time, heavy administrative responsibilities had the consequence that his scientific output was greatly diminished. Stokes was President of the Royal Society from 1885 to 1890. Through his position as Secretary and, later, President, Stokes was able to provide substantial assistance and support to a large number of younger scientists, as is evident in acknowledgments in the *Proceedings* and *Transactions* (see Chapter 8).

ACKNOWLEDGEMENTS

Thanks to Prof. Frédéric Dias for bringing the works of Olivier Darrigol to my attention at an early stage of this work.

CHAPTER 7

Stokes's Mathematical Work

RICHARD B. PARIS

Introduction

The collected mathematical and physical contributions of Sir George Gabriel Stokes occupy five volumes and amount to a total of 109 papers. Stokes's output covered a broad spectrum of physical problems, ranging across acoustics, wave propagation, gravitational attraction and thermodynamics, but the bulk of his work was in hydrodynamics and optics (he wrote over 60 papers in this latter field). His name is well known to generations of scientists and engineers, where in standard textbooks we encounter Stokes's law for the drag on a slowly moving sphere in a viscous medium, Stokes's theorem in vector calculus (although this was communicated to him by Lord Kelvin in a letter dated 2 July 1850 and set by Stokes as a question in the Smith's Prize Examination at Cambridge in February 1854), the Navier–Stokes equation in hydrodynamics, Stokes's conjecture on the wave of greatest height and, finally, the Stokes phenomenon in asymptotic analysis.

Apart from one or two brief mathematical notes concerned with certain differential operators repeatedly applied to a power of the variable, Stokes also wrote seven papers in mathematical analysis. It is with this work that we shall be concerned in this chapter.[1] In the majority of these papers, Stokes was dealing with important mathematical problems of the time arising from the needs of physical investigations and he clearly demonstrated much original thought and penetrative insight. Stokes's attitude to purely mathematical analysis, as opposed to mathematical physics, has been clearly summarized by G. H. Hardy,[2] who wrote in 1918: 'In the first place it must be remembered that Stokes was primarily a mathematical physicist. He was also a most acute pure mathematician; but he approached pure mathematics in the spirit in which a physicist approaches natural phenomena, not looking for natural difficulties, but trying to explain those which forced themselves upon his attention.'

Convergence of Series

The first purely mathematical paper Stokes wrote was entitled 'On the critical values of the sum of periodic series',[3] which was published in the *Transactions of the Cambridge Philosophical Society*

Paris, R. B., *Stokes's Mathematical Work*. In: *George Gabriel Stokes: Life, Science and Faith*, Mark McCartney, Andrew Whitaker, and Alastair Wood (Eds): Oxford University Press (2019). © Oxford University Press. DOI: 10.1093/oso/9780198822868.001.0007 in 1847. This justly famous memoir, which ran to over fifty pages, was written in a remarkably mature style (he was only 28 years old) and presented, in part, a general discussion of the possible modes of convergence both of series and integrals far in advance of current ideas of the time. Stokes's starting point was the Fourier sine series defined in the interval $[0, a]$ by

$$f(x) = \sum_{n=1}^{\infty} A_n \sin\left(\frac{\pi n x}{a}\right), \quad A_n = \frac{2}{a} \int_0^\infty f(x) \sin\left(\frac{\pi n x}{a}\right) dx;$$

an analogous treatment applies to Fourier cosine series. He examined the nature of the convergence of such series both when $f(x)$ is continuous in $[0, a]$ and in the neighbourhood of a finite discontinuity in $f(x)$ at an interior point $x = x_0$. He demonstrated that in the latter case the convergence of $f(x)$ at $x = x_0$ to the value $\frac{1}{2}\{f(x_0 + 0) + f(x_0 - 0)\}$ is non-uniform in the neighbourhood of $x = x_0$. The approach adopted by Stokes, which had been previously employed by Poisson, consisted of consideration of the above Fourier sine series as the limit of the series

$$\sum_{n=1}^{\infty} (1 - \epsilon)^n A_n \sin\left(\frac{\pi n x}{a}\right)$$

as $\epsilon \to 0+$. This was different from the approach used by Dirichlet in 1829,[4] which consisted of replacing the Fourier series truncated after n terms by an integral representation followed by determination of the limit of this integral as $n \to \infty$. In fact, from a remark made in a footnote[5] it is possible that Stokes was unaware of Dirichlet's paper when he first started his investigation, a fact possibly resulting from his greater familiarity with the work of the French analysts. He also went on to discuss the 'inverse' problem for continuous $f(x)$, showing how the nature of the resulting Fourier series could be deduced from the asymptotic structure of the Fourier coefficients A_n for large n. Part of this analysis is reproduced in the classic text by Whittaker and Watson.[6]

The second part of his 1847 memoir contains Stokes's introduction of the idea of uniform convergence. Although Weierstrass is rightly attributed with the discovery of uniform convergence and for fully realizing its far-reaching importance as one of the fundamental ideas of analysis, Stokes has priority of publication. He considered the sum

$$S_n(h) = a_1(h) + a_2(h) + \cdots + a_n(h)$$

whose terms depend on the positive variable h and which is supposed convergent as $n \to \infty$ for $h < h_0$. As is now well known, the value of this sum as both $n \to \infty$ and $h \to 0$ is not necessarily independent of the order in which these limits are taken. To illustrate this fact, Stokes cited the example where

$$a_n(h) = \frac{1}{n(n+1)} + \frac{2h}{[1 + (n-1)h](1 + nh)}$$

$$= \frac{1}{n} - \frac{1}{n+1} + \frac{2}{1 + (n-1)h} - \frac{2}{1 + nh}$$

upon partial fraction decomposition. This is a 'telescoping series' and so it is easy by a cancellation of terms to see that

$$S_n(h) = 3 - R_n(h), \quad R_n(h) := \frac{1}{n+1} + \frac{2}{1 + nh}$$

with the result that

$$S_1 := \lim_{n\to\infty}\lim_{h\to 0} S_n(h) = 1 \quad S_2 := \lim_{h\to 0}\lim_{n\to\infty} S_n(h) = 3.$$

In Stokes's words: 'The limit $[S_1]$ can never differ from $[S_2]$ unless the convergency of the [infinite] series becomes infinitely slow when h vanishes.' He defines the rather descriptive term *infinitely slowly convergent* to refer to the situation when the number n of terms that must be taken for the remainder $R_n(h)$ to be numerically less than a given arbitrarily small quantity ϵ increases beyond all limit as h vanishes. This is essentially equivalent to the modern definition of uniform convergence, though, as Hardy has pointed out,[2] what Stokes actually defined was *quasi-uniform convergence in the neighbourhood of a point*, rather than the usual sense of uniform convergence over an interval due to Weierstrass. This subtle distinction perhaps explains two mistakes made by Stokes. The first, one of omission rather than an actual error, was Stokes's failure to recognize the importance of uniform convergence to term-by-term integration of infinite series (indeed, Stokes quoted the false theorem that a convergent series may always be integrated term by term, appealing to an erroneous proof of Cauchy). And the second was that, although he correctly showed that a uniformly convergent sum of continuous functions implies continuity, his proof of the converse of this result was incorrect. Stokes also applied the idea of uniform convergence to integrals and took as one of his examples

$$\frac{2}{\pi}\int_0^\infty \frac{\sin au}{u}\,du = \begin{cases} 1 & a>0 \\ 0 & a=0, \\ -1 & a<0 \end{cases}$$

which he showed similarly to possess non-uniform convergence in the neighbourhood of $a = 0$. The memoir concluded with a discussion of the application of Fourier series to three physical examples.

It is not without interest to mention that in several places in the analysis of his 1847 paper, Stokes made use of what is today known as the Riemann–Lebesgue lemma (used by Riemann[7] in 1854). This asymptotic result states that if the integral of the function $f(x)$ over the interval $[a,b]$ exists and (if an improper integral) is absolutely convergent, then

$$\int_a^b f(x)\begin{Bmatrix} \sin \\ \cos \end{Bmatrix}\lambda x\,dx = o(1)$$

as $\lambda \to \infty$ (that is, the integral tends to zero); if, in addition, $f(x)$ has limited total fluctuation in (a, b) the order term can be strengthened to $O(\lambda^{-1})$. Stokes obtained this result (in the case of bounded functions) from an earlier paper of Hamilton in 1843 entitled 'On fluctuating functions'[8] and extended its validity to include the case when $f(x)$ becomes infinite. He also established a result equivalent to the sharper order estimate in a subsequent paper in 1850.

A Differential Equation for Bridge Deflection

A second paper that appeared in 1849 was concerned with the breaking of railway bridges[9] and appears to have been one of his few forays into 'industrial applied' mathematics of the time. We will give only a brief description here as this topic forms the subject of Chapter 9. This paper is of quite a different character to his others in that it was not driven by physics but by the solution of a practical engineering problem. With the advent of the railways in the first half of the nineteenth century, the effect of heavy moving loads over bridges and the associated enhanced stresses were of considerable concern. Having reduced the problem to its essentials, Stokes studied the second-order inhomogeneous differential equation

$$\frac{d^2 y}{dx^2} + \frac{\beta y}{x^2(1-x)^2} = \beta, \quad y(0) = 0, \ y'(0) = 0,$$

describing the deflection y of the bridge at a distance x (normalized to the length $2L$ of the bridge) from one end. The dimensionless parameter $\beta = gL^2 / (4V^2 y_0)$, where g is the gravitational acceleration, V the uniform horizontal speed of the train and y_0 is the central static deflection.

Stokes obtained the solution of this equation both numerically and in closed analytical form. Since typical values of β could be considerably greater than unity, he also carried out a perturbation expansion in inverse powers of β. From these different forms of solution, he was able to determine, in particular, how the deflection varied with the horizontal distance x traversed and the speed V. He established that the maximum deflection (which occurred for $\frac{1}{2} < x < 1$) was about $1.7 y_0$ for speeds corresponding to β somewhat greater than unity. In spite of this being Stokes's only applied investigation, he was invited some thirty years later to collaborate with engineers on a Royal Commission set up by the Board of Trade in July 1880 to consider and report on the effect of wind pressure on railway structures. This was as a consequence of the tragic Tay rail bridge collapse in Dundee in December 1879 in which many people lost their lives.

Asymptotic Analysis

The remaining five papers of Stokes's mathematical output lie in the domain of asymptotic analysis, to which Stokes made significant and far-reaching contributions. His third paper,[10] which appeared in 1850, contained a discussion of an asymptotic procedure for calculating the zeros of functions defined by integrals. A few years earlier, G. B. Airy had published an important investigation on the intensity of light in the neighbourhood of a caustic[11] in which the new wave theory of light applied to the familiar rainbow led to the intensity of illumination being proportional to the square of the integral

$$W(m) = \int_0^\infty \cos \frac{1}{2}\pi(w^3 - mw)\, dw, \tag{1}$$

where m is proportional to the perpendicular distance from the caustic ($m > 0$ towards the illuminated side of the rainbow). The location of the zeros of $W(m)$ correspond to the location of dark bands in a system of supernumerary rainbows of which up to thirty had been observed experimentally; see Fig. 7.1. Using the convergent power series expansion for $W(m)$ given by

$$W(m) = \frac{(\pi/2)^{-1/3}}{3} \Re\left\{ e^{-\pi i/6} \sum_{k=0}^\infty \frac{\Gamma\left(\frac{1}{3}k + \frac{1}{3}\right)}{k!} (ze^{\pi i/3})^k \right\}, \quad z = (\pi/2)^{2/3} m, \tag{2}$$

Airy succeeded in computing the value of the integral for $|m| \leq 5.6$, thereby obtaining the first two (positive) zeros. Further computation of the zeros by this method was impractical on account of the enormous accuracy required to overcome the cancellation of terms in the expansion. It must be remembered that all this was carried out by laborious hand-calculation using log tables (of limited accuracy). It can be seen from Fig. 7.1 that, being limited to $m \leq 5.6$, Airy just missed being able to determine the third zero of $W(m)$.

This was the position when Stokes took up the problem. He developed an asymptotic procedure for obtaining the value of $W(m)$ for large values of $m > 0$, thereby enabling the calculation of the zeros to be carried out with ease and with great accuracy. His approach, which he also applied to the Bessel function $J_0(x)$, consisted of showing that $W(m)$ satisfied the differential equation $u'' + zu/3 = 0$, where z is the variable defined in (2). This equation is of course, now known as Airy's equation and $W(m)$ can be expressed in terms of the well-known Airy function $\mathrm{Ai}(-3^{-1/3}z)$. He then gave the formal asymptotic solutions in the form

$$W(m) = Az^{-1/4}\exp[2i(z/3)^{3/2}]R_+(z) + Bz^{-1/4}\exp[-2i(z/3)^{3/2}]R_-(z), \tag{3}$$

where

$$R_\pm(z) = 1 \mp \frac{1 \cdot 5i}{144}(z/3)^{-3/2} + \frac{1 \cdot 5 \cdot 7 \cdot 11 i^2}{2!144^2}(z/3)^{-3} \mp \cdots$$

and A, B are constants to be determined. Although such a procedure today is routine, it would appear that this was not the case in 1850, for Stokes felt compelled to add in a footnote that he had obtained this manner of representation from Cauchy's analysis of the Fresnel integrals.[12]

The values of the constants A and B were then obtained by matching the above asymptotic solution to the large-m behaviour of $W(m)$ deduced directly from the integral (1). Stokes achieved this by exploiting Cauchy's theorem which enables the integration path to be suitably deformed in the complex plane of the integration variable. By writing the cosine as the real

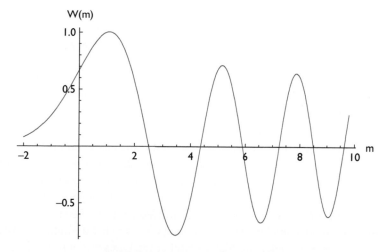

Fig. 7.1 The graph of the function $W(m)$.

part of its associated exponential and rotating the integration path through $-\pi/6$, he was able to transform the integral in (1) to $W(m) = (\pi/2)^{-1/3}\,\Re\{e^{-\pi i/6}\,I(m)\}$, where

$$I(m) = \int_0^\infty \exp(-x^3 + 3q^2 x)dx, \quad q = (\pi/2)^{1/3}(m/3)^{1/2}\,e^{\pi i/6}. \tag{4}$$

Then, recognizing that $x = q$ is a special point, he introduced the new variables $x = q + \tau$ followed by $\tau = (3q)^{-1/2}t$ to obtain the transformed integral

$$I(m) \;\;= e^{2q^3}\int_{-q}^\infty \exp(-\tau^3 - 3q\tau^2)\,d\tau$$

$$= \frac{e^{2q^3}}{\sqrt{3q}}\,\int_{-t_0 e^{\pi i/4}}^\infty \exp(-t^2 - (3q)^{-3/2}t^3)\,dt, \quad t_0 = 3^{1/2}\,|q|^{3/2},$$

where Cauchy's theorem has again been employed to replace the upper limit in the second integral by $+\infty$. Stokes then deformed the path of integration into the circular arc from $-t_0 e^{\pi i/4}$ to $-t_0$ and thence along the real t-axis to $+\infty$. By bounding the various contributions from different parts of this modified path, he was able to show that the dominant contribution to the integral arises from the segment $(-a, a)$ surrounding the origin $t = 0$, where $a \ll t_0$. Letting first $|q| \to \infty$ *followed* by $a \to \infty$, he finally obtained the desired result

$$I(m) \simeq \frac{e^{2q^3}}{\sqrt{3q}}\int_{-a}^a e^{-t^2}\,dt \simeq \sqrt{\frac{\pi}{3q}}\,e^{2q^3}. \tag{5}$$

This then yielded his asymptotic approximation for $W(m)$ in the elegant form

$$W(m) \sim \frac{(2/3)^{1/2}}{(m/3)^{1/4}}\cos[\pi(m/3)^{3/2} - \pi/4], \quad m \to +\infty. \tag{6}$$

Comparison of the above asymptotic form for $W(m)$ with the formal solution in (3) then yielded the values of the constants A and B to be

$$A = \frac{\pi^{1/6}}{2^{2/3}3^{1/4}}e^{-\pi i/4}, \quad B = \frac{\pi^{1/6}}{2^{2/3}3^{1/4}}e^{\pi i/4}. \tag{7}$$

By writing (3) in a suitable real form, Stokes subsequently was able to evaluate the function $W(m)$ for large positive m and gave a table of the first fifty zeros accurate to four decimal places together with the first ten zeros of the derivative $W'(m)$.

What Stokes had done here was effectively to employ the saddle-point method for the asymptotic approximation of Laplace-type integrals when the saddle is situated off the real axis. The Laplace-type integral $I(m)$ in (4) has the exponential factor associated with two saddle points given by $d/dx(x^3 - 3q^2 x) = 0$; that is, at the points $x = \pm q$. The relevant saddle is the one situated in the right-half x-plane at $x = q$. The first transformation of the variables then places the saddle $x = q$ at the origin in the τ-plane, with the deformation of the path along the real axis in the t-plane corresponding to a path which, in the neighbourhood of $t = 0$, coincides with the steepest descent path through the saddle. Thus, although nowhere did he mention the terms 'saddle' or 'paths of steepest descent', Stokes was effectively employing the saddle-point method in the complex plane more than a decade before Riemann, whose fragmentary manuscript[13] on this subject was dated October 1863. In a footnote, Stokes pointed out that the result in (6) could also be obtained directly from (1) by stationary phase arguments.[14]

The estimation of $I(m)$ for large positive m is today a standard exercise in asymptotics using the saddle-point method, or more precisely the so-called *method of steepest descents*. With $I(m)$ written in the alternative form

$$I(m) = \int_0^\infty e^{|q|^3 \psi(x)} dx, \quad \psi(x) = -x^3 + 3xe^{\pi i/3}, \tag{8}$$

saddle points of the integrand occur when $\psi'(x) = 0$; that is, at $x = \pm e^{\pi i/6}$. The paths of steepest descent and ascent through the relevant saddle $x_s = e^{\pi i/6}$ are determined by $\Im\{\psi(x) - \psi(x_s)\} = 0$ and are illustrated in Fig. 7.2(a). The path of steepest descent labelled AB passes to infinity in the directions $\arg x = 0$, $2\pi/3$; the other path indicated corresponds to the steepest ascent path (on which $\Re\psi(x)$ steadily increases away from x_s) that passes to infinity in the directions $\arg x = \pm\pi/3$. On the path AB, $\Im\psi(x)$ remains constant thereby eliminating the rapid oscillations in the integrand that would be present on other path choices. In addition, and most significantly, $\Re\psi(x)$ on the path AB decreases the most rapidly possible on either side of x_s; see Fig. 7.2(b) for an illustrative example. The integration path $[0,\infty)$ is then deformed (by appeal to Cauchy's theorem) first along the ray $\arg x = 2\pi/3$ out to infinity and then along the steepest descent path AB through x_s to $+\infty$. Since the contribution from the path OA is $O(q^{-3})$ as $|q| \to +\infty$, the main contribution to $I(m)$ can be shown to result from the neighbourhood of the saddle, where the integrand becomes increasingly sharply peaked as $|q|$ increases and can be locally approximated in terms of a Gaussian exponential. This is essentially what Stokes achieved with his sequence of transformations leading to his approximation in (5).

Stokes's discussion of the bright side of the caustic ($m > 0$) was entirely satisfactory. On the dark side of the caustic ($m < 0$), however, he encountered a major problem. With

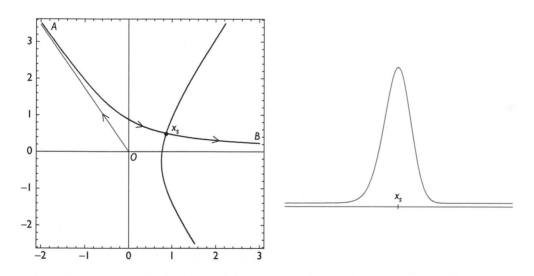

Fig. 7.2 (a) The deformed integration path OAB for the integral in (8). The path OA is the ray $\arg x = 2\pi/3$ and AB is the steepest descent path through the saddle $x_s = e^{\pi i/6}$. The other path is the steepest ascent path through x_s passing to infinity in the directions $\arg x = \pm\pi/3$. (b) The typical behaviour of $\Re\psi(x)$ for large $|q|$ *along* the path AB in the neighbourhood of x_s.

$z = -|z| = |z|e^{\pi i}$ (recall that z is defined in (2)), the leading terms of the formal asymptotic solution in (3) become

$$W(-m) \sim A z^{-1/4} \exp[2(|z|/3)^{3/2}] + B z^{-1/4} \exp[-2(|z|/3)^{3/2}].$$

Recognizing that A, B must change when $\arg z$ is increased by 2π (since $W(m)$ is an entire function of m whereas the asymptotic forms on the right-hand side of (3) contain fractional powers of m), Stokes knew from physical considerations that the constant A must be zero on $\arg z = \pi$ to remove the exponentially large term, with B remaining at its value given in (7). He states: 'This mode of determining the [constants] is anything but satisfactory. I have endeavoured in vain to deduce the leading term in [$W(m)$ for m] negative from the integral itself, whether in the original form [(1)] or in the altered form [(4)].'

The realization that the constant A (now known as a *Stokes multiplier*) must 'switch off' somewhere between $\arg z = 0$ and $\arg z = \pi$ is, of course, an example of the *Stokes phenomenon*. It took Stokes a further seventeen years to elucidate the nature of this phenomenon, which appears to have come only as the result of much effort and lucubration. This important discovery was published in 1857 in a paper entitled 'On the discontinuity of arbitrary constants which appear in divergent developments',[15] together with a sequel[16] in 1868 and a brief survey in what appears to have been his last paper[17] in 1902. In these papers, Stokes showed that the change in the constant A has to take place on the ray in the z-plane (now known as a *Stokes line*) where its associated exponential factor is maximally subdominant with respect to the other exponential factor. In the case of the function $W(m)$ in (1), this occurs on the ray $\arg z = \pi/3$, where the first exponential in (3) becomes $\exp[-2(|z|/3)^{3/2}]$ while the second exponential becomes $\exp[2(|z|/3)^{3/2}]$.

In his 1857 paper, Stokes verified this assertion with a numerical example applied to the Airy function $\mathrm{Ai}(z)$. The (formal) asymptotic solutions of the differential equation $u'' = zu$ satisfied by $\mathrm{Ai}(z)$ are

$$u_{\pm}(z) = \frac{z^{-1/4}}{2\sqrt{\pi}} \exp\left[\pm\frac{2}{3}z^{3/2}\right] \sum_{k=0}^{\infty} \frac{(\pm 1)^k c_k}{z^{3k/2}}, \quad c_k = \frac{\Gamma\left(3k + \frac{1}{2}\right)}{3^{2k}\sqrt{\pi}(2k)!}.$$

For $z > 0$ (the dark side of the rainbow), $\mathrm{Ai}(z) \sim u_{-}(z)$, which represents an exponentially decaying function corresponding physically to an evanescent wave. For $z < 0$ (on the bright side), it can be shown using the circuit relation satisfied by the Airy function that $\mathrm{Ai}(z) \sim u_{-}(z) + i u_{+}(z)$, which represents two exponential oscillatory terms corresponding physically to two waves. It is seen that the evanescent wave continues across $z = 0$ to become an oscillatory wave, but that somewhere in this transition the formal solution $u_{+}(z)$ makes its appearance.

Stokes resolved this difficulty by rotating about $z = 0$ in the complex z-plane (at large $|z|$). He argued that the birth of the term $u_{+}(z)$ should take place on the ray $\arg z = 2\pi/3$ (120°), where the exponential factor in $u_{-}(z)$ is $\exp\left[\frac{2}{3}|z|^{3/2}\right]$ and that in $u_{+}(z)$ is $\exp\left[-\frac{2}{3}|z|^{3/2}\right]$. This corresponds to where the difference between the two exponential factors is greatest and where the terms $(-1)^k c_k z^{-3k/2}$ in the dominant expansion $u_{-}(z)$ all have the same sign, so that the divergence of this series is most severe. The ray $\arg z = 2\pi/3$ is called a Stokes line for the function $\mathrm{Ai}(z)$ whereas the rays $\arg z = \pi/3$ and π are called anti-Stokes lines, where the exponentials in $u_{\pm}(z)$ are oscillatory and of the same order; see Fig. 7.3(a). To detect the birth of such exponentially small terms requires the calculation of the dominant series to at least a

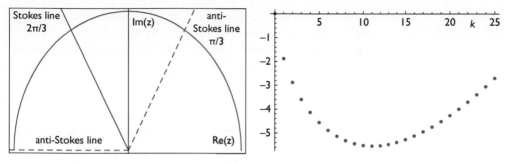

Fig. 7.3 (a) The Stokes (solid) and anti-Stokes (dashed) lines in the upper half-plane for the Airy function Ai(z). (b) Magnitude of the terms (on a $\log_{10}$ scale) in the dominant asymptotic series $u_-(z)$ against ordinal number k for Ai(z) when $z = 4e^{2\pi i/3}$. The optimal truncation index corresponds to $k = 11$.

comparable accuracy. Stokes achieved this by employing optimal truncation of the asymptotic expansion of the dominant function $u_-(z)$ to produce an exponentially small remainder; that is, truncation just before the least term (in modulus) of the expansion: see Fig. 7.3(b). It is clear that this truncation index depends on $|z|$ and must increase with z. Not content with this level of accuracy, he proceeded to 'resum' the divergent tail of the dominant asymptotic expansion by what is effectively an Euler transformation and further enhanced the accuracy of his calculations. It should be emphasized that all this was carried out well before both Poincaré's 1886 definition[18] of an asymptotic expansion and Stieltjes' work[19] on divergent series of the same year.

Stokes considered Ai(z) at 30° either side of the ray $\arg z = 2\pi/3$. He computed (by means of laborious hand-calculation) the function on these rays using the convergent series expansion of Ai(z) for a convenient value of $|z|$. In this way he was able to demonstrate numerically that the solution $u_+(z)$ is absent on $\arg z = 90°$ but is present on $\arg z = 150°$, thereby confirming his theory. He also realized that this phenomenon did not only apply to Ai(z) but that it had a pervasive nature applying to a wide variety of functions defined by integrals and solutions of differential equations.

Since Stokes's discovery of this phenomenon over 150 years ago, the conventional view has been that of a discontinuous change in the constant associated with subdominant asymptotic expansions. The fact that an analytic function can possess such a discontinuous representation has always been problematic. This 'hide-and-seek' nature of the multiplier, together with the vagueness associated with the precise location of these jumps in the complex plane, has resulted in the Stokes phenomenon being enshrouded in an element of mystery. Indeed, in his 1902 paper Stokes describes the process in the following highly imprecise terms: 'The inferior term enters as it were into a mist, is hidden for a little from view, and comes out with its coefficient changed.' This viewpoint, however, was completely overturned when, in 1989, Berry[20] argued that the coefficient of the subdominant expansion should be regarded not as a discontinuous constant, but, for fixed $|z|$, as a *continuous* function of arg z. He showed that, when viewed on an appropriately magnified scale, the change in the subdominant multiplier across a Stokes line is in fact continuous. For a wide class of functions, the functional form of the rapid but smooth transition possesses a universal structure described approximately by an error function, whose argument is an appropriate variable describing transition across the Stokes line; see articles by Berry and Howls,[21] and also Paris and Wood,[22] for more details.

In 1889, at the age of 70, Stokes briefly returned to consideration of an alternative method of determining the large-variable asymptotics of a function[23] in an attempt to simplify the computation of the constants in its formal asymptotic expansion. He approached this problem not from an integral definition of the function as he had done in 1857, but from its convergent power series expansion. He had observed that many of the functions arising in physical investigations possessed Maclaurin expansions whose coefficients involved quotients of factorial (or gamma) functions in the form

$$f(x) = \sum_{n=0}^{\infty} a_n x^n, \quad a_n = \frac{\prod_{r=1}^{p} \Gamma(n+a_r)}{\prod_{r=1}^{q} \Gamma(n+b_r)},$$

where the parameters $a_r, b_r > 0$ and it is supposed that the integers p and q satisfy $q \geq p+1$ so that the sum is uniformly and absolutely convergent for all finite x. Stokes was primarily interested in computing a function along a Stokes line, where all the terms in its asymptotic expansion are positive, and so he confined his attention to the behaviour of $f(x)$ as $x \to +\infty$.

The large-n behaviour of the coefficients a_n can be determined by use of Stirling's approximation for the gamma function given by

$$\Gamma(n+a) \sim \sqrt{2\pi}\, n^{n+a-\frac{1}{2}} e^{-n} \quad (n \to +\infty).$$

Some routine algebra then shows that $a_n \sim (2\pi)^{-\kappa/2} n^{\vartheta-\frac{1}{2}} (e/n)^{n\kappa}$ as $n \to \infty$, where the parameters κ and ϑ are defined by

$$\kappa = q - p, \quad \vartheta = \sum_{r=1}^{p} a_r - \sum_{r=1}^{q} b_r + \frac{1}{2}(1+\kappa).$$

It can then be established that the maximum value of the terms in the sum occurs for large x when $n = n_0 \sim x^{1/\kappa}$. Stokes applied the discrete analogue of Laplace's method (that is, the saddle-point method on the real axis) for the asymptotic approximation of integrals centred around $n = n_0$ to obtain the leading behaviour

$$f(x) \sim (2\pi)^{(1-\kappa)/2} \kappa^{-1/2} x^{\vartheta/\kappa} \exp[\kappa x^{1/\kappa}], \quad x \to +\infty;$$

see Paris[24] for a justification of this procedure.

In modern notation, the function $f(x)$ is proportional to the generalized hypergeometric function ${}_pF_q(a_1,\ldots,a_p;b_1,\ldots,b_q;x)$. It appears that this was the earliest attempt at the determination of the asymptotic behaviour of such functions, the later asymptotic theory of integral functions of hypergeometric type being developed by many authors, mainly in the first half of the twentieth century.

Conclusions

As we have shown, the relatively small number of purely mathematical papers compared to Stokes's total scientific output bears no relation to their importance or ramifications in mathematics today. It is also true that Stokes assumed considerable administrative duties, not the least of which was his appointment as Secretary of the Royal Society of London from 1854 to 1885, as described in Sloan Evans Despeaux's chapter (Chapter 8) in this volume. The result of all these responsibilities was a regrettable fall-off in his research output. However, Stokes's skills as an analyst are undeniable and, recalling Hardy's words cited at the beginning, one can only regret that his attention was not drawn to purely mathematical questions without being driven by an underlying physical problem.

CHAPTER 8

Stokes and the Royal Society

SLOAN EVANS DESPEAUX

Introduction

When reading tributes to George Gabriel Stokes from obituaries and appreciations, one notices that his contemporaries often referred to one facet of his career with something close to regret: Stokes's service to the Royal Society. Many of his colleagues felt that his service diverted valuable time away from original research. This chapter will explore Stokes's almost four decades of service to the Royal Society, first as Secretary and then as President. While his work as Secretary placed extreme demands on his time and energy, Stokes took on much of the heavy workload by choice. In his work as Secretary, Stokes acted as gatekeeper to Britain's most prestigious scientific publication, the *Philosophical Transactions*. Stokes broadened the role of his Secretaryship to include editor, mediator, and mentor. While he left many of the political machinations of the Royal Society to other officers, he found that he could not completely avoid all political turbulence during his term as President. For better or worse, the Royal Society provided the outlet for much of Stokes's commitment to science in Victorian Britain.

Master of 'Internal Scientific Work': Stokes as Royal Society Secretary

Two years after becoming Lucasian Professor of Mathematics at Cambridge, Stokes was elected a Fellow of the Royal Society. His election, on 6 June 1851, was the same day as that of his friend William Thomson, who would remain his ally throughout Stokes's life, and Thomas Huxley, who would later become his adversary.[1] By 1853, Stokes was chosen to sit on the Royal Society Council, and the next year he succeeded Cambridge mathematician S. H. Christie as Secretary. Stokes was elected 'A Secretary', that is, the Secretary who would deal with mathematics and physics. The year before, William Sharpey began his tenure as 'B Secretary', who covered the biological sciences.

The largest proportion of Stokes's workload as Secretary revolved around editing the Royal Society's journals, the *Philosophical Transactions* and the *Proceedings*. In particular, Stokes had

Evans Despeaux, S., *Stokes and the Royal Society*. In: *George Gabriel Stokes: Life, Science and Faith*, Mark McCartney, Andrew Whitaker, and Alastair Wood (Eds): Oxford University Press (2019). © Oxford University Press. DOI: 10.1093/oso/9780198822868.001.0008

Fig. 8.1 During an 1882 meeting of one of the Royal Society's many committees (specifically, a Transit of Venus Committee meeting), someone sketched this series of doodles on an ink blotter. G. G. Stokes is bottom right, with T. H. Huxley almost directly above him. © The Royal Society.

to manage the Royal Society's peer review process. While the Royal Society had been in the vanguard of British scientific societies in establishing a peer review process, its procedures were initially quite informal. However, by 1832, Royal Society President the Duke of Sussex was able to announce a Council resolution 'to allow no Paper to be printed in the Transactions of the Royal Society, unless a written Report of its fitness shall have been previously made by one or more Members of the Council, to whom it shall have been especially referred for examination'.[2] While this existence of discipline-specific Secretaries might suggest a natural division of labour in handling referees' reports on biological and physics papers, during Stokes's tenure as Secretary the workflow was quite different. Michael Foster, 'B Secretary' from 1881 to 1903, maintained that Stokes handled practically all of the 'internal scientific work, all that relates to the communications made to the Society, to the reading of papers, and to their publication in the *Transactions* or *Proceedings*', for not only the physical but also the biological communications. Foster, and the preceding 'B Secretaries' who had worked beside Stokes, handled the 'external scientific work, the negociations [*sic*] with the Government or with other bodies, home or foreign'.[3]

As the Secretary in charge of 'internal scientific work', Stokes 'made it his duty to make himself acquainted, so far as it was possible for him to do so, not only with the form but also with the substance of every paper which came in. He spared no pains in his efforts to secure that the form should be as good as possible under the circumstances.'[4] Specifically, he refereed many of the physics submissions himself.[5] Regardless of subject, Stokes acted as what Harrison has called a preliminary referee and a 'participating editor' who was very active in the revision of submitted papers. In this way the Royal Society's 'definition of the nature of the scientific enterprise thereby devolved upon the person of its long serving secretary [Stokes], to quite an extraordinary degree'.[6]

In his work as editor for the Royal Society, Stokes maintained a high standard for publication. His daughter, Isabella Lucy Humphry, recounted a conversation with Stokes in which he 'spoke of the misfortune which it would be to Science if…[the Royal Society] ever published anything for any reason than its first-rate excellence…he remarked that he gave up an immense amount of time to the improvement of hopeful work, but that he could not make bad work good'.[7] An aversion to controversy often motivated the Secretary: 'Stokes seemed to want the R.S. publications to contain only completed truths which could not be sullied by competitive gainsayings.'[8] Harrison has said that the maintained natural byproduct of this tendency was 'a host of emasculated papers being published without the opinions of the author or with a set of opinions provided by Stokes and lightly disguised as those of the author'.[9]

At the same time that the Royal Society's peer review process was becoming more formal, getting published in the *Philosophical Transactions* became more closely tied to being elected a Fellow. In 1839, new guidelines for Fellowship applications 'required a list of a potential Fellow's publications, and having a paper in the *Transactions* was considered an especially noteworthy qualification for fellowship'.[10] Therefore, Stokes acted as a gatekeeper for entry not only into Royal Society publications but also into the Fellowship itself.

Upon receiving a manuscript submitted for publication in the *Philosophical Transactions*, Stokes would select two or three referees. According to the 1832 peer review reforms, these referees came from the Council membership. While the pool from which referees could be drawn eventually widened to include Fellows,[11] Stokes still used a relatively small number of referees to evaluate submissions. For example, Harrison found that out of the 266 submissions made between 1859 and 1862, only seventeen referees (one of whom was Stokes) reviewed 166 of the manuscripts.[12] Referees knew the identity of the authors, because the Society wanted to enable the review of the prospective author's intellectual reputation as well as his social standing.[13] Referees also often knew who else was reviewing a given paper, and while the process was single-blind, established authors could often divine the identity of their referees. The peer review process was certainly not perfect and the 'unknown man's route to the pages of the Society's journals was far more steep and perilous, regardless of the apparent merit of his work vis à vis that of a well-known contributor'.[14]

Stokes did not directly pass on the referees' reports to the authors. In fact, the main purpose of these reports was 'not to direct the author about possible revisions; rather the reports were intended for the internal use of the Royal Society. Their function seems to have been to ensure that decisions about which researchers would have their papers in the *Transactions* were fair and thoughtful.'[15] Thus, not only did Stokes have to manage receiving the reports, arbitrating between the referees, and communicating them to the Council (which acted as the Committee of Papers that ultimately decided the fate of the submissions), but he also maintained a detailed correspondence with authors in which he transmitted selected referees' comments along with his own editorial directions.[16]

Case Study: The Disgruntled T. P. Kirkman

Stokes's correspondence with author Thomas Penyngton Kirkman provides a window into how tedious the Royal Society Secretary's work could be. After rejecting a career in his father's cotton business, Kirkman supported himself through private tutoring, earned a BA from Trinity College Dublin, and began a fifty-two-year career as rector of Croft, a parish in Lancashire. This parochial life enabled him to devote much time to mathematics, an avocation that began in response to a prize question Kirkman read in the *Lady's and Gentleman's Diary* in 1844.[17] After establishing his reputation as a mathematician, Kirkman focused once again on a prize question. In 1861, the Paris Académie des Sciences announced a Grand Prix de Mathématiques to be awarded to the mathematician who 'perfects in some important point the geometrical theory of polyhedra',[18] and this prompted Kirkman to consolidate his research in this area. Kirkman had earlier submitted an entry for the 1860 Grand Prix question concerning groups of substitutions. While the prize committee recognized innovations in his, and especially the memoirs of the other two contestants, Camille Jordan and Emile Mathieu, they believed that none of the submissions warranted the prize.[19] Kirkman wrote to Stokes that 'I am so annoyed at the manner in which the Académie had their competition in refusing to acknowledge any of their results, and so suspicious of their fairness to the Foreigner, that I have determined…not to send it [the memoir on polyhedra] to Paris, but to present it to the Royal Society'.[20] In this twenty-one-section article, 'On the Theory of Polyedra',[21] Kirkman tackled the enumeration and classification of polyhedra.

The referees assigned to Kirkman's massive treatise handled their task in a variety of ways. Thomas Archer Hirst, who would be elected to the Royal Society Council in 1864, declined to report on the memoir, citing a lack of expertise in the area as well as a lack of time.[22] William Spottiswoode, who would become President of the Royal Society in 1878, reported that, 'although I have not really mastered its contents, I have seen enough of it to recommend it for publication in the Philosophical Transactions'.[23] Unlike Spottiswoode, Arthur Cayley, an F.R.S. who had recently won the Society's Royal Medal, was not satisfied with the paper and criticized the quality of its contents:

> It is very probable that Mr. Kirkman's paper on Polyhedra contains results of great value and that he may have effected as he claims to have done the solution of the problem of the classification and enumeration of Polyhedra. But it appears to me on the first page (which is all that I have read) of the memoir that the author has not taken sufficient pains to present his results in a clear and intelligible form and that the memoir ought really to be rewritten…it would be a mere waste of time to attempt reading a paper characterised by such a want of precision of expression.[24]

Clearly unhappy with the latter referee's opinion as communicated to him by Stokes, Kirkman argued his case to the Secretary. He declared that except for one small detail,

> I have not been able to discover the slightest value in the remarks of your Referee…In all the work I have not been able to find an example of the carelessness of composition with which he charges me; and I consider it to be a simple piece of incivility only excusable on the ground of his not understanding the subject, when he says that I never used certain definitions…'for no other purpose than to save the author the trouble of defining &c.' There is not the least uncertainty that my work should be printed by the R.S. And indeed, if it was printed, I suspect that no few men in the world will read it.[25]

This response reveals that the identity of his referee must have been a mystery to Kirkman, because Cayley and Kirkman had been mathematical correspondents since the 1840s, when Cayley had very positively refereed a submission Kirkman had made to the *Cambridge and Dublin Mathematical Journal*.[26] A week after his frustrated letter to Stokes, Kirkman again engaged the Secretary with details of his disagreement with the referee. He clarified that the object of his quarrel was neither the Royal Society nor its Council and that he would not want the referee,

> wherever he may be, to know what I say of his criticisms. I am perfectly unconscious of having done or said anything that should have led him to treat me so harshly: and it is possible that the newness & difficulty of the subject may have so rebuffed him... But he has certainly failed completely to justify his own severity – and has not alluded to anything *out of the first page*.[27]

Cayley evaluated Kirkman's revisions to the paper but still considered 'the labour of going thro' it...far more than I am able to undertake'. However, he proposed the publication of the first two sections (which still occupied forty quarto pages) and noted that 'to save his rights of priority the author would probably wish and there would be no objection that the Introduction and Table of Contents of the whole memoir should be published with the first two sections'.[28]

The Royal Society published Kirkman's work in the manner Cayley suggested. In the published version's introduction in the *Philosophical Transactions*, Kirkman pointed out that besides the two sections 'here presented to the public...[m]uch remains of the entire work [sections 3 to 21], which is, however, completely written, and in the possession of the Royal Society'.[29] Here, Kirkman referred to the Society's practice of archiving work that it chose not to publish. While an author could ask to copy the submission, according to a 1776 statute 'the original copy of every paper read at the Society shall be considered as the property of the Society'.[30]

While the Royal Society never published the remaining sections of Kirkman's memoir, it did choose to publish some of the tables in its *Proceedings*.[31] By mid century, this secondary publication outlet for the Royal Society opened to include memoirs not selected for the *Philosophical Transactions*.[32] At this time many short papers entered the *Proceedings* without being refereed through what Moxham and Fyfe describe as 'long-standing, tacit, and social processes for winnowing papers—which relied on the judgement of the fellows acting as "communicators" and of the secretaries'.[33] While the Royal Society at least partially presented his work to the public, Kirkman never again published his work in Royal Society journals. He did recognize the role of Stokes in the affair, writing that 'the trouble you have had with me gives me a very lovely idea of patience and fortitude of a Sec^y of the Royal Society. You have no sinecure evidently'.[34]

A Pegasus harnessed to the Dust-Cart? Stokes's Extreme Service Load as Secretary

Secretaryship as sinecure was inconceivable to Stokes, who has been considered 'if possible, the most hard-working of all nineteenth-century Secretaries'.[35] Part of his heavy workload could be blamed on the increased number of submissions for publication to the Royal Society. 'Whereas between 1790 and 1799 only 319 papers were communicated to the Society, the number in the ten years 1850–59 was 672'.[36] In response to the growth, in 1860 the Royal Society Council almost doubled the annual salary of each secretary to £200.[37] This stipend was an important addition to Stokes's salary; he married Mary Susanna Robinson in 1857 amid worries about financially supporting his family.[38]

Another reason for Stokes's heavy workload as Secretary could also be attributed to the extreme care he put into his correspondence with authors. In the *Memoir and Scientific Correspondence of the Late Sir George Gabriel Stokes*, several of Stokes's contemporaries commented on his approach to correspondence. William Huggins, who served on the Royal Society Council during Stokes's tenure and who became the first twentieth-century Royal Society President, wrote:

> One of the most distinguishing characteristic qualities of Sir George was the generous way in which he was always ready to lay aside at once, for the moment, his own scientific work, and give his whole attention and full sympathy to any point of scientific theory of experiment about which his correspondent had sought his counsel. Notwithstanding the many heavy duties resting upon him, his reply came nearly always without delay by an early post.[39]

Similarly, telescope maker Howard Grubb, F.R.S, related that in his role as consultant on Grubb's lens-making, Stokes wrote

> in his spare moments, sometimes in railway trains, sometimes at the Royal Society and sometimes at home ... He was wonderfully painstaking in answering any queries, so much so that I sometimes hesitated to ask him even a simple question, fearing it would encroach upon his time, for he went so deeply and so minutely into every aspect of the question that in some cases I had as many as 5 postcards or letters from him in 24 hours each describing some new view of the particular subject I had enquired about.[40]

One byproduct of Stokes's excessive correspondence was his eventual resemblance to what Baldwin called a 'Victorian search engine'. The longer Stokes served as Secretary, 'the more familiar he became with most of the physics research in Great Britain. This expertise led to another kind of correspondence – researchers writing to Stokes to ask whether a particular phenomenon had been noticed before or whether he could recommend anyone expert in a particular subject'.[41] Baldwin conservatively estimates the size of the Stokes correspondence as '20,000 letters – and probably closer to 35,000. A printed version of the Stokes Correspondence would likely require over seventy volumes.'[42]

While Stokes was perhaps overworked, he was by no means inefficient. By 1878, he began using a typewriter, in order to facilitate making copies.[43] He jokingly wrote to Royal Society Councilman William Huggins that with this technological innovation, 'now I shall be able to read my own letters'.[44] Moreover, during almost all of Stokes's tenure, from 1861 to 1885, Stokes had the help of Assistant Secretary Walter White, 'an odd young man, of little formal education but with a great yearning for a literary life... Although he was an eccentric, his eccentricities clearly created no difficulties with his scientific superiors.'[45]

Stokes's colleagues still despaired that the Royal Society put an excessive service burden on Stokes's shoulders. Scottish mathematical physicist P. G. Tait made Stokes the subject of the fifth instalment of a *Nature* series on 'Scientific Worthies'. In this article, Tait complained that the 'Mole-eyed State' had the tendency 'to harness Pegasus to the dust-cart!' In particular, he complained that a 'genius like that of Stokes' [is] wasted on the drudgery of Secretary to the Commissioners for the University of Cambridge; or of a Lecturer in the School of Mines; or the exhausting labour and totally inadequate remuneration of a Secretary to the Royal Society!'[46] In November 1884, William Thomson tried to convince his friend Stokes to apply for the Cavendish Professorship in Cambridge because

> the income of the experimental physics chair is decidedly more than you have in the Lucasian and I thought possible the difference might amount to even a money compensation for giving up the Royal Society work. Thus I thought of the whole thing rather as freeing you from fatiguing or possibly irksome work.[47]

The effort that Stokes put into Society correspondence induced Michael Foster to remark that

> Correspondence, etc. ought not and certainly will not in the future be as great as Stokes has made it. It has been painful to see how his energy has been wasted in this way. Mr. Rix [the Assistant Secretary who succeeded White] is a very competent person, and can be entrusted with much more than he now has, and the council I think will distinctly approve of this kind of work being taken off the secretaries.[48]

Foster wrote these remarks to Lady Rayleigh in the hopes that she would convince her husband Lord Rayleigh to succeed Stokes in the Secretaryship. Rayleigh began his eleven-year tenure as Secretary in November 1885, and soon after, the Assistant Secretary took on more of the correspondence. 'The library, editorial responsibilities, and much else devolved gradually upon the minor officials, with the Secretaries exercising "knowledge and judgement" as Foster put it.'[49]

The end of the Secretaryship did not mark the end of Stokes's service to the Royal Society, but rather the beginning of his work as Society President.

George Gabriel Stokes, P.R.S.

Stokes was first approached about the Presidency in 1877, near the end of botanist Joseph Hooker's tenure as President. Stokes hesitated to give up his Secretary's stipend, which remained an important addition to his salary. Moreover, as he said to his father-in-law, astronomer Thomas Romney Robinson, 'I am naturally of rather a retiring character, and should feel not a little out of my element in being brought so prominently forward.'[50] Evidence of the 'retiring character' comes from the fact that during his tenure as Secretary, the Society minutes record only thirteen instances of him speaking.[51]

Stokes's daughter, Isabella Lucy Humphry, elaborated on qualities her father possessed that were perhaps at odds with the political challenges faced by a Royal Society President. She recollected that Stokes 'always expected all sorts of people to adhere to the truth, even to their own disadvantage. This was so much the case that regular scamps used to get money from him by their plausible tales.'[52] Furthermore, he 'was so totally unaware of small-mindedness that it made him quite sad to have it pointed out'. Once, when his daughter pointed out the anger a visitor had exhibited during a discussion about the suitability of paper for publication by the Royal Society,

> My father hitched his chair and got rather red, and seemed aware for the first time that this had been the case. He said, looking most uncomfortable, 'I thought that he seemed a little warm, but that he could not possibly be angry with me about what was purely a matter of business.' Then he became somewhat roused and said, 'It is most unjust!'[53]

Stokes avoided the Presidency in 1878 when William Spottiswoode succeeded Hooker. Spottiswoode, like Hooker, was a member of the X Club, described by Baldwin as 'a well-known society of nine British men of science who sought to promote scientific naturalism and increase the social status of science in their native country'.[54] A third X Club member, Thomas Huxley, followed Spottiswoode in the Presidency in 1883, but only served for two years owing to ill health. By 1885, the time had come for Stokes to accept the Presidency. His co-Secretary Michael Foster had forebodings about Stokes as President, referring to Stokes as 'Mumbo

Jumbo' in letters to Thomas Huxley and consulting Huxley in Royal Society matters even after Stokes had assumed the Presidency.[55] Foster was aligned with Huxley and other scientific naturalists, whose approach to science and religion was at odds with that of Stokes and his allied physicists from the North of Britain.[56]

Feelings of unease about Presidential fitness were mutual between Huxley and Stokes: Assistant Secretary Walter White wrote in his journal entry of 25 April 1871 of an earlier Royal Society Presidential transition '[t]hat Mr. Stokes dreads Huxley's being President so accepts [Astronomer Royal George Biddell] Airy'.[57] When White's nephew published the Assistant Secretary's private journal a quarter of a century later, Stokes was most unhappy that these views were made public. For both Stokes and Huxley, '[w]hat is sure is that their dread of each other's occupation in the Presidential chair was a matter that each for his own reasons wished to keep concealed'.[58]

It was Huxley's turn to express concealed disapproval when, in 1886, Stokes assumed the Presidency of the Victoria Institute, which sought to find commonalities between science and religion, and the next year became a Member of Parliament for Cambridge University. Huxley and his friends, including *Nature* editor Norman Lockyer, had a 'fear that the office of the President of the Royal Society, who was by this time generally supposed to be the leading spokesman of science to the public, might be associated with an ultra-conservative political and theological viewpoint'.[59] In an anonymous letter in *Nature*, Huxley wrote that Stokes had recently 'accepted the Presidency of a body of pronounced theological tendencies [the Victoria

Fig. 8.2 A wood engraving made during 1885, Stokes's first year as Royal Society President. Included in the group portrait are many of the men mentioned in this chapter including Thistleton Dyer (back row, third from left), Huggins (back row, fifth from left), Lockyer (back row, second from right), Stokes, Hooker and Huxley (front row, first, second, and fourth from left, respectively). Image courtesy of the Wellcome Collection.

Institute]; and he now accepts the nomination of a no less pronounced political party [the Conservatives], and has issued an address in which he promised to devote himself to certain party objects'.[60] While some applauded the fact that Stokes, as Lucasian Professor, M.P. for Cambridge, and President of the Royal Society, would be walking in the footsteps of Newton, Huxley maintained that

> the House of Commons of the end of the nineteenth century is a very different body from the House of Commons of the beginning of the eighteenth century. The position of an independent member has become impossible; and those who refer to Prof. Stokes's address will see that, whatever his first feelings may have been, he, now at any rate, does not propose to be anything but a staunch Conservative.[61]

Botanist, director of Kew Royal Botanic Gardens, F.R.S., and Joseph Hooker's son-in-law, William Thistelton-Dyer also wrote against Stokes in *Nature* that 'a President of the Royal Society owes it to himself and to his position to hold aloof from all influences that would impair his freedom, and as a consequence, that of the Society'.[62]

In the next week's edition of *Nature* after Huxley's letter, Stokes's champions, namely the physicist Balfour Stewart, F.R.S., and Alexander Williamson, Royal Society Foreign Secretary, defended his choice to serve as M.P., P.R.S., and President of the Victorian Institute concurrently. Stewart asked if 'the President [must] be precluded from going to church during his tenure of office? Unquestionably the going to church implies taking part in a public action about which the opinions of the Fellows could be divided'.[63]

Six days after Stewart's letter appeared in *Nature*, when Stokes presided over the Anniversary Meeting of the Royal Society, the President gave no hint of political strife. The most notable item in his brief address was his announcement of the splitting of the *Philosophical Transactions* into two series, one concerning mathematics, physics, and chemistry, and the other concerning biological sciences.[64]

Foster, writing an appreciation in Stokes's posthumous *Memoirs*, wrote that Huxley's letter in *Nature* 'had no effect on Stokes...he hardly seemed to realize the point of view of the writer of the letter; he fancied that it was dictated by other motives'.[65] Huxley revealed his authorship of the letter in *Nature* to Stokes in December 1887, writing that 'its anonymity is due to nothing but my strong desire to avoid the introduction of any personal irrelevancies into the discussion of a very grave question of principle'.[66] Stokes replied that he did not conceive of any incompatibility between P.R.S. and M.P. and that he had thought the *Nature* letter 'must have been written by some hot Gladstonian'. Huxley replied 'We must agree to differ',[67] and there the matter stood.

Stokes completed his five-year term as Royal Society President in 1890 and his stint as M.P. in 1892. Foster, writing after Stokes's death, described his Presidency as unmarked 'by any stirring or disturbing events...The circumstances of the Society during his term of office demanded no more than that he should "pursue the even tenour [*sic*] of his way."'[68] Stokes's daughter remarked that he was known for his regular attendance in person but not in voice in Parliament and that some regretted that 'his silent services in House were given at the cost of a great deal of time, which was taken from his own work, while his constant attendance involved considerable bodily fatigue'.[69]

In his last annual address as President to the Royal Society, Stokes reflected that

> I am deeply sensible of the kindness which I have always experienced from the Fellows, and of the indulgence with which they have overlooked my deficiencies, due, in part, to the pressure

Fig. 8.3 Royal Society portrait of George Gabriel Stokes commissioned by a committee of subscribers in 1891, the year after he ended his tenure as the Society's President. © The Royal Society.

of other work. It cannot be without a strong feeling of regret that I come to the close of an official connexion with the Society that has not extended over full half of my life. But I feel that it is time that I should make way for others, and that I should not wait for those infirmities which advancing years so often bring in their train.[70]

Stokes actually continued on the Council for two more years. He was succeeded in the Presidency by his lifelong friend William Thomson, who, on the Royal Society's behalf, awarded Stokes the Copley Medal (Fig. 8.4) in 1893.

Conclusion

In her account of the career of her father George Gabriel Stokes, Isabella Lucy Humphry wrote that

> As the years ran on, he was left very little time for himself; when his services were requisitioned to assist scientific investigations, the appropriateness was obvious; but Church reform, questions of belief, politics, University legislation, etc., all claimed his time and involved continual committees, and constant streams of letters which he always most dutifully answered, and to very dull people, often at great length.[71]

In trying to explain his motivations for such service, she noted that while he was 'very ambitious about everything he did, and desired to do it thoroughly well and to excel…he was not

Fig. 8.4 The Royal Society Copley Medal awarded to Stokes in 1893. Image courtesy Nick Lefebvre.

ambitious in the ordinary way and for the usual objects.... It was this mixture of intense but exalted ambition joined with great conscientiousness which spurred him on.'[72]

While some of his colleagues regarded Stokes as a victim of this conscientiousness, trapped in administrative tasks which stole valuable time away from his own original research, no one could fail to appreciate the help he gave to his fellow scientists, especially those wanting to publish in the *Philosophical Transactions*. Stokes's expertise and position in the Royal Society made him, as Wilson has described, 'a kind of bastion of orthodoxy in Victorian physical science... As none other did, Stokes, as guide and critic, functioned as almost an official interpreter of what constituted proper physical theory.'[73]

During his long tenure as Royal Society Secretary,[74] Stokes assumed duties—the 'internal scientific work'—that played to his strengths and eschewed the 'external scientific work' that could often involve political manoeuvring. He was not entirely successful in avoiding all things political, especially during his tenure as Royal Society President. However, his lasting Royal Society legacy lies on the pages of the *Philosophical Transactions* and the *Proceedings*, where he stood as guard and shepherd for three decades.

CHAPTER 9

Stokes and Engineering

The Analysis of the Structure of Railway Bridges and Their Collapse

ANDREW WHITAKER

Introduction: Railway Bridges and Their Collapse

Stokes made important contributions to the safety of railway bridges, both through his own analysis and through his contributions to inquiries. To appreciate the significance of his work it is helpful to have a general understanding of the state of railway bridge building at the time and its deep-seated defects.

David Deutsch[1] has pointed out that, when we admire centuries-old structures, we should not forget that we see only the ones that have survived. The great majority of structures built in those times fell down long ago, often soon after they were built. So for obvious reasons, the builders were inclined to stick rather closely to designs and techniques that had stood the test of time. Deutsch reassures us, though, that modern science enables today's designers and builders to construct buildings with practically full confidence that they will survive, even if both the design and the materials out of which they are constructed may be novel and indeed bold.

The pattern of design and construction of railway bridges in the last three-quarters of the nineteenth century presents in some aspects a rather different pattern.[2] While bridges had, of course, been constructed since the earliest times, the requirement to support massive locomotives travelling over them at great speeds provided a totally new challenge. These large metal bridges represented the forefront of technology for this period, rivalled only by ships such as Isambard Kingdom Brunel's *Great Eastern* (1858), the roof of St Pancras Station (1868), the Eiffel Tower (1887–9) and, later in the century the New York skyscrapers.

Much basic research was indeed carried out.[3] Charles Coulomb had already made important contributions in the 1770s, and in the nineteenth century many researchers were well known to mathematicians and physical scientists of today, including Lord Rayleigh, Macquorn Rankine, Benoît Clapeyron, James Clerk Maxwell, Gabriel Lamé, Fleeming Jenkin, Alfred Clebsch and Augustin-Louis Cauchy.

However the actual practitioners, being confident self-made men, took little notice of these academic publications. They did use the results of Rankine and, in particular, an equation of

Whitaker, A., *Stokes and Engineering: The Analysis of the Structure of Railway Bridges and Their Collapse*. In: *George Gabriel Stokes: Life, Science and Faith*, Mark McCartney, Andrew Whitaker, and Alastair Wood (Eds): Oxford University Press (2019). © Oxford University Press. DOI: 10.1093/oso/9780198822868.001.0009

Eaton Hodgkinson for the bending of a beam under a load (Hodgkinson, after a career in business, became a professor of engineering at the age of 58), but even here they were inclined to apply the formulae to situations well beyond the rather straightforward and limited ones for which they were derived.

In an extremely important thesis[2] published in 1977, together with a paper published with his supervisor[4] in the same year, Paul Sibly discussed the very limited role of research in the development of a sequence of types of bridge, each ending in disaster—the collapse of the Dee Bridge in 1847, the Tay Bridge in 1879, the Quebec Bridge of 1907 and the Tacoma Narrows Bridge of 1940. In his discussion of Sibly's work, Henry Petroski[5] adds the 1970 incident with the steel box-girder design at the Milford Haven Bridge, and Subramanian[6] has suggested that the failure of the I-35W Mississippi River Bridge in 2007 could be added to the sequence.

These authors admit that the very nearly thirty-year gap between each of these events may be a coincidence, but they nevertheless do not rule out the idea that a generation of thirty years is the time taken for the concerns and safeguards with respect to a particular type of bridge to be all but forgotten.

When a new bridge type is considered, engineers working on the new development will be expected to take special care with every aspect of the design. First-principles research may be carried out, probably by the associates of the engineer rather than the academic community, and a model may be built and tested. Possible failure modes may be considered and the bridge designed so as to avoid them. Close attention will be paid to the actual building of the bridge right through its construction. Financial considerations and even aesthetic arguments will be, at most, secondary considerations.

Assuming, though, that the first bridge of this type is constructed and operates satisfactorily for a period, there is a natural tendency for the same and other designers and builders to think that all is well with the general method, and thus to build further bridges of the type without much (if any) further thought. Even worse, almost certainly they will gradually extend the assumed permissible length of the bridges constructed—often by bold leaps rather than incremental steps, as Petroski[5] stresses.

Sibly[2] describes the situation as a design method, construction technique or material unwittingly applied outside its range of validity—usually some principle has been developed and then used for ever larger applications until finally it is applied to a structure of such scale that it has many different properties, and many different failure modes, in comparison with the original. As a particularly important example, for shorter bridges it may well be permissible to assume that the structure will not depart perceptibly from the main plane of the bridge, but for longer structures buckling may take place leading to torsional behaviour and failure.

In addition, as longer and longer bridges of a particular type have been constructed, often the design work has not been handled by the named experienced engineer but mostly by an assistant, who may have had little direct experience of the actual construction. Also, while the early bridges of a particular type may have been well supported financially, funding for subsequent longer (and sometimes more complicated) bridges of the same type may actually have been reduced because of complacency or just exposure to economic realities. Lastly, whereas for a novel bridge type those responsible may just be glad if the bridge stays up, as later workers become less concerned about stability there is a tendency to become rather too concerned about aesthetic issues. Overall it is scarcely surprising that in all the cases mentioned above, pride did indeed come before a fall!

We shall briefly describe how these factors are relevant for the cases discussed. The Dee and Tay disasters will only be sketched at this point, as we shall say much more about them later on in connection with the work of Stokes.

On the Dee case we will merely note the comment of Sibly that the root cause of the accident was the trussed girder principle, to be described later. In 1845 the use of this principle for a bridge of 30.5 m span must have seemed a reasonable step from the span of 25 m for the Stockton Bridge and the group of structures in the 20–25 m range. However these precedents had been extrapolated from earlier examples. Working backwards it would be found that when these structures were introduced to railways in the 1830s, the design basis was an experimental equation, Hodgkinson's formula, which was established for the use of trussed girders in factories around 1815 for applications where it was necessary to use beams longer than those obtainable from a single casting operation. By a series of small steps, short, single piece, simply supported beams designed to carry a static point load at mid-span became pre-stressed trussed girders carrying heavy fast-moving trains.

Moving on to the Tay Bridge, it may be remarked[7] that until the 1870s, of the two great railway routes to the North of Scotland, the eastern route had a distinct advantage over its western rival in directness and gradient until the deep and wide inlets of the Firths of Forth and Tay were reached, when passengers had to leave the comfort of their carriages and were tossed around in small ferries. It was inevitable that the company owning this railway would plan bridges across these two storm-lashed estuaries, and a contract for a bridge across the Tay was signed in 1871. However, owing to wrangles over contracts, the designers and builders were to be desperately short of money through the project.

According to Sibly,[2] Thomas Bouch's bridge was to provide 'another example of how a large and daring structure was built using design information which, although it had proved adequate for earlier smaller-scale designs, was wholly insufficient for the design of a much larger project'. Its collapse in 1879 was probably inevitable.

The Tay Bridge was rebuilt between 1883 and 1887, but there was probably even more interest in the possibility of the construction of a bridge across the Forth. Benjamin Baker[8] was a promising engineer, who had studied the effect of wind on bridges both in the UK and in America, and had given evidence on the topic to the Tay Inquiry.

Baker, together with John Fowler, was chosen to build the Forth Bridge which, it is generally agreed,[2] represented the highest pinnacle of nineteenth-century engineering.[9] It was a logical extension of existing principles, but Baker and Fowler also recognized the need to go back to first principles to incorporate the best of existing technology. A new grade of steel of high elastic limit was also developed to economize on dead weight. Sibly and Walker[4] report that there was 'much patient research, experimentation and testing before and during the construction period'. The construction period was from 1883 to 1890, and little expense was spared for the provision of a bridge that was required to give confidence to those travelling over it. While beauty must be in the eye of the beholder, many might agree that the bridge was reassuring rather than beautiful.

There could not have been a greater contrast, in every way, with the building of the Quebec Cantilever Bridge, a huge bridge designed to carry railway, tram and road traffic across the St Lawrence River. This was mooted as early as 1887, but because of continuing financial pressures construction did not commence until 1905. Apparently, though, Theodore Cooper, the main spirit behind the bridge and a respected consulting engineer from New York, who was in poor health, felt that economy was possible. Of the Forth Bridge he wrote that it was 'the clumsiest structure ever designed by man; the most awkward piece of engineering in my opinion that was ever constructed. An American would have taken that bridge with the amount of money appropriated and would have turned back 50% to the owners instead of collecting, when the bridge was done, nearly 40% in excess of the estimate.'

The complacency is obvious, and it scarcely requires hindsight to study a description of the building of the bridge and to suspect that disaster was forthcoming. Cooper himself remained 725 km away from the bridge and made only extremely occasional visits; the managers at the bridge were not incompetent but certainly did not have the experience or knowledge to take a major part in such a massive project.

The builders neglected the fact that changes had been made since the 1898 proposals, so they used weights that were too low for the bridge components, as a result of which those near the main towers were considerably overloaded and the stress level was extremely high. (Cooper came to understand this but rather reluctantly did not take any action.) British engineers remarked that the bridge appeared spidery and insubstantial compared to the Forth Bridge, the tension members were weak, the material that was used had lower characteristic strength, and the joints were also inadequate. Whereas the elements in the Forth Bridge were splayed to allow for the effect of wind, those in the Quebec Bridge were strictly in the vertical plane. It should also be mentioned that the design had no provision for preventing progressive collapse—the failure of a single member would be fatal.

In July 1907, as one of the cantilever arms was being erected it was found to be severely distorted. Cooper was unconcerned and still did not visit the site. However in August one member buckled. Cooper was informed of the development, though meanwhile work continued. Then on 27 August the buckling developed in an alarming fashion. Again informed of the problem, Cooper realized that the bridge was in dire distress, and immediately sent a telegram ordering work to be suspended. Sadly before it arrived the whole structure folded up with the loss of 75 men killed.

The Commission of Inquiry reported that the main cause of the failure of the bridge was failure of compressive load-bearing members in the anchor arm due to defective design, but subsidiary causes were the weakness of project management and the absence of a first-class engineer on the site. Given such a person, they pointed out that the mistakes made could have been spotted and rectified. Rather amazingly they said that 'No one connected with the design fully appreciated the magnitude of the work nor the insufficiency of the data upon which they were depending.' It may be added that the designers carried out no experimental work of their own, preferring to rely on idealized theoretical analysis.

Much the same general remarks might be made about the Tacoma Narrows Bridge. As a long-span suspension bridge this could be compared with the Brooklyn Bridge.[5,10] The latter was planned by John Roebling, and work began in 1869, though the bridge was only completed by his son and daughter-in-law after his death. It opened in 1883, and originally carried railway traffic as well as horse-drawn vehicles. Roebling had had previous success building suspension bridges, including the Niagara Bridge, and had studied in depth the quite large number of failures of such bridges. As a result he kept the roadway of the Brooklyn Bridge from being destroyed in the wind by increasing its weight and tying it down, and he prevented the deck from being deflected by making it extremely stiff.

The Tacoma Narrows Bridge had been mooted as early as 1933 to carry road traffic across the Puget Sound at Tacoma in Washington state, but finance was difficult to obtain and the bridge could be at best marginally economic. In 1937 the consulting engineer was able to reduce the cost by 10 per cent of the original $3 million, but in doing so reduced the dead weight by around 30 per cent. The bridge was also exceptionally narrow, and it should have been natural to inquire whether the lightness and narrowness would lead to new and unexpected modes of behaviour and perhaps failure. Designing and moving as far as a signed contract took only six months from June 1938, and the bridge was completed by July 1940.

There was no time for research before building started, which was unfortunate as, after work had started, Professor Burt Farquharson of the University of Washington put great effort into analysis of the bridge, using a scale model, mathematical analysis and direct observation. He was able to demonstrate all the problems the bridge would in fact experience.

From the start the bridge demonstrated substantial oscillations in its plane; this behaviour was found to be amusing and even enjoyable and the bridge became known as 'Galloping Gertie'. However on 7 November 1940 under a steady wind these vertical movements became so pronounced that the bridge was closed to traffic. Then at 10 a.m. a torsional movement developed and half an hour later the deck began to break up; then most of the central span of the bridge dropped away and the towers bent towards shore. Farquharson came off the bridge only at the last moment, narrowly escaping going down with 'his' bridge.

The actual collapse was not due to resonance, as often stated, but to aerodynamic effects, which were only explained fifteen years or so later by Theodore von Kármán. However in more general terms, as Sibly explains,[2] the bridge, as perhaps not really understood by its creators, was very daring, not because it introduced new principles to bridge design but because it stretched and extended existing design principles well beyond anything justified by theory or experience. It might also be suggested that the thin and light nature of the bridge was, to an extent, justified as 'elegance'.

The next category of bridge to be discussed briefly is the box girder bridge,[11,12] which is built from prefabricated steel girders in the shape of hollow boxes. The first bridge of this type was actually the Britannia Bridge of 1850, which will be mentioned in the following section, but the method was not used again for the best part of a century. It was developed in particular in World War II, being known in Britain as the Bailey bridge. The type of bridge became popular in the roadbuilding expansion of the 1960s, but around 1970, four collapsed—the Fourth Danube Bridge in Vienna, the Milford Haven Bridge in Wales, the West Gate Bridge in Melbourne and the Koblenz Bridge in Germany; the last three caused a total of 52 deaths.

The various reasons for collapse given in inquiries included faults in fabrication and erection, inadequate design, failure of site organization and incorrect sequencing in construction. Yet again in the UK a Committee of Inquiry was set up under Alec Merrison, and in this case announced strict regulations for the building of this type of bridge.

Finally the collapse of the I-35 Mississippi Bridge in 2007 is said[6] to be a result of a fault of the original design proposal in the form of undersized and inadequate gusset plates at joint locations.

The analysis of failure has already been stressed, and in his book *Design Paradigms*[5] Petroski argues how fundamental to successful design is the study of failure. He remarks that it is only by thinking in terms of obviating failure that we achieve successful designs. We learn much more from failures than from successes; indeed, as Petroski stresses, the history of engineering is full of examples of dramatic failures that were once considered confident extrapolations of successful designs, and it was the failures that ultimately revealed the latent flaws in design logic which until then had been masked by large factors of safety that in time gradually became relaxed.

Christopher Alexander[13] argues that we can never state a design problem except in terms of the errors we have observed in past solutions to past problems. Even if we attempt to design something entirely new, the best way we can state the problem is to anticipate how it could go wrong by scanning all the ways in which other things have gone wrong in the past. The ideally successful design anticipates all relevant ways in which failure can occur. Understanding failure should play a key role in error-free design of all kinds. A similar comment has been made by Rolt[7] in his study of British railway accidents: 'The accidents have made British railways the safest in the world.'

Thus it is quite clear how essential the various inquiries and Royal Commissions have been in the maintenance of high levels of safety in the railway system. Whereas the railway companies would have been inclined to take the line of least resistance and merely to avoid the areas which were suspected of causing the collapses, the statutory bodies were obliged to study the problems from a much more fundamental point of view, thus, it would be hoped, eliminating a much more general cause of future catastrophes.

It is here that the work of Stokes was so important. After both the Dee and the Tay collapses, his work was crucial not only in analysing the direct causes of the accidents but in participating in wide-ranging studies of railways, railway bridges and general safety matters.

It is interesting to consider the wide range of personal attributes and abilities that Stokes was able to demonstrate in this work. First there was his willingness to put his academic work on one side and devote considerable time and effort to the solution of problems of general interest and importance. Then there was, of course, his mathematical ability, but also his experience in applying his mathematical results to real-life problems. Lastly there was an ability to study complicated problems and arguments in a way which was based on science, but also on general common sense and a desire to get to the bottom of fundamental issues.

One last point will be raised here—the importance of dynamic effects. It is very natural to take—initially at least—a static approach, treating the problem of a stationary engine and carriages on the bridge. Of course the actual problem is the much more complicated one with the train often travelling at a high speed, causing motion and vibration in the bridge, which itself acts back on the train. Little attention was paid to dynamic effects until the work of Stokes and his fellow Cambridge professor Robert Willis in the late 1840s, work which was described by Charlton[3] as 'perhaps the most significant research of the century into dynamics of structures'.

One extremely simple example of a dynamic effect that had been studied, of course, was the well-known resonance effect on a suspension bridge caused by soldiers marching with a particular frequency of step matching that of a resonant frequency of the bridge. An example often given[3] is the collapse of the Broughton Suspension Bridge in 1831, and this led to the British Army order for soldiers to 'break step' when crossing such a bridge. Injuries but no deaths occurred at Broughton, but in a similar incident at Angers in 1850 over two hundred soldiers were killed.[9,14]

Treatment of more complicated treatment of dynamic effects had, though, as already said, to wait for Stokes and his followers.

The Dee Bridge and Its Collapse

The Chester and Holyhead Railway Company was set up in 1844 to construct a railway line principally for the conveyance of the Irish Mail from Chester (to which they had already been brought from London) on to Holyhead to be taken from there over the Irish Sea to Dublin.

Robert Stephenson[15,16] (1803–59) was appointed Chief Engineer. He was, of course, the famous son of the equally famous father, George (1781–1848). Both men were probably by preference mechanical engineers, enjoying constructing railway engines—Robert's name being used for the celebrated company[17] which built engines from 1823 to 1960—but both also became practically by default civil engineers, responsible for building railway lines all over the country and beyond.

Robert refused a knighthood in 1850,[16] but he became a Fellow of the Royal Society in 1849, subsequently sitting on the Council of the Society. He was elected Member of Parliament for Whitby in 1847, and was awarded an Honorary DCL degree from Oxford University in 1857.[15] He followed his father in becoming regarded as a 'Grand Old Man' but came closest to losing this respect in 1847 with the collapse of one of his bridges that was part of the Chester and Holyhead line and ran over the River Dee.

It will be helpful to give some account of the state of railway building at the time. The first period of railway building was up to 1840, and during this period it was broadly the case that railway lines and particularly stations were built with great care and considerable expense. As a result fares were generally quite high, usage of the trains comparatively light and finances rather unsatisfactory. From 1840 to 1843 there was an economic depression and very little work was performed on the railway network.

Then in 1843 the depression ended and the age of *Railway Mania* began.[2,5,15] Enormous efforts were made to generate finance, and by 1846 Parliament had granted powers for many thousands of kilometres of railway to be built, which were to cost many tens of millions of pounds sterling. However the numbers of expected users of the new lines needed to be high for finances to be satisfactory, so fares needed to be quite low and thus structures had to be built generally as cheaply as possible. In addition many engineers had moved to other jobs during the economic depression, or were busy debating with the necessary Parliamentary committees who would give or refuse the required permission for development. It is scarcely surprising that much of the work done was fairly haphazard and when the Royal Commission reported after the Dee failure, a great deal of analysis of different types of problem was required.

When Stephenson planned the Chester to Holyhead line, he would have realized that by far the most significant problem would be the bridging of the Menai Strait, which separates the island of Anglesey from the mainland of Wales. Because of the importance of the Strait for navigation, the Admiralty ruled out an arch bridge. Stephenson's solution was the Britannia Bridge,[5,15,18,19] which Petroski[5] says was 'generally regarded as the most significant construction project under way, in the British Isles, if not the world, in the late 1840s'. There is a well-known picture by John Lucas of a number of (also well-known) engineers surveying the plans, with the bridge itself in the background (Fig. 9.1). Engineers including Brunel visited the site and gave Stephenson advice and support.

The bridge consisted of huge wrought-iron rectangular tubes through which the trains would travel. Because of the audaciousness of this idea, a large amount of practical research and testing was commissioned by Stephenson, which was carried out by William Fairbairn and Eaton Hodgkinson. Still the cost of this research came to only 0.67 per cent of the total cost of the bridge, which eventually came into commission in March 1850. A smaller version of the idea on the same railway line across the Conway River had been completed two years earlier.

In contrast to the significant amount of research on the tubular bridge, virtually no original thought was assumed to be needed for the crossing of the River Dee, for which the bridge was *only* [one might say Petroski's italics] an extension of many bridges of the trussed girder type

Fig. 9.1 'Conference of Engineers at the Menai Straits Preparatory to Floating one of the Tubes of the Britannia Bridge' by James Scott, after John Lucas. Robert Stephenson is seated at the centre of the image. © National Portrait Gallery, London.

previously built, quite a number of them designed by Stephenson himself. Sibly[2] suggests that probably his assistant Charles Wild carried out most of the design under minimal supervision.

Trussed girder bridges use both cast and wrought iron. Cast iron is iron that is smelted and cast into moulds; it behaves well under compression, so it is suitable for arches, but it is brittle and is weak under tension. Wrought iron, on the other hand is worked with tools while still hot; it is strong under tension, so would be expected to behave better than cast iron in bridges formed from girders. However in the first half of the nineteenth century, wrought iron was considerably more expensive than cast iron (though this was changing; the price of wrought iron fell by a factor of three between 1806 and 1831[5]).

Hodgkinson's formula mentioned above gave the amount of cast iron that would be required for a particular length of girder (and, as implied above, in ideal circumstances). However the point of the trussed girder was that wrought iron tie-bars would run the entire length of each girder. Initially the thinking was that the cast iron should be strong enough to hold the bridge up on its own; the wrought iron was there merely as a safeguard if the cast iron should fail. However, as implied in the previous section, and in typical fashion, bridge designers became so convinced of the beneficial effects of the wrought iron that they came to include its strength in the design.[4] It was admittedly impossible to calculate the interaction of the beam/truss interaction, but it was assumed to be positive! Thus the amount of cast iron could, it seemed, be significantly reduced.

For the Dee Bridge a crossing of 76 metres was required, and the bridge was a three-span structure with two masonry piers. With his confidence that the wrought iron was working with the cast iron, Stephenson was happy with a remarkably low calculated safety factor for the cast iron alone of around 1.5. This is defined as the ratio of the load the bridge can withstand to the maximum load expected. Most trussed girder bridges that had been built had safety factors of around 3 or 4, but many of Stephenson's bridges of this type had the longest spans and the smallest safety factors.[2,5]

The bridge was completed in September 1846, inspected by the Board of Trade Inspector-General, Major-General C.W. Pasley, in October, and opened for traffic on 4 November. As a test, heavy-laden wagons drawn by three 30-ton locomotives were driven over the bridge with no ill-effects. The Chester to Holyhead line was not yet completed, of course—as we have seen the Britannia Bridge would not be open for four years, but local traffic and traffic in connection with construction of the line was using the bridge, and the Shrewsbury and Chester Railway were given running powers over the bridge.

It is clear that Stephenson had absolutely no idea that the safety of the bridge was in any way in doubt. At about this time, one of the bridges on the Great Western line had burnt down when hot coals from an engine landed in it. As a result, on 24 May, after six trains had traversed the Dee Bridge without problem, Stephenson, who happened to be in Chester on business, ordered 5 inches or 18 tons of ballast to be laid on the bridge.

The very next train to attempt to cross the bridge was the 6.20 from Chester to Shrewsbury consisting of a locomotive and five flimsy carriages. This had reached the last of the three spans when one of the girders directly underneath it broke into three pieces. The driver felt the bridge 'sinking' under him and, opening his throttle, managed to get to the other side. Unfortunately the locomotive parted company with the rest of the train, which fell into the river. Five people died, though sixteen passengers survived, most of them being injured (Fig. 9.2).

This was the first serious structural collapse on the railway network[2] and it caused a 'tremendous furore in engineering circles'.[16] Pasley was dismissed on 29 May, but even before then, Captain J. L. A. Simmons of the Royal Engineers, who was an Inspector of Railways, had been called in; on 27 May he inspected the bridge and on the following day he examined the broken parts and undertook a number of tests, including running a locomotive gently onto the bridge and checking for distortion and vibration—both of which were substantial. Simmons wrote an initial report; James Walker, a civil engineer, was also asked to go to Chester, and Walker and Simmons prepared a further joint report, which was ready as early as 15 June. Both reports are included in a document prepared for the House of Commons.[18]

While the report raised several issues, including the distortion and vibration mentioned above, their fundamental point was that neither the cast iron nor the wrought iron was sufficient on its own to ensure the stability of the bridge. Had Stephenson been able to guarantee that the two would act together, the bridge would have been secure, but Walker and Simmons stressed that this was certainly not the case. They stated that if the principle of the bridge design had been to utilize such joint action, 'we do not approve it'.[18] We may note Rolt's remark that, in the view of Robertson, the engineer of the Shrewsbury and Chester Railway, in fact the wrought iron tie-bars actually weakened the girders by imposing severe local stresses.[16]

Parenthetically we may note two other suggestions for at least part responsibility for the collapse of the bridge.[19,20] The first is the possibility of metal fatigue. This was only very vaguely understood at the time, but it was known that the substantial oscillation of the bridge could lead to components failing well before their rated strength would predict.

SCENE OF THE LATE RAILWAY ACCIDENT, AT CHESTER.—DILAPIDATED SPAN OF THE DEE BRIDGE.

Fig. 9.2 A 'Scene of the late railway accident, at Chester' showing a 'dilapidated span of the Dee bridge'. Taken from the *Illustrated London News*, 12 June 1847.

The second relates to complacency creeping in to a seemingly successful design, and artistic issues becoming more important than stability, as mentioned in the previous section. In this case cavetto mouldings, popular in contemporary carpentry designs, were used for the corners of the cast iron girders. Lewis and Gagg[19] showed that these acted as stress raisers and the fracture of the girders is likely to have begun at these points. As the title of their paper indicates, they suggest that this artistic flourish weakened the design 'catastrophically'.

To these possibilities, of course, must be added the near certainty that the laying of the ballast must have acted as the rather substantial 'straw that broke the camel's back'.

Meanwhile the inquest had begun in Chester Town Hall.[16] Locally there was intense anger against Stephenson, stoked up by Robertson, and Sir Edward Walker, foreman of the Coroner's jury, was determined to get a verdict of manslaughter or even murder against Stephenson. Manslaughter was actually not out of the question.

Stephenson was prepared to admit his fault, but the solicitor to the Chester and Holyhead Railway insisted that he claimed instead that the locomotive had derailed, thus causing the damage to the bridge, and in this he received the backing of many of the most famous builders of railway bridges. This argument was obviously incorrect, but a visit to the inquest by Walker and Simmons persuaded the jury to return a verdict of accidental death. However they added a remark that the bridge had been unsafe, that the many bridges of the same type must be potentially unsafe, and that a Government inquiry into the matter should be held.

Before acting on this suggestion, though, on 23 July the Government Commissioners of Railways sent a circular[15] to all the railway companies asking them to report back on all the iron bridges on their lines and requesting that they take all measures necessary to add to their stability, in the meantime ordering speeds on these bridges to be reduced. The Commissioners must have been shocked by the replies because on 29 June they issued a minute[18] effectively repudiating any responsibility for the iron bridges in the country on the grounds that their inspectors had had the opportunity of only a superficial observation of the bridges. Responsibility for the design of the bridges, they insisted, must remain the responsibility of the companies.[15]

From this point of view the dismissal of Pasley seems alarmist. He had been an excellent military man, Director of Field Instruction in the Army for twenty-nine years until his Board of Trade appointment, and was a Fellow of the Royal Society.[21]

Also on 29 June the Commissioners wrote to the Government[18] calling for a Commission, composed of practical men and engineers, to be set up to investigate the use of iron, in particular cast iron, in railway bridges. They pointed out that there was great difference of opinion among engineers about the form and dimensions of girders required, stressing the importance of studying the effect of violent concussions and vibrations, and also of recognizing the fact that the bridges were repeatedly crossed by loads of extraordinary weight and at great velocities.[15]

The Commission[15,22] was appointed on 27 August 1847 and consisted of highly distinguished men. Lord Wrottesley was a well-known astronomer, later to be President of the Royal Society. George Rennie and William Cubitt were excellent engineers, both Fellows of the Royal Society, as was Eaton Hodgkinson, who was mentioned in the previous section; Cubitt was also later knighted. Robert Willis was Jacksonian Professor of Natural Philosophy at Cambridge University; he performed important work in engineering, but he also became well known later for studies in speech and architecture. Two Captains in the Royal Engineers were also appointed, Henry James as a member of the Commission and Douglas Galton as an (in fact highly active) Secretary. James was, and Galton would become, Fellows of the Royal Society and Galton would be knighted.

As stressed in the previous section, the paucity of research in earlier years meant that the Commission had a vast amount of work to perform. Their final report came to around 450 pages. A large number of founders and bridge builders were interviewed by the Commission, including Stephenson, Brunel and Fairbairn. They provided a remarkably complete account of the beliefs and practices of those who provided the iron, those who designed the bridges and those who built them.

Hodgkinson presented a very substantial report on an enormous number of experiments he carried out in Manchester and Lambeth, in which he studied the resistance of bars of iron, mostly cast iron but a few wrought iron, to violent concussion in the form of blows from heavy balls. He also studied the elasticity of the forms of iron by experiments on the extension and compression of the metal. In addition Hodgkinson provided an account of his experiments in connection with the design of the Britannia and other bridges, which was, of course, still ongoing.

This was all extremely useful information, of course, but more novel was the work carried out by Willis together with James and Galton, mostly at Portsmouth Dockyard, where James was Director of Works, but also at Cambridge and on the rail network at Ewell and Godstone, both of which are south of London. These experiments studied the effect of heavy bodies moving with high velocity over bridges. The report of the Commission stressed that direct scientific investigation of such cases had never been carried out, and that it was extremely important that it should now take place!

Typically in the Portsmouth experiments a railway track ran down an inclined plane of length around 30 feet, which was joined by a suitable arc, so as to avoid jerking, to a horizontal rail, the behaviour of which could be carefully monitored. A loaded railway car weighing between half a ton and two tons was propelled down the inclined rail and the behaviour of the horizontal rail was studied. Mass and velocity on the rail were in many cases increased to breaking point. It was found that when the railway car was moving the deflection of the rail in all cases increased compared to when the same car was stationary. The increase depended of course on the various quantities involved but was often as higher as a factor of two or in extreme cases even higher. (It is interesting that the official report of the Commissioners (p. xiii) remarks that most engineers expected the deflection caused by a moving weight to be **less** than that caused in the static case.)

Willis analysed the problem theoretically, but he reached a complicated equation and at this point he was unable to continue. Fortunately he was, of course, well acquainted with Stokes, at that stage a fellow of Pembroke College, though by the time the report was presented he had been appointed Lucasian Professor. Stokes was willing to take on the mathematical problem, and his results[23] were published as a paper in the *Cambridge Philosophical Transactions*; the report of the Commissioners merely made reference to this paper. This work constituted the first significant theoretical study of the dynamic effect of a moving railway vehicle on a bridge, and the work will be discussed in the next section. A substantial account of this paper as well as the work of Willis has been presented by Todhunter[24] in his magisterial account of the theory of elasticity and the strength of materials through the ages.

Stokes and the Effect of Moving Railway Engines on Bridges

Willis's analysis, which was taken over by Stokes, was presented as follows.

Initially in order to make the problem at all tractable the inertia of the bridge was neglected while the moving body was treated as a heavy particle. (It may be objected that this was rather dramatically unlike actual engines and bridges, but at least it was hoped that it would represent the results of the Portsmouth experiments.) The coordinates of the moving body were denoted by x' and y', with x' being measured from one end of the bridge and y' downwards, and y' is considered small.

Also, M was the mass of the body, V its speed when it enters the bridge, $2c$ the length of the bridge, S the deflection of the bridge produced by a particle placed at rest at the centre of the bridge and R the reaction between the moving body and the bridge. Since the deflection was small, R could be treated as vertical and the horizontal component of the velocity of the body would remain constant as V.

The bridge was treated as an elastic bar which was supported at its ends and which sags under its own stiffness. Then the depth that a weight will sink when placed at a particular position on the bridge will be proportional to $\left(x'^2\right)\left(2c-x'\right)^2$, assuming that the curvature is proportional to the bending moment. Then since the mass of the bridge is being neglected, the relation between the depth y' that the moving body will have sunk at any time and R will be the same as if R were a weight at a distance x' from the end of the bridge, so we may write $y' = CR\{(x'^2)(2c-x')^2\}$, where C is a constant. When R is equal to Mg and x' is equal to c, then y' must be equal to S, we must have C equal to $S/(mgc^4)$.

Thus we may write

$$d^2 y' / dt^2 = g - R / m = g - (gc^4 y') / \{Sx'^2 (2c - x')^2\} \tag{1}$$

and since $dx'/dt = V$, we obtain

$$d^2 y' / dx'^2 = g / V^2 - (gc^4 y') / \{V^2 Sx'^2 (2c - x')^2\}. \tag{2}$$

With $a = g / V^2$ and $b = (gc^4) / (V^2 S)$, the equation that Willis presented to Stokes was:

$$d^2 y' / dx'^2 = a - by' / \{x'^2 (2c - x')^2\}, \tag{3}$$

the particular conditions being that when $x' = 0$, both y' and dy'/dx must equal zero.

The solution of this differential equation is also noted by Richard Paris in Chapter 7. Stokes simplified the equation by writing

$$x' = 2cx; \; y' = 16c^4 ay / b; \text{ and } b = 4c^2 \beta$$

and finally obtaining

$$d^2 y / dx^2 = \beta - \{(\beta y) / (x - x^2)^2\}. \tag{4}$$

In the solution of this equation, the general size of β, which varies inversely as the square of the horizontal motion of the body, will be found to be crucial. To be precise, $\beta = (gc^2) / (4V^2 S)$.

In his study of this equation, Stokes utilized, with great skill, a considerable range of mathematical techniques that included series expansions but also more direct methods of solution. With supreme care he chose particular forms of expansion and particular approximations to provide convenient expressions for the various quantities that concerned him and for different ranges of experimental conditions, in particular different ranges of β. He also spent enormous effort on providing numerical solutions for the important theoretical results, working to four or five places of decimals.

To solve Equation (4), he initially utilized a convergent series. With $y = (x - x^2)^2 z$, he defined the infinite series:

$$z = A_0 + A_1 x + A_2 x^2 = \Sigma A_i x^i \tag{5}$$

where i takes values from 0 to infinity, and obtained the result that:

$$A_i = \{2 A_{i-1} - A_{i-2}\} \left\{ 1 - \frac{\beta}{(i+1)(i+2) + \beta} \right\} \tag{6}$$

which converges for the appropriate range of x.

It seems that he only subsequently realized that a solution could be found in terms of definite integrals:

$$y = \frac{\beta}{m-n} \{ x^m (1-x)^n \int_0^x x^n (1-x)^m \, dx - x^n (1-x)^m \int_0^x x^m (1-x)^n \, dx \} \tag{7}$$

where m and n are the roots of the equation

$$z^2 - z + \beta = 0 \tag{8}$$

The arbitrary constants of the complete integral vanish because of the particular conditions given above.

Initially Stokes worked with relatively small values of β, between 5/36 and 5/4. He found that in these cases the trajectory of the body is decidedly unsymmetrical with respect to the centre of the bridge. The tangent to the trajectory is close to horizontal at the near end of the bridge, the minimum trajectory occurs substantially past midway, and the gradient is positive and large—indeed quite close to vertical—towards the far end.

He then demonstrated great mathematical dexterity to produce convenient expressions for two centrally important quantities—the depression of the track at any point, and the tendency of the bridge to rupture—both relative to the effects produced by the body stationary at the midpoint of the bridge.

The numerical values of both quantities were low at the near end of the bridge but rose gradually and then, towards the far end, extremely quickly before decaying at the far end. Though the high values reached demonstrated that the various approximations made in the analysis must be illegitimate by this stage, it will be noted that, until that point, both results are broadly in line with the Portsmouth experiments, for which these values of β would be appropriate.

However Willis then informed Stokes that for practical bridges, values of β would almost invariably be much higher. For example values of β for the Ewell and Godstone Bridges would lie between roughly 30 and 600 depending on the value of V. Stokes therefore calculated the values of the same quantities. For these values of β, both the relative values of depression of the track and tendency to fracture relative to the static values were reduced compared to the values for smaller values of β. Nevertheless there were considerable portions of the bridge where these quantities were clearly greater than unity. As β increased, the trajectory of the body became closer and closer to being symmetrical.

Also in this addition to his paper, Stokes included an attempt to address the opposite case to that of the mass of the body being much greater than that of the bridge: that where the mass of the body may be neglected in comparison with that of the bridge. He was stimulated to carry out this work by being informed of some experiments of Willis in which a balanced lever was connected with the centre of the bar, so as to increase its inertia without increasing its weight. In some cases the deflection produced by a given weight was increased, but in others it was decreased. Stokes attempted an approximate solution of the problem taking into account the inertia of the bridge.

He found that when each force acting on the bridge is replaced by a uniformly distributed force which produces the same mean deflection as that produced by the force alone, and when the difference between the pressure exerted by the travelling mass on the bridge and its weight is neglected, the equation may be solved in finite terms.

In this case the solutions depend on a quantity q given by

$$q^2 = \frac{252 g c^2}{31 V^2 S_1} \tag{9}$$

where S_1 is the central static deflection that would be produced by a mass equal to that of the bridge.

Stokes discussed this as follows. Suppose the travelling load is removed and the bridge is depressed slightly, then the period of an oscillation would be given by

$$P = 2\pi \sqrt{\frac{31S_1}{63g}} \tag{10}$$

Thus:

$$q = \frac{4\pi c}{PV} = 2\pi \frac{\tau}{P} \tag{11}$$

where τ is the time that the mass takes to cross the bridge. Stokes sums up this result by saying that the deflection depends on the ratio of the periodic time of the fundamental mode of vibration of the structure to the time taken for passage of the load over the structure. Then the ratio of the central deflection to that for the static case is given by $25/8q$.

Stokes provided tables and graphs of his results, and paid particular attention to two important cases. For the Ewell Bridge and if the velocity of the body is around 30 miles per hour, the ratio is equal to around 0.04. For the earlier assumption that the mass of the body is much greater than that of bridge we would have β around 127 and the ratio around 0.008. Thus the effect of the first assumption is smaller than that of the second assumption, but both are small. For the Britannia Bridge, one might surmise that the massive weight of the tubes might cause the inertia of the bridge to be important, but one finds in fact that, again for a velocity of 30 miles per hour, the relevant ratio is still only around 0.43.

Stokes sums up his results as follows. The problem, he reports, has been worked approximately only for two extreme cases in which the mass of the moving body is infinitely greater or infinitely smaller than that of the bridge, but the causes of the increase in deflection for these two extreme cases are totally distinct. In the first case the increase in deflection is a direct result of the difference between the pressure on the bridge and the weight of the body, and it may be regarded as depending on the centrifugal force. In the second case it depends on the manner in which the force, regarded as a function of time, is applied to the bridge.

In practical cases, Stokes says, the masses of body and bridge will usually be comparable with each other, and thus the effects are combined in the final results. However, if it is found that each effect, considered separately, is too small to be of practical concern, we may conclude without much fear of error that the actual effect will also be insignificant. Stokes had shown that, in the two calculations, the most important terms in the increases of deflection were proportional to $1/\beta$ and $25/8q$ respectively, which in practical cases are small. Stokes suggests that the magnitudes of these fractions will enable us to judge the amount of the actual effect.

Willis,[22] in fact, had prepared tables of values obtained merely by combining the results of the two effects, and had found reasonable agreement with the Portsmouth experiment and others carried out for the project.

In 1848, Homersham Cox[25] had written an interesting paper on broadly the same topic as that of Stokes's paper. He had based his treatment around the use of *vis viva*, equivalent in today's terms to half the kinetic energy. Essentially he used the Principle of Conservation of Energy and, relating work done and extension produced, he claimed to deduce that the maximum static deflection which a travelling load may produce cannot be more than doubled when motion takes place at any velocity. This contradicts, of course, both the experiments of Willis and his colleagues, and also Stokes's analysis.

In his paper of 1849, Stokes[23] remarked that Cox treats the subject in 'a very original and striking manner' but nevertheless severely criticized this claim. He pointed out that, among the sources of labouring force that may be used to deflect the bridge, Cox had omitted the *vis viva* due to the horizontal component of the motion of the body. But Stokes showed that inclusion of this is essential. A portion of this *vis viva* is converted into labouring force and is used to deflect the bridge. For during the first part of the motion, the horizontal component of the reaction of the bridge against the body impels the body forwards and thus increases the *vis viva* due to horizontal motion; the labouring force which produces this extension is derived from the bridge, and so the bridge is deflected less than if the horizontal component of the motion had remained unchanged. But during the latter part of the motion, the horizontal component of the reaction acts backwards, and a portion of the *vis viva* due to the horizontal motion of the body is continuously converted into labouring force which is stored up in the bridge.

Because of the asymmetry of this motion, the direction of the motion is more inclined to the vertical over the second part of the motion than for the first part, and also the reaction is greater. For both these reasons, more *vis viva* of the horizontal component of motion is lost in the second part of the motion than is generated in the first part. Therefore overall the bridge is deflected more than if the horizontal component of the motion had remained unchanged throughout.

Stokes pointed out that, although the change of the horizontal component of the velocity is small (it is, of course, neglected in his own treatment), Cox cannot ignore it because he must deal with and compare changes in the squares of the horizontal and vertical components of the velocity.

Charlton[3] reports that in the 1850s and 1860s, several authors—Phillips, Mallet, Winkler, Renaudot and Kopytowski—analysed the same problem, but says that all this work is of trivial historical interest compared to that of Stokes. Stokes had made the first systematic attempt at the dynamical problem of a railway bridge, it was an intense and highly skilled treatment, and deservedly it is cited right up to the present day.[26,27,28]

This work was not superseded until that of Adhémar Saint-Venant, whose work on elasticity, according to Todhunter[24] makes him the equal of Cauchy and Poisson. It was included in his 1883 'translation' of a book of Alfred Clebsch,[29] originally published in 1862. Clebsch himself had died in 1872. Saint-Venant took the opportunity to add substantial 'notes' to the effect that the translation was three times as long as the original book. Included in this addition is a very substantial treatment of Stokes's problem, in which Saint-Venant was able to retain the masses of both beam and travelling object to a suitable approximation. In the appropriate limits he obtained Stokes's own solutions.

The Tay Bridge: Triumph and Disaster

The collapse of the Tay Bridge[2,7,9,30,31] on 28 December 1879 (Fig. 9.3) sent shock waves through complacent Victorian society, and particularly in Scotland it was to be remembered with sadness and horror for decades to come. It was to finish the career and in effect the life of the creator of the bridge, Thomas Bouch.

Born in Carlisle, Bouch had first come to the east of Scotland thirty years earlier when he was appointed at the age of 26 as manager of the Edinburgh and Northern Railway (as it was at the time). He already had a good deal of practical experience in the construction of railways, though in truth he had only patchy knowledge of engineering and no capacity at all for mathematical calculations. What he did have was enormous determination and ambition. During

Fig. 9.3 The Tay Rail Bridge before and after collapse. On Monday 29 December 1879, the day after the disaster, the Dundee Evening Telegraph, stated 'The Tay Bridge—a structure in which the most timid and sceptical were beginning to have confidence—has fallen a victim to the terrific gale which raged with inconceivable fury from dark till a late hour last night.'

his spell of only two years in Edinburgh he had designed and brought to completion 'flying bridges' for the Tay and Forth estuaries; using these, wagons could be run straight from railway line to flying bridge and off again at the other side.

This achievement gave him enough renown to be able to leave the employment of the Railway in 1851 and set himself up as a consulting engineer—in effect as a catch-all railway designer. By this time the main railway lines were mostly in existence, lines which would receive an enormous amount of traffic in the form of heavy engines, and which had to be built with the appropriate strength and resilience; they were necessarily expensive but they had been able to recoup their costs by their heavy passenger numbers.

From the 1850s it was the time of the small community which wanted the glory and convenience of their own railway. A track and bridges of only modest strength would be required for only light engines and loads travelling at low speeds, and a low price was essential. For such groups, Bouch was most decidedly their man! Even at the Inquiry after the Tay Bridge disaster, when asked how many bridges he had built, he retained enough pride to reply: 'I do not suppose anybody has built more.'

He built railways in many parts of northern England and Scotland, eschewing the heavy plate girder bridges of Brunel and Stephenson, and making a trademark of lattice girders supported on slender cast-iron girders. Everything he produced was reasonably durable and cheap—indeed to one client it seemed too cheap. When his design for the Hownes Gill viaduct was submitted, the management of the railway involved asked Stephenson for a second opinion. Stephenson felt that the bridge would be unstable and suggested a range of strengthening measures to ensure stability.[31]

Bouch was decidedly not a member of the 'establishment', the select group of engineers based around Westminster who had built major bridges across the world, and who were inclined to look down on the semi-qualified engineer from the North, who was content to take work wherever it happened to come from. When the chips were down—as they certainly were after the collapse of the Tay Bridge—it would not be surprising to detect a measure of closing ranks or even scapegoating.

While Bouch had been content to build up a substantial bank balance with relatively humdrum jobs, ever since his very first days in Edinburgh he had nursed a dream on a completely different scale—bridges across the Firths of Tay and Forth. To his employers at the time, it seemed an impossibility. They shuddered at the immense length that the bridges would need to be in conditions so different from those of an inland quiescent river. Most of all they were deterred by the massive finance that would be required!

Over the coming years, the question was raised again several times both by Bouch himself and by some, though by no means all, the wealthier citizens of Dundee, who saw the commercial advantage of a bridge across the Tay. They expected the presence of a Tay bridge would drastically increase the goods and mineral traffic to Dundee, and the general prosperity of the city. The North British Railway, the body that would build the bridge, was in a fairly parlous state at the time, but expected that both the amount of traffic from Edinburgh to Dundee and Aberdeen, and its share of it, would increase enormously, and also that the Fife coalfields would be opened up. All this would indeed happen when the bridge was built.

By 1869 the North British was in full financial support of a bridge, and a subscription list was set up with great success. Royal Assent was attained in July 1870, and not surprisingly Bouch was appointed as designer. Much will be said of the merits and otherwise of Bouch's work, but, as an echo of the arguments of Sibly,[2] it might immediately be suggested that, though Bouch was proud to boast of the length of the bridge he was to build, he never really faced up to the different order of challenges compared to virtually all his previous work.

From the start, prices were forced down. When arguing for the bridge, Bouch had actually claimed that he could build it for £180,000 (perhaps £20 million in today's terms), and in the end a contract was struck for £229,000 with the lowest bidder. When this bidder was forced to pull out, the new contractor, Charles de Bergue, signed for even less, £217,000, the specified time for completion being three years. In 1873, de Bergue died, and it soon became clear that the company was in dire straits, solely because of the Tay Bridge contract—clearly de Bergue had put in a wildly unrealistic tender in order to win the contract. The firm that took over the contract, Hopkins, Gilkes and Company, took on not only the workforce but also the work arrangements of de Bergue, but in accepting the latter they inevitably did considerable damage to the project, as we shall see.

In the remainder of this section, a general account of the design, construction and maintenance of the bridge will be given, leaving to the following section the discussion of the problems associated with the wind, including the advice obtained by Bouch, the policies he followed and the discussion of Stokes and others after the collapse.

In Bouch's original plans, the bridge was to be a latticed girder bridge on tall brick piers carrying a single railway track on a wrought iron superstructure. There were to be eighty-five spans, ranging in length from 28 feet to 285 feet. While the shore-to shore distance was just over a mile, the oblique direction of the course of the bridge meant that its total length was nearly two miles.

The City of Perth, which was situated further up the Tay, insisted on maintaining navigation out to sea, and so for the central fourteen spans of the bridge, 2,800 feet in all, there was additional clearance. Total clearance for this section was 88 feet, and on this section the rail line ran through the box girders rather than on top of them. This section was called the High Girders.

While early progress on the construction of the bridge was good, 1873 was a year of crisis (or, as Sibly put it, as with so many cut price projects, the optimism of the initial proposals was not borne out in practice[2]). Apart from the stress of changing the contractor, there was a major crisis regarding the foundations of the bridge. The original borings had indicated that there was a rock bed around twenty feet under the gravel surface layer right across the river. This was entirely wrong. In many places it was forty feet under the gravel and one sounding went down 157 feet and still found no rock!

Bouch was forced to adapt his design. First the bases of the piers had to be increased in area to reduce the pressure acting on them. For the same reason the masonry towers of the original plan were replaced by iron columns, which Bouch suggested would be stronger and could be erected in half the time at less cost. Overall the pressure on the base was reduced from 6¼ to 2¾ tons per square foot. In fact the iron towers were more expensive than the masonry ones and so a further change was made: the number of piers in the High Girders was reduced from fourteen to thirteen, but the length of each was increased and the total length became 3,149 feet. Finally, and against all engineering advice, Bouch reduced the number of iron columns at each pier in the High Girders from eight to a hexagonal arrangement of six.

At the Court of Inquiry, Henry Rothery, its Chairman, concluded that the bridge was badly designed and he held Bouch wholly responsible for this.[32] He included the lack of proper study of the foundations, for which he blamed Bouch—had it been possible to build brick piers, he said, doubtless the bridge would still have been standing. He criticized the inadequate size of the base and the hexagonal form of the columns; the shape of the caissons which supported the piers; the lack of a continuous girder at the top of the columns; and the lack of a spigot in the lower columns.

Rothery compared the design of the Tay Bridge with that of the Belah Viaduct, which Bouch had designed in 1861, and which was not finally demolished until 1963. Despite the Belah Viaduct being described as the lightest and cheapest of its kind ever erected, all the unsatisfactory features of the Tay Bridge that Rothery pointed out had in fact been handled wholly satisfactorily on the Belah Viaduct. When asked, in particular, why Bouch had not used the type of ties used at Belah for the Tay Bridge, he could only reply that 'They were so much more expensive; this was a saving of money.' It is tragic to realize that this penny-pinching was to have such a catastrophic result.

It is one thing to point out obvious faults in the design of the Tay Bridge; something else to suggest how it might have been designed genuinely taking on board the novel challenge in a positive way. This may actually be seen in the differences in the design and construction of the first and second Tay bridges,[9,33] the latter taking place between 1882 and 1887. The short period in which the first bridge had served railway and community so successfully meant that it was regarded by all as essential to replace the failed bridge as soon as possible, and penny-pinching was forgotten. While the first bridge had, in the end, cost £350,000, the second was to cost £670,000.

The girders of the shorter spans were re-used, being shifted sideways onto the new piers, which were erected 60 feet to the west of the old ones. Thus the new bridge used the same spans as the old, and the stumps of Bouch's bridge acted as breakwaters for the new bridge. There were, though, many changes, and in particular there were two tracks of railway, and in important areas iron was replaced by steel. Bouch's girders were used for the exterior sides and two new sets were used for the inner sides. The new piers were more massive than the old ones with a gate-like fully enclosed structure. For the centre of the river, there were thirteen new girders of wrought iron. The overall height was reduced and the whole structure is more solid than that of Bouch, but also obviously much safer.

We now turn to the construction of the original bridge. Rothery reported that the bridge was badly constructed and that this was principally the responsibility of Bouch, though the contractors also bore some responsibility.

Certainly there were several aspects of the construction that were totally appalling. Sibly suggests that there was an attempt to recoup the money lost in the enforced change of design by the use of poor quality materials, the iron in particular being especially poor, and impossible to use for decent quality casting for the piers.[2]

While they were in charge, de Bergue's had established their own foundry, despite the fact that there were many foundries in Dundee that could have provided the necessary columns cheaply and quickly. The replacement contractors kept the foundry going, and it soon became a byword for its dreadful conditions and dubious product. Those nominally in charge were totally ignorant of foundry work, and those in control by default had little incentive towards good work. Many columns produced were so faulty that they had to be scrapped, but many more, little if any more regular and crack-free, passed inspection and went on the bridge.

Much used at the foundry was *Beaumont Egg*, which is a corruption of 'Beaumontage', named after the French geologist Elie de Beaumont.[9] This is a mixture of rosin, lamp black, beeswax and iron borings, which could be used to fill holes in the iron columns. It would be melted in and it would harden just like the metal from which it would be indistinguishable except that it could be picked out with the point of a knife. Prebble has suggested that: 'Often it seemed that all that held the bridge together had been this magic, malleable substance.'[30]

The greatest defect was in the columns, in particular in the cast-iron lug ties to which the ties and struts were attached, and in failing to provide not only that they should be cast with the columns but with the holes ready-made. It is to this, to the casting of the holes in the flanges of the 18-inch columns, and to failing to ensure that these holes were made properly cylindrical, and that the bolts fitted them accurately, that the weakness of the piers—and the fall of the structure—was, according to Rothery, mainly due.

Having discussed design and construction, we must refer to the inspection of Major-General C. S. Hutchinson on February 1878. Hutchinson spent three days studying the bridge, going as far as sending six coupled ballast engines over the bridge at speeds up to 40 miles per hour, while he was hauled up inside one of the piers under the High Girders. He was generally satisfied with the bridge and its behaviour, but made two important points. First he stipulated 25 miles per hour as the maximum speed at which a train should cross the bridge. Secondly he said that he would like to see the effect of a high wind when the train was crossing. It should be mentioned that the first point was apparently routinely ignored by the train drivers, who delighted in racing the ferries, while the second was never followed up.

The coming into service of the bridge was greeted with great acclamation, Queen Victoria paying a visit and knighting Thomas Bouch.

However the maintenance of the bridge was as deficient as its construction. Bouch was nominally in charge and accepted a salary of £100 for this task. Rothery stated that the bridge was badly maintained, and that Bouch was principally, if not totally, responsible.

In fact Bouch left in sole charge of the bridge Henry Noble, who was an expert on bricks but knew absolutely nothing about iron! He initially had a team of seven men, which was gradually reduced by the board to three; wages were also decreased.[31] Noble was an extremely conscientious man, who even took on the diving work though he was totally unskilled. His main aim seemed to be to save money and to avoid reporting any problems to Bouch.

When he found large numbers of nuts and bolts, used to secure the timber baulks carrying the rails to the transverse way beams, on an exposed sandbank, he merely replaced them. Finding slits in the iron columns, he strapped them with bands of iron. Cracks in the masonry were also just strapped. He found large numbers of gibs and cotters loose, and wedged them with slivers of iron he bought locally, in fact fixing them in distorted positions.

When the bridge was repainted in spring 1879, the painters noticed a variety of dangerous signs: slack tie-bars painted over and ignored; rivet heads found and corresponding holes ignored; loose fitting of girders; extreme vibrations both lateral and vertical when trains passed. Again little of this got even as far as Noble.

As is well known, the end came on 28 December 1879. On an evening of intense wind, the 5.27 train from Burntisland, bringing many passengers who had travelled from Edinburgh over the ferry, passed on to the bridge and entered the High Girders, which at that point collapsed and the train together with seventy-five people fell into the River Tay.

The following days were, of course, highly traumatic and it may be the case that, in Scotland in particular, certainly in Dundee, deep memories still remain. Here we wish only to discuss the intensity of the weather that night. Some inhabitants of Dundee reported it as undoubtedly windy, but only of such a fierceness to be expected every few years or even several times a year.[30] Yet stories of the engine-shed foreman fearing that the gale would rip off the big doors, the wind driving three loaded coal wagons out of a siding, and the grand glass roof of the Tay Bridge Station being practically destroyed would suggest that the storm was of a greater magnitude altogether.

Wind and the Tay Bridge

As has been said, Henry Rothery was Chairman of the Court of Inquiry and he ended up issuing a report on his own, the other two members, Colonel William Yolland, Chief Inspector of

Railways, and William Barlow, President of the Institute of Civil Engineers, presenting their own report. Perhaps it was because he was not a civil engineer but a lawyer and Wreck Commissioner that Rothery felt free in his report to put virtually all the blame for the disaster onto Bouch. The other two members were far more circumspect, particularly understandably in the case of Yolland, as we shall see, and they reacted quite negatively to the suggestion that Rothery effectively spoke for them as well.[31]

In his report, Rothery commented that in France and the United States, 55 and 50 lb/sq ft respectively were allowed for wind pressure, while even two British witnesses to the Court of Inquiry, including Benjamin Baker, recommended around 30 lb/sq ft, which, as we shall see, are much higher figures than any considered by Bouch. Yet it must be said that at that time, and obviously highly regrettably, there was practically no communication between British builders and those from other countries on these matters. Indeed there was much extremely loose talk about wind pressures before the building of the bridge, and after its collapse many were 'wise after the event'.

In his evidence to the Court of Inquiry—where Rothery said that he had given perhaps more attention to the motion of fluids that any person in the country—and later as a member of the Wind Pressure (Railway Structures) Commission,[34] Stokes was able to cut through the rhetoric and to present firm scientific facts and recommendations. These have played a substantial part in safeguarding railways and bridges ever since.

Sibly[2] stresses that though at that time engineers had developed some empirical understanding of wind forces, there was no reputable scientific method for estimating the effects of the wind until the development of wind tunnel models at the end of the century. So wind loading calculations were not normally carried out for structures completed before the Tay Bridge accident, though horizontal bracing was incorporated to assist the distribution of load. In addition drag and lift forces were not understood at all.

Bouch, in fact, had been extremely exercised as to the necessity of allowing for wind pressures, and unfortunately had been led to the belief that the wind would not generate significant forces. What data was available seemed highly unreliable.

What would have been regarded as the 'official' position was that of John Smeaton, from 1759. He claimed that a wind pressure of around 12 lb/sq. ft corresponded to a storm or tempest. It was true that he considered much higher pressures—up to 50 lb/sq. ft, but this corresponded to a hurricane that tore up trees and carried buildings before it. In any case, by the time of the building of the Tay Bridge, Smeaton's figures were looked on as highly conservative. For example, Eaton Clark, in his 1850 book on the Britannia and Conway tubular bridges, suggested around 20 lb/sq. ft as a maximum value, though the opinion of William Rankine in 1862 was that design wind load should refer to the maximum recorded wind gust recorded which was 50 lb/sq. ft.

Justifiably confused by all this contradictory 'information', in October 1869 Bouch wrote to Colonel Yolland in his capacity of Chief Inspector of Railways.[31] He asked 'In calculating the strains of malleable iron girders . . . is it necessary to take the pressure of the wind into account for spans not exceeding 200 feet, the girders being open lattice work?' He gave his own opinion that one and a quarter tons per foot run was sufficient for spans over 100 feet, and that it was not necessary to take the force of the wind into account. He stated that he asked for this information so that he could act in accordance with the views of the Board of Trade.

Yolland replied that 'A ton and a quarter foot run will be sufficient for spans over 100 feet, and we do not take the force of the wind into account when open lattice girders are used for spans not exceeding 200 feet.' It is hoped that he remembered this advice when he met in judgement on Bouch roughly a decade later.

Still uncertain, Bouch referred to a letter he had received from Sir George Airy, the Astronomer Royal, who was responsible for the maintenance of meteorological records. This letter had actually been sent in connection with Bouch's design for the Forth Bridge. A committee of experts called in by the North British Railway, which included William Barlow, had been suspicious of Smeaton's data and asked for Airy's opinion.

At least in retrospect, Airy's letter was unhelpful and misleading. He stated that upon very limited surfaces and for very limited times, the wind pressure may be as high as 40 lb/sq. ft, and he admitted that in Scotland it might be even higher. But he qualified this by adding that the times involved would be momentary, and the high pressures little more than 'irregular whirlings', which would extend for a distance far smaller than the length of the proposed bridge. In practice he suggested that the greatest wind pressure over the entire length of the bridge would be 10 lb/sq. ft.

Incidentally Airy's evidence given to the Court of Inquiry was also equally unhelpful. It appeared that he knew nothing about what was the crucial point—given the maximum value of a gust of wind, what was the length of time for which a gust might last and the physical dimensions over which it might act? His expertise was limited to the values produced by certain meteorological instruments which themselves did not give information on gusts. It turned out that this was just because of a lack of interest; once the Court of Inquiry requested the design of such instruments, the omission was quickly put right.

Faced with this contrary information, Bouch decided to make no particular provision for wind pressure–though his assistant, Allan Stewart, reported that he had arranged for some relatively small provision. Bouch's design was examined and approved by five eminent engineers, of whom T. E. Harrison reported that 'the bridge was sufficiently strong in all its parts' and J. M. Hepper said that 'I can therefore have no hesitation in stating my complete conviction of the efficacy of the design in every particular, whether in regard to the foundations or the superstructure.'

We add one significant further remark concerning wind. Sibly remarks on the discovery in 1881 that the wind pressure increases parabolically with height. Moving from a height of 6.1 m to one of 30.5 m, wind pressure might increase by 88 per cent.

Before moving on to work of Stokes himself, we briefly mention two people who, it has been suggested, may share some of the blame for the disaster with Bouch. Sibly says that Airy's statement was based on scanty information and casually glossed over uncertainties and inaccuracies. He remarks that his comments may seem unduly critical of one man, and admit that his opinions were endorsed by the best talents of the day.

Not so forgiving are Koerte[9] and Hammond[35] with their comments on Hutchinson. They feel that his words on wind did no more than give a free pass to Bouch on the topic. Also they feel that his failure to spot what they feel much have been grave failures in the bridge construction that should have been obvious from initiation, were inexcusable. They suggest that his responsibility for the collapse should be regarded at least as high as that of Bouch.

We now turn to the evidence of Stokes to the Court of Inquiry.[36] When, as already stated, Rothery described him as the person who had perhaps given more attention to the motion of fluids than anyone else in the country, Stokes agreed but commented that his knowledge of the various related experiments was not quite as high.

He was particularly requested to discuss the relation between the velocity and the pressure of the wind. He first stressed that there were two measures of pressure which were unfortunately very often confused with potentially tragic consequences. The first is the usual hydrostatic pressure, which Stokes calls 'standard'; he says that this could be measured by presenting to the wind the open mouth of a tube connected to a liquid gauge.

The second method is to measure the pressure on a plate, which will almost certainly give a very different result. The wind will strike the plate, but then divides and goes left and right, up and down. While the pressure in the centre of the plate will be the same as the hydrostatic pressure, it will diminish towards the sides, and towards the edge it may be negative. Thus Stokes says that the front of the plate may experience a pressure less than the standard. However, at the back of the plate there is a lowering of pressure; he suggests that the air passing the edges breaks into eddies which mix with the still air behind and drag it along, thus producing a partial vacuum. The increase in pressure in front and the decrease behind combine, with the result that the pressure sustained by the plate is higher than the standard pressure by a factor Stokes estimates as 80 per cent. He was asked to comment on how the size of the plate would affect this result: for relatively small plates, he thought pressure would be independent of area, but suggested that experiments on larger plates would be well worthwhile.

However Stokes says that it is far more common to measure the force of a wind by its velocity rather than its pressure. This is done by a Robinson's cup anemometer, which registers the number of rotations that the cups make round an axis in a given time. He suggested that it was best to take the velocity of the cups to be a factor of around 2.4 times that of the wind. This would often give a velocity in excess of 100 miles per hour, so it was essential to be able to relate this to the pressure of the wind.

Stokes reported that experiment had shown that a wind velocity of 20 miles per hour corresponds to a standard pressure of 1 lb/sq. ft, and his theory agreed with this value; since pressure varied as the square of the velocity, a velocity of 100 miles per hour would give a standard pressure of 25 lb/sq. ft, and a plate pressure of around 45 lb/sq. ft.

Much of Stokes's questioning related to the earlier advice and evidence to the Court of Inquiry of Airy. While Airy had recognized pressures of this order, or even higher in Scotland, his belief had been that, because of the limited times and distances over which such pressures exist (momentary whirlings, as he put it), the greatest mean pressure would be around 10 lb/sq. ft. Stokes described this figure as 'low.'

There was an interesting altercation when the counsel for Bouch twice suggested to Stokes that it was quite reasonable for any engineer to be influenced by the opinion of Airy, the Astronomer Royal, but twice objections were made by the counsels for the Board of Trade and the railway company. Eventually the question was allowed and Stokes could only reply that 'Everybody knows that the Astronomer Royal's opinion upon such a subject is a very valuable one, and entitled to all consideration.' It seems clear that Bouch's concerns about the wind were hardly given a fair hearing. Airy was actually recalled by the Court and found considerable difficulty in reconciling his earlier remarks with his present full agreement with Stokes.

Stokes was pressed very strongly by the counsel for Bouch over whether such violent gusts might be only momentary or how far they might be extended. He replied that evidence from anemometers was inconclusive—in fact anemometers at the time could give no information on gusts, but that common experience would tell us that a very heavy gust would very often blow for two or three minutes. Though he considered that a heavy gust might be confined to a 'narrow track', he explained that 'narrow' meant having a breadth of a few hundred yards.

Under further questioning he said that a storm with a velocity even of 60 miles per hour, which would thus travel half a mile in half a minute, might also have a lateral extent of half a mile. Overall Airy's approach was broadly untenable.

Sibly commented that in his testimony Stokes had been able to provide little useful information.[2] While that seems something of an exaggeration, it may also be suggested that it is not really the point. While he had provided what information was available, both on experimental procedure and on the behaviour of the wind, he refrained from ill-informed guesswork. He made clear suggestions for where obtaining further information was crucial and suggested the types of experiment that might provide this.

As a result of the general confusion exhibited at the Tay Bridge Court of Inquiry, and in particular the suggestions of Stokes, the Board of Trade set up a Commission to study the whole question of wind pressure and its effect on railway structures[34]. Yolland and Barlow from the Court of Inquiry moved to this Commission, and they were joined by Stokes himself and engineers John Hawkshaw and W. G. Armstrong. It was quite clear that Stokes's evidence at the Court of Inquiry had been much appreciated. Indeed, while there was a wealth of knowledge of railways and bridges on the Commission, an understanding of the nature and effects of the wind would be provided almost entirely by Stokes.

The first question investigated was the highest pressures of wind in the country. To this end, statements were obtained from those observatories and stations where the velocity or the pressure of the wind was measured; and lithographed copies of wind diagrams taken by self-registering apparatus were obtained from Bidston near Liverpool, and from Glasgow and Greenwich. At some of the stations, wind pressures were measured directly by self-registering pressure anemometers, while at others only the velocity of the wind was measured by rotating anemometers, the velocity of the wind being taken as three times that of the revolving cups.

For some stations the Commission noted that the only information provided was the run in miles of the wind during each hour. While there could be a clear relationship between this value and the greatest pressure experienced during the same hour, it was discovered by analysis of stations were both were recorded, that, for the case of the high winds which were the subject of the study, a suitable relationship was $P = V^2/100$ where P is the maximum pressure in the hour and V is the maximum run of wind, both given in the usual units.

Maximum wind pressures differed quite markedly at different stations. For example, at Glasgow the maximum was 47 lb/sq. ft, which was quite typical, while at Bidston they were as high as 90 lb/sq. ft. The group were clear that the exceptional results at Bidston were the result of the conformation of the ground rather than instrumental malfunction. However they wondered whether such high pressures were rather local, though they were not able to find any evidence on the suggestion.

To back up the evidence from meteorological stations, the Commission studied the occurrences of collapse of buildings, tall chimneys and so on and also the cases where railway carriages had been overturned by the force of the wind.

Their recommendations were as follows:

(1) A maximum wind pressure of 56 lb/sq. ft should be assumed. It will be remembered that this is quite close to the values assumed in France and the United States.
(2) If the train passes through the girders, the full pressure should be applied to the surface of the main girder only, but if the train is higher than the top of the main girder, the surface area of the train should be added.

(3) The Commissioners considered the case where a bridge is of lattice or open form. Quite contrary to the fairly casual remarks before the building of the Tay Bridge, their requirements were tough. On the windward girder, the full 56 lb/sq. ft should be applied to the area from rails to the top of the train and in addition to the area of the girder below the rails or above the train. Also a substantial force should be assumed on the leeward girder, the details depending on the surface area of the open spaces.

(4) Arches and piers should be considered in general conformity to the rules above.

(5) The margin of safety should be a factor of four, although where wind is opposed only by gravity this could be reduced to a factor of two.

Additional points were that if trains ran on the top of girders, parapets should be built; there was a recognition that the velocity of wind increased with height, as already stated; that even given the wind pressures implied at Bidston, bridges would be safe; and that modifications of the rules would be allowed for structures of small altitude or in sheltered locations.

This set of rules were well thought out and sensible. It was a great pity that the railway industry or failing that the government had not carried out this work many years before!

Lastly it may be mentioned that Stokes himself put considerable effort into analysing the way in which anemometers worked. Experiments had actually been carried out in 1872 at Kew; it had been thought that whirling anemometers could help to provide stable results, and three different types of cup were used. However the results were not thought to be interesting enough to publish. Stokes however was given the results and decided that they were of considerable interest. He published a substantial account of the theory of the experiment and also detailed tables of the results and analysis.[37] Earlier he had published some further discussion[38] of a paper on the same topic published by Robinson in Dublin.

· ·

ACKNOWLEDGEMENT

I would like to thank Annette Ruehlmann of the Institution of Civil Engineers for kindly and speedily providing me with material relating to the Court of Inquiry into the Tay Bridge disaster.

CHAPTER 10

Faith and Thought

Stokes as a Religious Man of Science

STUART MATHIESON

George Gabriel Stokes was one of Victorian Britain's most celebrated physicists, and one of its most prominent religious scientists. Stokes worked during a period of religious upheaval. Geology, biology, physics, philosophy, biblical criticism, and moral concerns all appeared to pose a threat to Christianity. The natural sciences were professionalizing, leaving little room for the gentleman amateurs, often clergymen, who had previously undertaken much research. Many of this new class of professional scientists were also keen to see their disciplines secularized. Stokes, by contrast, saw a harmonious relationship between his scientific work and his faith. He therefore offers an opportunity to explore the moral and intellectual threats to Christianity in the mid-nineteenth century, particularly as he was one of the most public advocates of conditional immortality and opponents of the doctrine of eternal punishment. Stokes's work on natural theology also gives a particularly clear example of how evangelicals could interpret their scientific work and their faith in mutually reinforcing ways, and demonstrates how William Paley's teleological argument could survive threats from geology and Darwin. Further, his role with the Victoria Institute, alongside his work for the Royal Society and evangelical organizations, shows how Stokes acted as a leader for Victorian Britain's religious scientists, and how scientists of faith continued to operate in a professionalized, secularizing environment.

Faith in Crisis? The Nineteenth-Century Context

The middle of the nineteenth century was a challenging period for Christianity in Britain. A series of scientific discoveries, new approaches to the Bible, and moral concerns all combined to create a crisis of faith. The 1851 census had included, for the first time, a survey of religious observance, which suggested that only around half of the population went to church on a Sunday.[1] Evidence from geology indicated that the earth was much older than was supposed by those, such as the seventeenth-century archbishop James Ussher, who had used the Bible to estimate its age at less than six thousand years.[2] Instead, works by James Hutton (1726–97),

Mathieson, S., *Faith and Thought: Stokes as a Religious Man of Science*. In: *George Gabriel Stokes: Life, Science and Faith*, Mark McCartney, Andrew Whitaker, and Alastair Wood (Eds): Oxford University Press (2019).
© Oxford University Press. DOI: 10.1093/oso/9780198822868.001.0010

Charles Lyell (1795–1875), and Hugh Miller (1802–56) suggested that the earth had been formed by gradual processes acting over an almost indefinite period of time. This theory, uniformitarianism, contrasted with the previously popular idea of catastrophism, which argued that the earth had been shaped by sudden changes or catastrophes, such as the Noachian deluge.[3]

Charles Darwin's theory of evolution, published in *On the Origin of Species* (1859), relied on a similarly ancient earth for the mechanism of natural selection to create separate species. Indeed, Darwin had been heavily influenced by Lyell's *Principles of Geology* (1830–3) during his voyage on the *Beagle*, when he collected the evidence that would inform his theory.[4] Although Darwin did not explicitly state that humans were descended from other animals until his *Descent of Man* (1871), the implication was immediately clear. Evidence from two disciplines now combined to argue that the earth was much older than suggested, and that both animal species and geological features were a result of gradual, mechanical changes over time rather than interventions by God.

Darwin is often misconstrued as having dealt a decisive blow to religious orthodoxy. Yet many religious people accepted the biological principle of his theory with little difficulty, even if the idea of human descent proved more troublesome.[5] At least as problematic as *Origin of Species*, however, was a book published just four months later. *Essays and Reviews* (1860) was a collection of essays written by six Church of England clerics and one layman, which questioned the reliability of the Bible as a historical document and cast doubt on the possibility of miracles.[6] This was the culmination of an intellectual trend, higher criticism, that had begun at Tübingen University in Germany, in the work of biblical scholars such as Friedrich Schleiermacher (1768–1834), Ludwig Feuerbach (1804–72), and David Strauss (1808–74). The higher critics advocated treating the Bible as they would any other historical document, using the same techniques as historians and literary critics; works by Strauss and Feuerbach were translated into English by the novelist George Eliot (1819–80) in 1846 and 1854 respectively.[7]

The publication of *Essays and Reviews* and, in particular, of biblical criticism from members of England's established Church, caused a scandal.[8] Two of the authors, Rowland Williams and Henry Bristow Wilson, were tried for heresy in ecclesiastical court.[9] Criticism was one internal threat to Christianity, but it was not the only one. Victorian Britain had a culture of popular piety and a strong emphasis on morality. However, just as some critics had examined the Bible and found its literal interpretation to be untenable, so had the moral implications of some Christian doctrines become similarly challenging.[10] Of particular concern was the principle of eternal punishment, something that was difficult to reconcile with the idea of a just, loving God.[11] Eternal punishment troubled many Victorian thinkers; Stokes was deeply concerned by it, and Charles Darwin, who called it a 'damnable doctrine', identified it as one of the main reasons behind his own loss of faith.[12]

While Darwin's role as the slayer of Victorian religious belief has been overstated, his work was appropriated by others who saw it as ammunition to be used in a struggle for cultural and intellectual authority. Chief among those were the biologist Thomas Henry Huxley (1825–95), who became known as 'Darwin's bulldog' for his tenacious defence of evolutionary theories, and the Irish physicist John Tyndall (1820–93). Both were members of the X Club, an informal dining group and talking shop that was dedicated to the professionalization and secularization of the natural sciences and represented a rising class of professional scientists.[13] The natural sciences had traditionally been the preserve of the gentleman amateur, particularly Anglican clerics, who had the benefits of a university education and a measure of free time to pursue their interests. The new generation of naturalists typified by the X Club saw science as the preserve of professional specialists rather than polymath priests, and the controversy around Darwin's theory as an opportunity to hasten the process of professionalization.[14] After an Oxford Union debate on Darwin in 1860, Huxley and Bishop Samuel Wilberforce, renowned as one of Britain's finest public speakers, had a stormy exchange in an event that has passed into folklore. After dismissing the scientific evidence for evolution Wilberforce trained his sights on Huxley, who had suggested that humans too had evolved, asking whether it was through his grandmother or grandfather that the biologist claimed to be descended from a monkey. While the outcome of the debate was inconclusive, Wilberforce's scathing assessment of evolutionary theory was held up as evidence of the church overstepping its intellectual authority, and an example of why the sciences must be pursued freely, without clerical interference.[15] This, then, was the religious and intellectual environment in which Stokes, as one of Victorian Britain's most prominent religious scientists, operated.

A Son of the Manse: Early Religious Life

Much of Stokes's approach to religion can be traced back to his early family life. His father, the Rev. Gabriel Stokes, was a Church of Ireland minister at Skreen, a small village in County Sligo, on Ireland's rural west coast. George Gabriel was the youngest of eight children, and his three elder brothers, John, William, and Henry, followed their father into the church.[16] Stokes's mother, Elizabeth, and her husband were both serious, taciturn, but good-natured parents, dedicated to providing a modest but happy life for their family, and ensuring that their sons were well educated. Rev. Stokes was a member of the evangelical wing of the church, and emphasized careful study of the Bible, conscientiousness, and moral uprightness. These virtues were not long in filtering down to his youngest son; Stokes, frustrated at learning to read from a children's book, asked his mother if he might instead learn from the Psalms. It is not difficult to deduce from where Stokes acquired his seriousness, sense of duty, and laconic manner of speech. Similarly, he inherited his father's straightforwardly evangelical theology. Like most protestants, Stokes was committed to the principle of *sola scriptura*, the doctrine that the Bible represents the complete and authoritative guide to the Christian faith. Stokes had a keen interest in the New Testament, which he often read in Greek.[17] He was also, as were most evangelicals, committed to social action, particularly efforts to communicate and spread his faith, joining several evangelical organizations such as the Church Missionary Society.

While the most striking feature of Stokes's religion was his evangelicalism, Anglicanism was also an important part of his religious identity. Following his father and brothers, he remained a member of the established church for the rest of his life. He served as a warden for his church in Cambridge[18] and was a lay member of the Church of England's synodical council, Convocation.[19] In 1849 his friend William Thomson, later Lord Kelvin, suggested that he apply for the recently vacated chair of mathematics at Glasgow. Despite seeking out testimonials from colleagues, Stokes expressed some reservation, for professorships at Glasgow required a statement of faith in Scotland's established, Presbyterian church. The Rev. John Whitley Stokes, a rector in County Tyrone, convinced his youngest brother that he should not make such a statement unless he was 'prepared to become a thorough Presbyterian'. This was, Stokes wrote to Thomson, something 'which certainly I do not mean to become'.[20] Undeterred, Thomson wrote back, reassuring Stokes that he had conferred with other Glasgow professors, and that a test of faith need not present an obstacle to an Episcopalian, as Anglicans were known north of the Tweed. In fact, Thomson cheerfully described his own ecumenical approach as 'the habit of regularly conforming to the Episcopal Church, & not appearing more than once or twice or three times in the course of a session at an Established Church', hoping to further persuade Stokes that he could continue to practise as an Anglican if he took up the post.[21] However, Stokes's mind was set; he was unwilling to make a half-hearted statement of faith, and thereby insult both his own church and the Church of Scotland, and did not apply for the post.

Stokes's early interest in biblical study, his moral earnestness, and his precocious mathematical ability meant that he had an unusual concern with the doctrine of eternal punishment. He recollected that, as a young child, he was 'so horrified at the idea of endless torments that I wished there was no God and no future state, lest I should fall into them'. Because he was naturally inclined toward mathematics, Stokes suggested that he 'took in the idea of infinite duration more readily than most children would have done'.[22] While he had little trouble in grasping the idea of infinity, Stokes found the idea of punishment meted out to immortal souls for an infinite duration difficult to reconcile with the idea of a just, merciful, and benevolent God with which he had been raised.

A 'Damnable Doctrine': Eternal Punishment and Conditional Immortality

While the concept of eternal punishment troubled Stokes, it did not lead, as it had with Darwin and many others, to a loss of faith. For Victorian evangelicals, God's defining characteristics were love, justice, and mercy. The doctrine of eternal punishment held that, after death, the immortal souls of the unsaved would be condemned to unending torment. This seemed at odds with the work of a loving, merciful God, and engendered fierce debate in evangelical circles. Indeed, the subject was so controversial that the Anglican theologian Frederick Denison Maurice (Fig. 10.1) was dismissed from his post at King's College, London for questioning eternal punishment in his *Theological Essays* (1853).[23] Similarly, Henry Bristow Wilson's contribution to *Essays and Reviews* saw him tried for heresy. In his essay, *Séances Historiques de Genève—The National Church*, Bristow rejected the idea of eternal punishment and the existence of hell. The debate was not confined to theologians: even former prime minister William Ewart Gladstone voiced his concerns over the doctrine in his *Studies, Subsidiary to the Works of Bishop Butler* (1896).

Stokes took a characteristically evangelical, studious approach to the problem of eternal punishment, and in 1851, once again troubled by it, he turned to scripture for guidance. Here he was in good company. Throughout the nineteenth century, protestant scholars combed the Bible, searching for a definitive answer. Richard Whately (1787–1863), archbishop of Dublin, argued in 1829 that there was no scriptural evidence to suggest that souls were immortal, nor that immortality was the natural state of the human soul.[24] John William Colenso (1814–83), the bishop of Natal, wrote a controversial commentary on Paul's Epistle to the Romans in 1861.[25] Colenso caused outrage when his commentary rejected several aspects of Christian doctrine as unscriptural, including eternal punishment.[26] Frederick Farrar (1831–1903), canon of Westminster Abbey, made both the moral and the scriptural case against eternal torment.

Fig. 10.1 Frederick Denison Maurice (1805–72), was dismissed from his position at King's College, London, after writing his *Theological Essays* (1853), one of which questioned eternal punishment.

There were many 'of the highest intellect…among our most eminent literary and scientific men', Farrar argued, who found eternal punishment 'one of their most insuperable difficulties in the way of accepting the Christian faith'. Farrar considered it his 'duty to show that those torments have been described in a manner unauthorized by Scripture'.[27] Since the doctrine of *sola scriptura* held that the Bible was the final authority on matters of faith, many evangelicals argued that the lack of scriptural evidence for immortal souls and eternal torment meant that these concepts should not be considered a canonical part of their religious views.

Stokes, having combed scripture for what he hoped would be the authoritative truth, agreed with this assessment, and was relieved to find nothing in the Bible that required belief in eternal punishment. However, it was not enough for Stokes to have settled the matter to his own satisfaction. As a committed evangelical, he felt a responsibility to discuss eternal punishment further, and to share it with others, particularly those who considered it an impediment to personally accepting Christianity. Stokes was a prolific writer of letters, and he wrote frequently to friends and acquaintances on this topic, urging them to read works by Henry Constable, a Church of Ireland minister in Cork, which argued against the possibility of eternal punishment. Indeed, Stokes sent a copy of Constable's *Future Punishment* (1868) to several correspondents, particularly fellow mathematical physicists who had graduated from Cambridge, such as William Walton (1813–1901), and Peter Guthrie Tait (1831–1901), a physicist who was professor of natural philosophy at Edinburgh.[28] Stokes also corresponded with several authors of books on eternal punishment. First, he exchanged letters with the Rev. Henry Smith-Warleigh, who wrote a series of tracts on the subject, including one that suggested that the Church of England had abandoned it as part of its official doctrine, which he sent to Stokes. Through Smith-Warleigh, he began corresponding with the Congregationalist minister Edward White (1819–98). White was the author of *Life in Christ: Four Discoveries upon the Scripture Doctrine that*

Immortality is the Peculiar Privilege of the Regenerate (1846), a book based on a series of lectures on the immortality of the soul. While questioning the doctrines of natural immortality and eternal punishment caused controversy within the Church of England, White's position in another denomination insulated him from the charges of heterodoxy and heresy that he might have attracted as an Anglican. Stokes was impressed by the book, and in particular White's version of the argument that eternal punishment not only lacked support from scripture but that it was in fact contrary to what the Bible actually taught. White instead insisted that the Bible taught the doctrine of conditional immortality. This meant that, instead of souls being naturally immortal, and therefore subject to eternal punishment, immortality was conditional upon the person being saved, by undergoing a conversion experience and becoming a born-again Christian. Those who were not saved would have their souls annihilated and cease to exist following Christ's return and judgement. Further, White argued that the notion of natural immortality was not only unscriptural; it was not Christian. Rather, it came from 'Platonists, and Pharisees'. By this, White meant that the notion of the immortality of the soul came from some interpretations of Judaism, and from Greek philosophy, which had intermingled with Christian doctrine through the work of early theologians such as Origen, Augustine, and Thomas Aquinas. The Greek philosopher Plato had popularized the idea that souls were immortal throughout the Hellenized world, which included the eastern provinces of the Roman Empire in which Christianity first flourished. The concept of immortal souls was also popular in several pagan religions, and White argued that it had been absorbed into Christian thinking without careful examination of the scriptures.[29]

White's interpretation of conditional immortality appears to have energized Stokes in his evangelical efforts. Not only could he point to a lack of evidence for eternal punishment, he had an alternative teaching that was supported by scripture. While it was Stokes's scientific mind that allowed him to grasp the implications of eternal punishment, it was as an evangelical that he attempted to find a solution, and here he had one that appealed both his moral sense and to his reverence for scriptural authority. He continued writing to friends and colleagues on the topic, particularly after becoming involved with the Christian Evidence Society (CES). The CES was an evangelical organization that was formed in London in 1870 to arrest the rising tide of 'infidelity' in British society. This it hoped to achieve through the use of apologetics, reasoned arguments defending the Christian faith, in the form of lectures and tracts. The CES attracted religiously minded scientists such as the chemist John Hall Gladstone (1827–1902), who were keen to demonstrate that science did not threaten their faith, nor did their faith impede their own scientific work.[30] Instead, they hoped to show that the two operated in harmony, indeed, actually reinforcing each other. Stokes, as a religious scientist who was firmly convinced of the harmonious relationship between his scientific work and his faith, was naturally drawn to such an organization. However, while the CES identified several threats to faith, including scientific threats, Stokes remained convinced that eternal punishment was the most dangerous. He wrote to the CES's secretary, the Rev. P. Barker, in 1877, urging the organization to make conditional immortality part of its official doctrine. A London conference on conditional immortality, attended by Constable, White, and the Rev. Samuel Minton (1820–94), another frequent correspondent of Stokes, had taken place in May 1876. The organizer of that conference, the Scottish Congregationalist minister William Leask, published a successful journal advocating conditional immortality, *The Rainbow*.[31] Yet, while conditional immortality was clearly generating a great deal of interest in evangelical circles, Stokes was unsuccessful in convincing the CES to adopt it.[32]

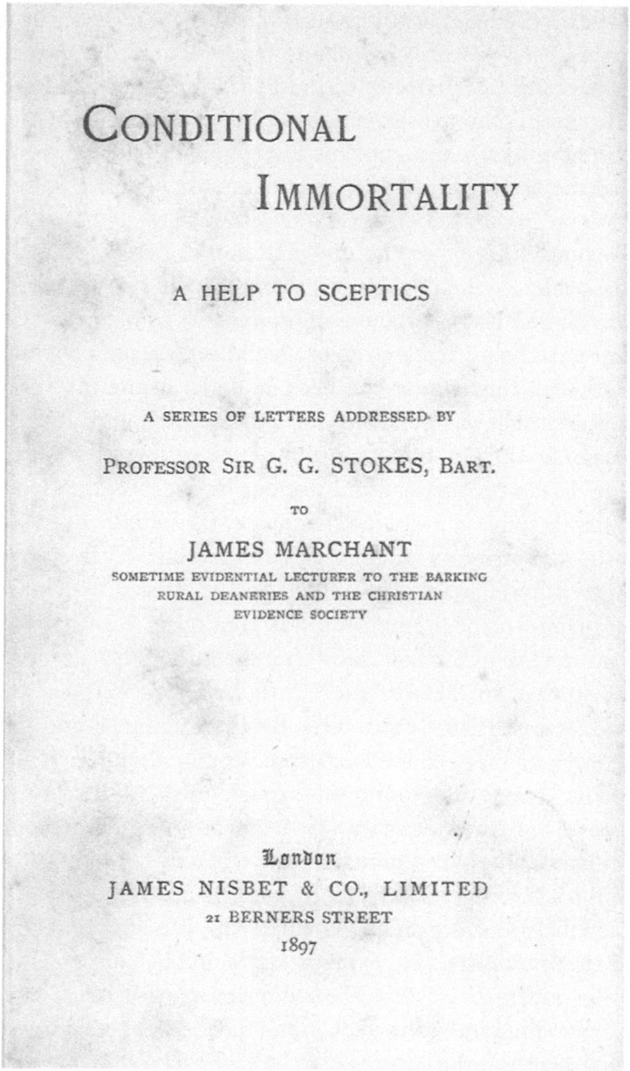
CONDITIONAL
IMMORTALITY

A HELP TO SCEPTICS

A SERIES OF LETTERS ADDRESSED BY

PROFESSOR SIR G. G. STOKES, BART.

TO

JAMES MARCHANT

SOMETIME EVIDENTIAL LECTURER TO THE BARKING RURAL DEANERIES AND THE CHRISTIAN EVIDENCE SOCIETY

London
JAMES NISBET & CO., LIMITED
21 BERNERS STREET
1897

Fig. 10.2 *Conditional Immortality* (1897) was a book of letters rather than lectures, with the effect that Stokes's writing voice carried both a directness and a sense of familiarity sometimes absent from his other works aimed at the general public.

Nevertheless, the CES did present Stokes with an opportunity to circulate his views more widely. Through it, he met James Marchant, a lecturer on apologetics, and the two began corresponding. Eventually, Marchant compiled the letters that Stokes had sent him on conditional immortality and edited them into a book, *Conditional Immortality; a Help to Sceptics* (Fig. 10.2), published in 1897. In *Conditional Immortality*, Stokes's views on the human soul were given particularly clear expression, and so it offers an opportunity to fully explore those views. Marchant exercised a great deal of restraint in editing the letters, retaining the original letter format as far as possible. The result was an epistolary book rather than a lecture series, with the effect that Stokes's writing voice carried both a directness and a sense of familiarity sometimes absent from his other works aimed at the general public. Stokes began by arguing that the notion of eternal torment presented 'a stumbling-block to atheists' and to the 'sceptically inclined'.[33] Here, again, Stokes was first and foremost writing as an evangelical. His task was to evangelize, to write, to explain, and to help convert. Using his abilities as a writer and thinker, he hoped to remove what he saw as the most obvious impediment to atheists becoming Christians. While he understood that eternal torment presented a moral concern, Stokes remained convinced that the answer lay in scripture. Eternal torment was flawed, because it was a result of mixing 'the clear teaching of Scripture, that the perdition of the lost is final and irreparable' with what Stokes described as 'the dogma of the immortality of the soul'.[34] Stokes clarified exactly what he meant by immortality of the soul: 'that man is by nature, by the very condition of his creation, an immortal being; that mortality affects his body only, but that his soul cannot but live for ever'. Although this dogma had been 'adopted in the supposed interest of the Christian faith', Stokes argued, it was 'nowhere taught in Holy Scripture'. Rather, it was 'opposed to the teaching of Scripture, and is a root-error from which springs a whole crop of theological difficulties, obscuring the fair proportions and glorious prospects of the Gospel of Christ'.[35]

Stokes accepted that the Bible did discuss the future state of human souls. It was 'clear as clear can be', he wrote, that scripture taught about a future life. Yet this was not the same as teaching natural immortality. Humans were created, Stokes insisted, 'with a view to immortality', but the 'actual attainment of immortality was contingent on [their] obedience'.[36] The disobedience of Adam and Eve meant that their 'whole being became forfeited to sin'. The result was that, as Paul had written, 'in Adam all die'.[37] This, then, Stokes argued, was the fate of the human soul: designed for immortality, but as a result of their sinful nature, not guaranteed it. It was not God's desire that humans be allowed to become immortal sinners. Nevertheless, humans had still been offered the chance of immortality, through Jesus; in Adam, all might die, but 'in Christ will all be made Alive'.[38] However, as with the initial promise of immortality, Stokes argued that this too was contingent. Through Christ humans were only 'brought within the possibility of immortality'; it was not guaranteed. While the gospels taught that everyone would be resurrected, Stokes insisted that resurrection would not be the same for all. Those who had been saved would indeed have eternal life, while the wicked would be 'consigned to a second death, from which there is no resurrection'.[39] This allowed for those parts of the Bible that 'represent the perdition of the lost as final and irremediable, and those other passages which picture to us a state in which all will once more be very good' to be 'reconciled with one another without the slightest strain in either case'.[40] Indeed, Stokes identified forty passages in the New Testament that mentioned eternal life, and argued that the context of almost all of them made clear that immortality was 'not the common lot of all men, but is peculiar to the redeemed'.[41]

Stokes also recognized that, while he as an evangelical considered the Bible to be authoritative, this would not be true for everyone. When debating eternal punishment, 'arguments for or against it involve a reference to the teaching of Scripture, and infidels would not admit the authority of Scripture'.[42] Nevertheless, if eternal punishment was to be used as an argument against Christianity, Stokes believed that it was important to demonstrate that it was not an essential part of that faith. 'There is not a syllable in the Liturgy', he insisted, 'or Articles of the Church of England affirming it'.[43] Similarly, Stokes recognized that some Christians, particularly Catholics, attached at least as much importance to tradition and the Church's own teaching as they did to the Bible. Here too, he had a solution. Stokes pointed to early Christian leaders such as Irenaeus and Theophilus, who he argued 'held that eternal living existence was for the saved only'.[44] Stokes acknowledged that one early father of the church, Justin Martyr, had mentioned eternal torment, but insisted that Justin was actually an example of 'how the doctrine of endless torments crept in' to Christian theology.[45] Justin, born in Samaria, had been a Platonist, and accordingly believed that the human soul was immortal and indestructible. Despite his converting to Christianity in later life, Stokes believed that Justin retained his Platonic conception of the soul. The notion of an immortal soul contrasted with the materialism of some other Greek philosophies, such as Epicureanism, and Stokes argued that this meant that Platonism had been 'received with undue favour by early Christians, especially in Greek-speaking countries'.[46]

'Design Implies a Designer': Cambridge, Paley, and Natural Theology

If Skreen was where Stokes's religious views first took root, it was at Cambridge that they flourished. When Stokes arrived in Cambridge as an undergraduate in 1837 it was a hotbed of evangelicalism, standing in marked contrast to the high church Oxford Movement, which emphasized ritual, doctrine, and the authority of the church.[47] Stokes entered Pembroke College to read mathematics, having already displayed a prodigious talent for algebra and arithmetic as a schoolboy. Indeed, he would pass the Mathematical Tripos in 1841 as senior wrangler, the highest scoring student at any Cambridge college. He was subsequently elected a fellow of Pembroke, and then offered the prestigious Lucasian professorship in 1849, a chair previously held by luminaries such as Isaac Newton (1643–1727), the Astronomer Royal George Biddell Airy (1801–92), and the computing pioneer Charles Babbage (1791–1871).[48]

While Stokes had distinguished himself in mathematical examinations, Cambridge undergraduates also had to read outside their field. Two items on the list of required reading were from the same author, William Paley (1743–1805). Paley (Fig. 10.3) had like Stokes been senior

Fig. 10.3 William Paley (1743–1805), the Anglican theologian whose work, particularly *Natural Theology* (1802), heavily influenced Stokes.

wrangler, yet it was his books on moral philosophy and Christian apologetics that were studied at Cambridge. Paley's *Moral Philosophy* (1785), which espoused a utilitarian morality, had become unfashionable by the time Stokes read it, but his *Evidences of Christianity* (1794) made a decisive impact. *Evidences of Christianity* was a work of apologetics laying out what Paley considered the most compelling evidence for the existence of God and was a response to sceptical philosophers such as David Hume (1711–76), who had questioned the possibility of miracles and the evidence for a divinely designed universe. Paley's best-known work, *Natural Theology* (1802), developed his ideas from *Evidences of Christianity* into a comprehensive argument for the existence of God from the evidence of the natural world.[49] Both *Evidences of Christianity* and *Natural Theology* had impressed Charles Darwin during his days as a Cambridge undergraduate. As Darwin recalled in his autobiography,

> I am convinced that I could have written out the whole of the Evidences with perfect correctness, but not of course in the clear language of Paley. The logic of this book and as I may add of his Natural Theology gave me as much delight as did Euclid.[50]

Stokes was similarly impressed by Paley's work, which informed his own views on the harmonious relationship between scientific inquiry and religious faith for the rest of his life. Indeed, almost forty years after he had graduated from Cambridge, Stokes still described his vision of science and religion as 'very homely' and said that he understood a designed universe 'much in the same way that was mentioned long ago by Paley in his *Natural Theology*'.[51] Paley's work was not entirely original, but its strength lay in its clarity of expression. In *Natural Theology*, Paley laid out his version of what has become one of the best-known arguments for the existence of God, the teleological argument, or argument from design. Teleology is the practice of explaining an object or phenomenon in terms of its purpose. An object that has a purpose has usually been designed with that purpose in mind, and for something to have been designed, it must have had a designer. The presence of clearly designed objects in the natural world would thus imply the existence of an intelligent designer who had in some way created them. Paley's most striking argument used the analogy of a watchmaker. Paley argued that if, when out walking, he hit his foot on a stone and wondered whence it came, it would be satisfactory to say that so far as could be supposed it had been there forever. If he struck his foot upon a watch, however, this would not seem a sufficient explanation. An examination of the watch, Paley argued, would conclude that it had a purpose. Even without knowing how a watch operated, or never having seen one before, the conclusion would be 'inevitable, that the watch must have had a maker: that there must have existed, at some time, and at some place or other, an artificer or artificers who formed it'.[52] Paley did not stop at merely laying out the argument in a clear fashion. He turned to examples from the natural world, such as plants and the human body, that he considered analogous to a watch. Since these bore the clear hallmarks of design, Paley argued, they were compelling evidence of a designer, and thus of God. For Stokes, the most important conclusion to be drawn from Paley's work was that scripture and science, properly undertaken, would always be in harmony. By examining the book of nature, scientists would inevitably find evidence of design.

As a set text, Paley's work was widely read by nineteenth-century Britain's intelligentsia. Paley, and natural theology more broadly, formed part of its 'common intellectual context'.[53] Indeed, during the 1830s, a series of eight tracts on natural theology had been commissioned by the Earl of Bridgewater; the last of these *Bridgewater Treatises* was published in 1840, while Stokes was still an undergraduate. Historians have debated the extent to which the *Bridgewater Treatises* followed Paley, but they certainly bore the marks of his influence, even if they departed from his approach.[54] Stokes was not unusual, then, in his espousal of natural theology. Nevertheless, natural theology became increasingly uncommon as the nineteenth century progressed, particularly after the publication of *Origin of Species* and *Essays and Reviews*; by the 1880s, Stokes was Britain's best-known scientific advocate of natural theology. In 1880, Stokes agreed to give a lecture to the Victoria Institute, an organization established in 1865 to investigate the relationship between science and religion. In his lecture, Stokes expressed clearly his view that natural theology offered the best explanation of how that relationship should be understood. Stokes argued that 'truth of one kind, rightly pursued, cannot conflict with our reception of truth of another kind, though from the imperfection of our knowledge and of our faculties temporary difficulties may arise'.[55] Stokes then followed Paley's formula by drawing on an example from the natural world, the human eye. As an expert in optics, Stokes was well placed to explain how an understanding of the physical laws governing human vision might initially undermine faith in a creator who had given sight as a gift to humans. Indeed, other physicists such as Tyndall had used those laws to argue for a mechanical, materialist universe. Nevertheless, Stokes argued, the notion of a creator who worked by establishing laws that governed the operation of the universe, rather than by fiat, could provide a compelling reason to believe in God. Just as a watch was a clear example of a designed object, and design implied the existence a designer, an orderly universe operating according to fixed laws implied the existence of a lawmaker. What initially seemed to place science and faith in conflict was thus a result of considering only one perspective; viewed through the prism of natural theology both could be easily reconciled.

For the rest of his career, Stokes was the obvious choice whenever a prestigious scientist was needed to deliver a lecture on natural theology. In 1883, he travelled to Aberdeen to become the inaugural Burnett Lecturer. John Burnett, an Aberdeen merchant, had bequeathed part of his fortune in 1784 to establish an essay prize for writers on natural theology, awarded at forty-year intervals. In 1881, Burnett's trustees decided that it would be more effective to endow a lectureship on the relationship between religion and the academic fields of history, archaeology, and the natural sciences, tenable for three years. Stokes was chosen to deliver the first series of Burnett Lectures, and chose the properties of light as his topic. While the lectures were aimed at a popular audience, Stokes was determined that they would be intellectually rigorous. He assumed 'on the part of the reader a knowledge of the rectilinear propagation of light in the same medium, of the laws of reflection and refraction, of the compound character and of the decomposition of white light', although he generously did not assume 'that he is acquainted with the phenomena of interference, or diffraction, or double refraction, or polarization'. Indeed, Stokes demonstrated a remarkable gift for understatement when he noted that 'some acquaintance with these subjects will make the lectures much more easy to follow'. Those sufficiently equipped to follow Stokes were treated to a comprehensive overview of the physical properties of light, before he highlighted the two key points that he felt should be taken away from this study, 'the importance of the ends, and the simplicity of the means'. By this Stokes clarified that he meant evidence of design, a conclusion that he found 'irresistible' when the eye was considered as a specialized device.[56] Here, Stokes once again drew on Paley's argument, using his authoritative knowledge of optics to suggest that the human eye was designed to obey rules laid down by a designer.

In 1891, Stokes returned to Scotland, this time Edinburgh, to deliver another series of lectures of natural theology. When Adam, Lord Gifford, died in 1887, he left provision in his will for a lecture series, divided between the four Scottish universities, on natural theology.

In Gifford's will, natural theology was defined rather strictly; lecturers were expressly forbidden from turning to revelation or scripture and restricted instead to the natural world. As Stokes put it,

> From the words of the bequest by which this Lectureship was established, it would seem that the idea of the Founder was that it was possible to establish the existence of a God, and to frame a perfect system of rules of duty, simply by the exercise of man's intellectual powers, exerted in a manner perfectly analogous to that by which the physical sciences have attained their present development.[57]

While Stokes had already demonstrated his ability to lecture on natural theology, as a committed evangelical he found the inability to refer to revelation or the authority of scripture restrictive. In his few previous theological writings, Stokes wrote, he had 'gone on the basis of accepting a supernatural revelation, more especially on that of accepting the resurrection of Jesus of Nazareth as a supernatural historical fact'. Stokes was being rather modest when he claimed to have 'never written on, and I may add I have never specially studied, natural theology', but it reveals the simplicity with which he understood the term.[58] A straightforward reading of Paley's watchmaker analogy, coupled with knowledge of the physical world derived from contemporary science, was sufficient to provide a compelling argument for design. Where his Burnett Lectures had focused specifically on light, Stokes used the Gifford Lectures to discuss a broader range of topics. The subjects that Stokes chose to discuss offer a remarkable insight into how he related his understanding of the natural world to his faith. Shorn of the ability to draw on scripture, Stokes turned first to science; of the twenty lectures he delivered, ten were primarily concerned with scientific topics. Stokes clearly believed that the human eye was the most effective way to explain his version of the design argument, basing four lectures around it. He also drew on other areas of physics, such as light, laws of motion, and astronomy. Here too, Stokes reiterated his view that the natural world revealed observable laws, and that these were evidence of a lawgiver. When Stokes departed from the natural sciences, he covered territory with which he was familiar, and about which he was passionate. Five lectures discussed the human soul, immortality, and future states, while three discussed a sense of duty. This was, again, an opportunity for Stokes to link one of his favourite topics with the design argument. If the human body was clearly designed, he reasoned, it was no great leap to assume that an innate sense of morality was built in to that design. Indeed, Stokes argued that it was possible to 'look on this innate sense of right and wrong as the will of God written upon the heart' and thus 'some rules of guidance may be obtained even without having recourse to a supernatural revelation'.[59]

Most of Stokes's lectures trod a familiar path through subjects he had covered in detail elsewhere: evidence for design in the natural world, particularly in his own field of physics, conditional immortality, and morality. Yet the Gifford Lectures allowed Stokes the opportunity to develop an idea at which he had previously only hinted. Stokes was concerned that the natural sciences were being commandeered by secular scientists to serve their own ends. In biology, Huxley was the obvious antagonist, while in Stokes's own field, it was John Tyndall. In his 1874 address to the British Association for the Advancement of Science meeting in Belfast, Tyndall had made vociferous calls for the emancipation of science from religion, and advanced a materialist understanding of the universe.[60] Materialism is the philosophical position that matter is the only fundamental substance, and that all phenomena, including the human mind, could be explained in terms of the interaction of matter; this was a mechanical, deterministic view that left little room for God. Tyndall drew on examples from physics, particularly the law of conservation of energy, to support his argument. Stokes, as a religious physicist, was perturbed that his discipline, which he believed provided compelling evidence of God's work in nature, might be used to advance materialism. Eventually, he was able to use his Gifford Lectures to suggest an alternative viewpoint, which he called 'directionism'.[61] Stokes chose directionism as a name to reflect its hypothesis that there was a force that directed the life and energy of living beings. He compared the two positions by considering their explanation of a living organism. According to materialism, Stokes argued, 'a living organism is a highly elaborate machine, constructed out of ponderable matter, and growing and performing all the functions belonging to its life under the influence of the physical forces and nothing more'. Under directionism, by contrast, while the organism was subject to the physical forces, it was also subject to 'an influence of some kind, not recognisable by our senses, nor observed in life-less matter'. The role of this influence, Stokes claimed, was 'not counteracting the action of the physical forces, but guiding them into a determined channel'.[62] Since the directing force existed outside the material realm, conservation of energy did not necessarily apply to it. For Stokes, this was doubly useful. First, it provided an alternative to materialism that allowed for the action of laws on the human body, in an analogue of how God might theoretically interact with the natural world. Further, since it removed the threat from the law of conservation of energy, it made possible the conditional immortality of the human soul.

Figurehead of the Scientific Establishment: The Victoria Institute

On 24 May 1865, a circular was distributed to newspapers and prominent individuals, announcing the formation of 'a new Philosophical Society for Great Britain'. The members of this society, the Victoria Institute, were to be 'professedly Christians' and its object would be 'to defend revealed truth from "the opposition of science, falsely so-called"'.[63] By 'science, falsely so-called' the Victoria Institute meant those modern intellectual trends, such as geology, biology, or biblical criticism, that they considered a threat to Christianity. Although headed by Anthony Ashley-Cooper (1801–85), the 'Evangelical Earl' of Shaftesbury, the Victoria Institute was the brainchild of James Reddie (1819–71), a civil servant at the Admiralty. Reddie had hovered on the periphery of London's intellectual circles for some time, although his idiosyncratic views, particularly his disavowal of Newtonian physics, meant that he was never likely to join the scientific mainstream. Convinced that he was being deliberately ignored by intellectual society because of his 'heretical' opinions on gravitation and the cosmos, Reddie had watched with alarm as Victorian Britain's cultural and intellectual authority began shifting away from churchmen and toward the professional scientific class from which he felt excluded. As the furore over *Origin of Species* and *Essays and Reviews* swept Britain, Reddie turned to a circle of evangelical acquaintances, and expressed his concern that orthodox Christian theology was in danger of losing its voice in the public sphere.

Indeed, Reddie was not the only person concerned that modern intellectual trends threatened Christianity. In 1864, the two contributors to *Essays and Reviews* who had been convicted of heresy, Rowland Williams and Henry Bristow Wilson, successfully appealed the decision. John William Colenso, who had followed his *Epistle to the Romans* with another work of biblical criticism, was at the same time engaged in a lengthy legal battle against attempts to depose him from his bishopric. With the Church of England in uproar, thousands of clerics, led by Wilberforce, sent a declaration to Convocation, affirming that the Bible was the inspired word of God. Following this assault on liberal theology, on 21 April 1864 a group of chemists followed suit against science, sending their own declaration to Convocation. The 'Students of the natural and physical scientists', as they styled themselves, argued that any conflict between science and religion was only apparent, and hoped to see an extra article of faith to that effect added to the Church of England's doctrine, to which students at Oxford and Cambridge were required to subscribe.[64] Stokes was invited to sign the declaration, but declined, noting that he did not think that there was any possible conflict, apparent or not. Nevertheless, these two declarations demonstrate the fractious relationship between modern intellectual trends and religion in the mid-1860s, and the charged intellectual atmosphere into which the Victoria Institute arrived.

Reddie assembled a cadre of likeminded middle-class gentlemen, principally Anglican clergy and retired military officers, to form a committee. Its members were mostly veterans of other evangelical organizations, and all shared Reddie's deep concern about the range of intellectual threats to Christianity. The committee drew up a constitution, invited Shaftesbury to become their president, and began soliciting funds and membership. After installing himself as honorary secretary, Reddie began using the Victoria Institute's funds to publish thousands of copies of *Scientia Scientarium*, a manifesto in which he combined the Victoria Institute's constitution and mission statement with some of his more esoteric views. Most important in *Scientia Scientiarum* were the seven objects codifying the Victoria Institute's *raisons d'être* (Fig. 10.4). According to the first of these, it was to 'investigate fully and impartially the most important questions of Philosophy and Science, but more especially those that bear upon the great truths revealed in Holy Scripture, with the view of defending these truths against the oppositions of Science, falsely so called'.[65] This echoed Paul's injunction to Timothy that he avoid 'profane and vain babblings, and oppositions of science falsely so called,' and offers an indication of the Victoria Institute's methodology.[66] Officially, it was to examine philosophical and scientific developments that related to religion. Indeed, every issue of the society's journal included a note reminding the reader that, as an organization formed to investigate, 'it must not be held to endorse the various views expressed at its meetings'. Nevertheless, this suggested that the Victoria Institute believed that whatever conclusions they reached through their investigation would coincide with biblical teaching. There were thus two sciences, a 'true' science, which accorded with scripture, and 'science, falsely so-called' or pseudoscience, which questioned it. Labelling certain scientific concepts as pseudoscience undermined the claim that they represented authoritative knowledge. In this way, otherwise plausible arguments could be delegitimized, and even dismissed, by attacking their epistemological foundations rather than their propositions. For example, if evolution could be demonstrated to contradict scripture, which was by definition true, it could be dubbed pseudoscience, and thus illegitimate. Indeed, since the categorization of an opposing argument as pseudoscience implies that one's own argument is scientific, the Victoria Institute need not even necessarily prove the scientific credentials of their own views. The Victoria Institute was thus another venue in which the struggle for cultural and intellectual authority would play out; in a certain sense, it represented an evangelical analogue to the X Club.

Despite promising beginnings, and an enthusiastic response to calls for membership, early meetings of the Victoria Institute saw regular clashes between members. Reddie often

The Victoria Institute,
or
Philosophical Society of Great Britain,

10, ADELPHI TERRACE, STRAND, LONDON, W.C.

OBJECTS.

THIS SOCIETY has been founded for the purpose of promoting the following Objects, which are of the highest importance both to Religion and Science, and such as have not been attempted to be attained by any previously-existing scientific Society, viz. :—

First.—To investigate fully and impartially the most important questions of Philosophy and Science, but more especially those that bear upon the great truths revealed in Holy Scripture, with the view of reconciling any apparent discrepancies between Christianity and Science.

Second.—To associate men of Science and authors who have already been engaged in such investigations, and all others who may be interested in them, in order to strengthen their efforts by association, and by bringing together the results of such labours, after full discussion, in the printed Transactions of an Institution ; to give greater force and influence to proofs and arguments which might be little known, or even disregarded, if put forward merely by individuals.

Third.—To consider the mutual bearings of the various scientific conclusions arrived at in the several distinct branches into which Science is now divided, in order to get rid of contradictions and conflicting hypotheses, and thus promote the real advancement of true Science ; and to examine and discuss all supposed scientific results with reference to final causes, and the more comprehensive and fundamental principles of Philosophy proper, based upon faith in the existence of one Eternal God, who in His wisdom created all things very good.

Fourth.—To publish Papers read before the Society in furtherance of the above objects, along with full reports of the discussions thereon, in the form of a Journal, or as the Transactions of the Institute.

Fifth.—When subjects have been fully discussed, to make the results known by means of Lectures of a more popular kind, and to publish such Lectures.

Sixth.—To publish English translations of important foreign works of real scientific and philosophical value, especially those bearing upon the relation between the Scriptures and Science ; and to co-operate with other philosophical societies at home and abroad, which are now or may hereafter be formed, in the interest of Scriptural truth and of real Science, and generally in furtherance of the objects of this Society.

Seventh.—To found a Library and Reading Rooms for the use of the Members of the Institute, combining the principal advantages of a Literary Club.

Fig. 10.4 The objects of the Victoria Institute, an organization that investigated the relationship between science and religion, and of which Stokes was president from 1886 to 1903.

abused his position as honorary secretary to interrupt lectures with his own thoughts, and to redact and edit papers to his own liking before they were published in the society's journal. Indeed, Reddie's pugnacious approach led two chemists, John Hall Gladstone and George Warington, to abandon the Victoria Institute as a lost cause; they would later find the CES a more congenial environment. With membership declining every year after 1867, it seemed that Reddie, who had almost single-handedly created the Victoria Institute in just a few years, would just as quickly destroy it, leaving it collateral damage in his war against Victorian Britain's scientific establishment. However, Reddie died in 1871, after a short illness, and was replaced as honorary secretary by Captain Francis W. H. Petrie (1831–1900). Unlike Reddie, Petrie had no interest in lecturing and was instead an incredibly capable administrator. He worked tirelessly to rehabilitate the Victoria Institute's image and began a relentless campaign of letter writing in an attempt to attract prestigious and credible members, and to reassure them that the Victoria Institute was a serious forum for mature reflections on science and faith.

Petrie hoped to convince Victorian Britain's best-known religious scientists to join the Victoria Institute, believing that their presence would lend prestige and scientific legitimacy that had been lacking during the Reddie years. In August 1871, Petrie wrote to Stokes, sending information about membership and commending Stokes for his work with the British Association for the Advancement of Science.[67] Petrie's subtle hint was not enough to entice Stokes into joining, and so he persisted, inviting Stokes to give a lecture in an upcoming season and appealing to his sense of responsibility as one of the 'leaders of the scientific world'.[68] When Stokes again declined, Petrie expressed regret, but asked him to maintain an interest in the Victoria Institute and its work.[69] Stokes was not the only Cambridge physicist that Petrie encouraged to join the Victoria Institute; his colleague James Clerk Maxwell declined an invitation in 1875, explaining that he considered attempts to harmonize science and faith to be a personal matter.[70] Having been rebuffed by Clerk Maxwell, Petrie began again to badger Stokes. Petrie attempted a variety of approaches; framing the Victoria Institute as a forum for the philosophy of science, appealing again to Stokes's sense of duty, sending pamphlets and papers that he felt might interest Stokes, repeatedly inviting him to become an honorary correspondent, even asking his opinion of the suitability other speakers, before Stokes finally relented.[71]

Stokes's first paper, on natural theology, was a success. Finally, the Victoria Institute had a first-rate scientist offering his considered opinion on the relationship between science and his faith. He returned to deliver another well-received paper in 1883, again arguing that there was no opposition between science and religion.[72] When the Earl of Shaftesbury died in 1885, Stokes was considered the natural choice to replace him as the Victoria Institute's president. In contrast to Shaftesbury, who had taken on little more than a ceremonial role presiding over the Institute's annual general meetings, Stokes became involved in the society's administration. This was, again, Stokes's sense of duty manifesting itself. In addition to having served as president of the British Association in 1869, Stokes had also been a secretary to the Royal Society since 1854. Reflecting on Stokes's tendency to become overinvolved in mundane administrative work, his colleague Michael Foster wrote that 'It has been painful to see how his energy has been wasted in this way'.[73] Stokes resigned as secretary in order to take up the Royal Society's presidency from 1885 to 1890, meaning he was concurrently president of both it and the Victoria Institute. During that period, Stokes was also elected as an MP for his university and was a committed member of both the Church Missionary Society and the British and Foreign Bible Society, the latter of which he was also vice-president until becoming its president in 1891.

Despite so many demands on his time, Stokes continued to take an active role in the operation of the Victoria Institute. He frequently corresponded with Petrie, who often wrote asking him to deliver papers and urging him to bring illustrious scientific colleagues along to meetings. After assuming the Victoria Institute's presidency, Stokes tended to deliver the address at its annual general meeting. His ability and his stature within the academic community meant that few members of the Victoria Institute could engage meaningfully with his ideas; polite admiration was the most common reaction. Indeed, it became tradition during the Stokes years that annual addresses would not have any subsequent discussion, presumably to spare the blushes of members ill-equipped to debate Stokes.[74] Stokes's earliest papers had been relatively straightforward works of natural theology, treating the relationship between science and revelation in its broadest context. However, his later addresses, as with his Burnett and Gifford lectures, drew heavily from his areas of expertise in physics. In a lecture on human sight, Stokes recycled his approach from the Gifford lectures, and studiously avoided any direct mention of revelation or teleology. Yet he phrased his lecture in such a way that the inference to design was, as with Paley's watch, unmistakeable. Lord Halsbury, chairing the meeting, noted that 'the first impressions given to us must be, I think, that we are, indeed, "fearfully and wonderfully made"', and that 'the inference drawn by Sir George Stokes is, I think, one that everybody will draw for themselves from his mere statement of facts'.[75]

Stokes did not ignore Petrie's request to bring scientific colleagues to meetings of the Victoria Institute. Indeed, he appears to have taken a leaf out of Petrie's book, badgering William Thomson, now Lord Kelvin, into accepting an honorary position, and eventually delivering a paper, at the society's annual meeting in 1897. Kelvin joined Stokes for several annual addresses, and at the 1898 address both were joined by Lord Lister, the antiseptic pioneer. Lister had succeeded Kelvin, who in turn had succeeded Stokes, as president of the Royal Society. Having three successive presidents of the Royal Society at a lecture was surely a high point in the Victoria Institute's attempt to obtain scientific legitimacy. Indeed, as the three presidents preceding Stokes—T. H. Huxley, William Spottiswoode, and Joseph Dalton Hooker—had all been members of the X Club, it appears that scientists of faith had experienced a revival in fortunes, and retained their cultural and intellectual authority well into the twentieth century.

Conclusion

The Victoria Institute, at its most successful, functioned as a safe haven for religiously inclined scholars, particularly gentleman amateurs and clerics, who found themselves increasingly unwelcome in the academic mainstream. Under Stokes, the perfect embodiment of the religious man of science, it was a confident, prestigious body, where matters of faith and thought could be articulated to an educated, sympathetic audience. Stokes was a figurehead of Victorian Britain's scientific establishment, serving as the president of several esteemed societies. Yet he was also an earnest evangelical who dutifully threw himself into the administration of groups as diverse as the Royal Society and the British and Foreign Bible Society. Through his work, those concerned about the moral implications of eternal punishment could have their fears assuaged. Scientists of faith also had, in Stokes, an impeccable example of how their religion and scientific work might be harmonized in natural theology. In the Victoria Institute over which he presided, they also had a respectable forum in which that natural theology could be discussed sympathetically.

CHAPTER 11

The Scientific Legacy of George Gabriel Stokes

ANDREW FOWLER

In deciding how to write this chapter, one might first consider what a legacy actually is: an instruction of procedure to be followed after a person's life has ended. Of course, scientists do not generally draw up such programmes, and just as the lineage of individuals explodes down the generations,[1] so also do the scientific ideas which can be traced back to a single author. Bearing this in mind, I have chosen to construct this chapter as a series of themes and variations, rather than (for example) a series of symphonically growing movements. In keeping with this intent, I make no effort to do what is probably in any case impossible and provide a complete review of the areas in which Stokes's influence is felt, but rather give a personal tour of a number of subjects where his concerns are still relevant.

A good place to begin is Lord Rayleigh's obituary of Stokes, published in the *Proceedings of the Royal Society*.[2] In this marvellous eulogy, Rayleigh gives an intimate description of Stokes's upbringing in Skreen, Co. Sligo, 'where chickens cost sixpence and eggs were five or six a penny', before proceeding to a discussion of some of Stokes's main achievements. Rayleigh being perhaps more a physicist (we will get to that below), he focuses on Stokes's work in optics, commenting later that a 'complete estimate of Stokes' position in scientific history would need a consideration of his more purely mathematical writings…but this would demand much space and another pen'. If nothing else, this is a good get-out clause. If *Rayleigh* is not up to the task of describing his work on asymptotic expansions, for example, then the present writer need feel no qualms in submitting a comparable sense of incapability, particularly with regard to the work in optics.

Stokes in the Nineteenth Century, and Beyond

To begin, let us place Stokes in his context. In present-day terms, he would be classed as an applied mathematician. But the term does not really make sense in the nineteenth century: the barriers that now separate (physical) applied mathematicians from other types of

Fowler A., *The Scientific Legacy of George Gabriel Stokes*. In: *George Gabriel Stokes: Life, Science and Faith*, Mark McCartney, Andrew Whitaker, and Alastair Wood (Eds): Oxford University Press (2019). © Oxford University Press. DOI: 10.1093/oso/9780198822868.001.0011 mathematicians—analysts of various types (functional, numerical), not to mention actual pure mathematicians—did not really exist in those days. But already, in the pantheon of great Victorian mathematical physicists, including Kelvin, Reynolds, Maxwell and Rayleigh as well as Stokes (three of them born in Ireland, by the way), one can see the evolution of a style of applied mathematics which has come down to us as the British school of physical applied mathematics.

In the early twentieth century, the epitome of this developing style was perhaps enshrined in the work of G. I. Taylor at Cambridge, and his ethos that where possible one should do experimental work. This doctrine was further established with the experimental laboratory in DAMTP, and the Cambridge school of applied mathematics is widely seen, particularly in GFD (geophysical fluid dynamics) circles, as the world-leading institution for the subject; to some, fluid mechanics *is* applied mathematics. But since the mid-twentieth century, further strings have been added to the bow. It is perhaps the case that this coincides with the development since the 1950s of nonlinear techniques of perturbation theory, as well as efficient methods of direct numerical computation, but it is also due to the spread of the application of mathematics to an ever-increasing number of fields: obviously, finance, biology, industry; but also many others: materials science, geoscience, physical chemistry, zoology, to name but a few.

And if we go back, we see the embryo of this development in the work of Rayleigh, Kelvin and Stokes, in particular; all three of them experimentalists as well as theoreticians, and all interested, as generally the case then (but not, mostly, now) in a wide variety of subjects. Other chapters in this book have elucidated this in some detail.

It is worth remembering that, apart obviously from the lack of computers, not to mention a few other useful paraphernalia such as cars, Xerox machines, electricity, email, and so on, the analytic techniques available to Stokes and his contemporaries were piecemeal and spasmodic. Boundary layers do not appear until Prandtl produces them in the 1900s; turbulence is not made explicit until Reynolds's experiments in the 1880s; nevertheless, Stokes and his contemporaries do wondrous things with the tools at their command, often using a superior form of intuition which almost defies reason.[3] A consequence of this is that the prose style of Stokes's papers (as also those of his contemporaries) is verbose; he cannot actually provide explicit approximations, other than via Taylor series, but he must explain verbally the intricacies of his thinking.

The Navier–Stokes Equations

Every student who has taken a fluid mechanics course will be familiar with the derivation of the properties of the stress tensor $\sigma = \sigma_{ij}$, both discussed by Stokes.[4] First the argument, based on consideration of an infinitesimal tetrahedron of fluid, that the stress tensor is indeed a tensor (which is to say, that under an orthogonal transformation of axes $\mathbf{e}_i$ to $\mathbf{e}'_i = l_{ij}\mathbf{e}_j$ (we use the Einstein summation convention), the stress tensor is indeed a tensor, i.e. $\sigma'_{ij} = l_{ip}l_{jq}\sigma_{pq}$); and then, by using torque balance, that the stress tensor is symmetric.

One can even wheedle out the Newtonian relation for viscosity, by assuming that σ_{ij} is a linear function of the strain rate tensor $\dot{\varepsilon}_{ij} = \frac{1}{2}\left(\frac{\partial u_i}{\partial x_j} + \frac{\partial u_j}{\partial x_i}\right)$. This is done, for example, in

Batchelor's 1967 book,[5] and leads, via some dealings with fourth-order tensors, to the classical Newtonian viscous relation for the stress tensor, and thus the Navier–Stokes equations as we now know them.

The Continuum Approximation

Even before all that, however, there is the continuum approximation. This is at once the most intuitive assumption, but also the most infuriating. It is found in every fluid mechanics text, and is introduced in section 1 of Stokes's major paper on fluid flow.[4] He defines the velocity of a fluid at a point as a local average of the motion of the molecules, and chooses to 'regard it as varying continuously with the coordinates' (we would say, being continuously differentiable). He then proceeds more or less deductively to establish the form of the equations.

One particular legacy of this approach is the programme of rational mechanics, championed in particular by Truesdell,[6] which aims to put continuum mechanics (including fluid mechanics and solid mechanics) on an axiomatic footing: there are postulates and laws, and the science is built somewhat independently of pragmatic issues. This approach to continuum mechanics has particularly found application in solid mechanics, where the formulation of appropriate constitutive laws for finite amplitude deformation is a complex matter. The same is true for viscoelastic fluids, for the same reason: constitutive laws for the relation between stress, strain, and rate of strain typically involve material derivatives of the stress tensor, but the usual definition of the material derivative does not yield a representation which is *frame-indifferent*. This key property thus becomes one of the pillars on which a successful theory must be built. And then, there is no unique adjustment of the material derivative of the stress, and so the theories which are developed do become more axiomatic in nature.

It is in the character of the subject, as well as the character of its exponents, which tends to follow the heritage of nineteenth-century German science, that this approach to continuum mechanics is long on formality and proof, and shorter on actual practical results, and its influence has disseminated into a wide number of different areas. As examples, we mention the approach taken by Hutter in his treatise on glaciology,[7] Bear and Bachmat in their book on transport in porous media,[8] Hills, Loper and Roberts in their study of solidification in binary alloys,[9] and Drew and Wood in modelling two-phase flow.[10]

The modelling of two-phase flow, in particular, has been a spectacular example of the kind of formal theory-building which is involved in this axiomatic approach. The reason for this is that, just as fluid mechanics aspires to create describing equations for fields such as density, velocity and temperature, which are not the primary describing constituents of the flow (these are the molecular motions), so in two-phase flow, such as that of air–water mixtures, or foams, slurries, suspensions, fluidized beds and the like, one wants to write equations for idealized quantities such as density and velocity, which equivalently are taken to be means. An axiomatic approach such as *mixture theory* (as used by Hills et al.[9] for example) posits the describing variables, a number of conservation laws, very general types of constitutive laws, and then focuses on frame indifference, entropy inequalities, and eventually other mathematical issues such as existence and uniqueness of the describing model.

At this point, axiomatic continuum mechanics meets its intellectual brother, the field of applied mathematics devoted to functional analysis. It allows continuum mechanics to function as a stand-alone science: the 'application' consists of showing that the model is well posed, and practical applications are consigned to computations. Two-phase flow modelling, in particular, has been obsessively concerned with the issue of writing down equations, often whole books of them, with very little effort to analyse the resulting models, and this remains true today.

It is appropriate to mention the fact that the three-dimensional Navier–Stokes 'regularity problem' has excited much interest since it was formally laid out by the work of Leray.[11] In two dimensions, the vortex stretching term $(\omega . \nabla)\mathbf{u}$ in the vorticity equation vanishes because the vorticity ω is perpendicular to the plane in which the velocity $\mathbf{u}$ lies. The absence of this term allows one to bound the vorticity from above. From this, it is easy to show that smooth solutions exist. In three dimensions, the presence of vortex stretching is a formidable technical obstacle. Leray showed that weak solutions are available in time-averaged form and we are restricted to these if we wish to have explicit estimates for the solutions. A necessary and sufficient condition for the existence and uniqueness of strong (i.e. smooth) solutions requires a bound on the energy dissipation rate $\nu \int_v |\omega|^2 dV$ for arbitrarily long times, which is a result that remains elusive. Instead, Leray showed that the time average of this quantity is bounded up to any positive time T. There have been many attempts to circumvent this problem but all require some form of approximation.

Boltzmann and Beyond

One of the curious features in the axiomatic approach to continuum mechanics (examples are in the book by Hutter[7] and the paper by Hills et al.[9]) is the proposal of an entropy inequality (entropy must increase following the fluid). This proposal stems from the parallel description of matter which is embodied in the laws of thermodynamics, and becomes involved in fluid mechanics when the energy equation is considered; but one does not usually see entropy production considered when writing the equations of fluid mechanics.

Thermodynamics has its mysterious origins in the nineteenth century, and has been a source of bemusement to many applied mathematicians ever since. Truesdell had little patience with this, and provided a similar axiomatic approach.[12] An alternative standpoint was adopted by Maxwell and Boltzmann, who aimed to provide an explicit description of the statistical properties of the molecular motion from first principles, and this led to the renowned Boltzmann equation, which describes the evolution of the distribution of particle velocities in space and time, based on the free motion of particles subject only to inter-particle collisions.[13]

It is a remarkable fact that the Boltzmann equation allows one to *deduce* the three conservation principles of mass, momentum and energy, and includes naturally the prescription of the stress tensor and the heat flux vector, the former precisely in keeping with Stokes's representation. It even allows the deduction (via the Chapman–Enskog expansion[14]) of the form of the linear viscous stress/strain-rate equation, and Fourier's law of heat conduction.

It is even more remarkable that the same equation allows one to prove that in the approach to the equilibrium velocity distribution (the Maxwellian), a quantity—which one can define to be the entropy—increases to its unique maximum (this is a consequence of Boltzmann's H-theorem). The Boltzmann equation thus provides a statistical mechanical underpinning of fluid mechanics and thermodynamics; it even has other ramifications as we discuss below.

Of course the Navier–Stokes equations require boundary conditions. Stokes comes first[15] to the stress conditions at an interface, and for a free surface provides the kinematic boundary condition and the normal stress condition, allowing also for a capillary attraction, i.e. a surface tension term. This latter boundary condition is something of an anomaly in fluid mechanics. One commonly thinks of it as arising through a force balance, with the surface possessing a tension, but is perhaps better to think of it as arising through the thermodynamic constraint of locally minimizing the (Helmholtz) free energy of the system, which itself is a consequence of maximising entropy. The capillary effect arises through the surface energy associated with the interface.

The idea that one replaces a force balance condition with an energy minimization criterion carries a certain philosophical baggage with it. Because surface tension is associated with a force per unit length (of a line drawn in the interface surface), it is not an immediately accessible concept. One starts to think, what actually *is* a force? And then we think, how do we measure a force? One way you measure a force of an object is by extending an elastic spring, but really this measure involves the equation of the work done by the force to the stored elastic energy of the spring. The more one thinks about it, one realizes that force is not really a primary quantity, but rather energy. Equally, pressure is a force per unit area, but also an energy per unit volume. And indeed, it is common in soil mechanics, for example, to think of the driving force for fluid flow (such as the pressure in Darcy's law) as the gradient of a hydraulic potential, comprising gravitational, capillary, osmotic, and perhaps other effects.[16] More generally, material flow is driven down chemical potential gradients. Thus thermodynamics is intimately bound up with fluid dynamics, even if this is not commonly recognized.

The No-Slip Condition

Next, Stokes considers the boundary condition between a fluid and a solid. To us, now, the condition of no slip at a solid wall is a matter of uncritical acceptance. But in the nineteenth century, this was not the case. After some consideration, Stokes takes the view that no slip is the appropriate assumption. But other possibilities had been suggested: notably Navier had much earlier suggested a slip condition in which the slip velocity was proportional to the shear stress,

$$\beta u = \mu \frac{\partial u}{\partial n}, \tag{1}$$

where μ is the viscosity. The quantity μ / β is a length, known as the slip length. At the end of his (edited) two-volume compendium on fluid mechanics, Goldstein provides an illuminating commentary on the nineteenth century debate on the fluid/solid boundary condition.[17] Ultimately, the no-slip condition was accepted on the pragmatic basis that it agreed with experimental results.

A curious variant of this condition occurs in glaciology, in the study of the flow of glaciers and ice sheets, a topic which has risen from obscurity to recent prominence due to its relevance to studies of climate change and sea level rise. Ice is a polycrystalline solid, but nevertheless is able to deform under the action of stress due to various solid-state processes such as dislocation creep or grain boundary creep. For example, dislocations in the crystalline structure are places of weakness which migrate under the action of an applied stress, leading over long timescales to an effective creep of the material as the grains deform. Insofar as ice is a polycrystalline rock, it shares this property with other (silicate) rocks which form the Earth's crust and mantle. Results of deformation of crustal rocks are readily visible in folds in rocks, but rock motion extends throughout the Earth's mantle, to a depth of 3,000 km, and is the cause of the gradual drifting of the continents over time scales of hundreds of millions of years. It is commonly known, for example, that the rise of the Himalayas is due to the collision of India, as it migrated northward, with Asia. Indeed it is also commonly thought that the resulting increase of continental erosion by mildly acidic (H_2CO_3) rain caused the long-term decline in CO_2 which cooled the planet, and in the process formed the Antarctic and Greenland ice sheets. That was not known in Stokes's lifetime, and the formulation of the theory of continental drift by Wegener[18] (whose book was originally published in 1915) and Holmes's later explanation of it in the 1930s as being due to convection[19] also comes later. But the establishment of the motion of ice sheets and continents as being due to solid state creep processes is just another example of the application of the Navier–Stokes equations.

Because ice is a solid it can of course melt, and it is commonly the case that it does so at the base of a glacier. Indeed, some glaciers (known as temperate glaciers) are at the melting point throughout, and thus contain small quantities of liquid water. At the bed of a glacier, this water can form a thin film which lubricates the interface and allows the ice to slide over the bed, thus providing for non-zero slip. However, although the lubricating layer provides for an effectively frictionless contact, the resulting ice flow is resisted by a non-zero basal stress, which is due to the small scale roughness of the bed: a stress is needed to drag the ice over the rough bed.

Much attention has been devoted to providing theoretical expressions for the resulting *sliding law*, but generally they have the form

$$\tau_b = f(u_b, N), \tag{2}$$

where τ_b is the basal shear stress and u_b is the basal velocity. The quantity $N = p_i - p_w$ is the *effective pressure*, a quantity familiar from soil mechanics, and equal to the difference between the ice overburden pressure p_i and the subglacial water pressure p_w, since, apart from the lubricating water film alluded to above, basal water is produced, which flows in its own subglacial hydraulic system, with its own self-determining water pressure (and the determination of this latter quantity is the subject of another whole area of study and current interest). These and other curiosities of ice sheet flow are discussed by Cuffey and Paterson.[20]

Stokes Flow

Ice sheet and mantle flow are two (extreme) examples of *Stokes flow*, in which the inertial acceleration terms are negligible. In modern terms, this is because the Reynolds number is small. Stokes introduces the slow flow approximation in a paper which is largely concerned with the effect of viscous drag on pendulums,[21] and the early part of the paper is taken up with the effect of viscosity on an oscillating sphere or cylinder. Later, he derives the classic formula $6\pi\mu U a$ for the drag on a sphere, and also shows that the solution method cannot be made to work for flow past a cylinder. For the cylinder he could not make any headway, not surprisingly as the full resolution of the matter had to await the invention of matched asymptotic expansions a century later. Oseen's approximation (introduced after Stokes's death) provides a solution which works, but it is not until the paper by Proudman and Pearson that the matter is fully resolved.[22]

The construction of matched asymptotic expansions at that time saw efforts to put the procedure on a secure footing, with the work of Kaplun and Lagerstrom. Kaplun's early death brought this effort to a premature close, and the mystery of why asymptotic expansions are so effective remains, even though they are commonly used by applied mathematicians: they share this elusiveness with the use of anaesthesia in surgery. Somewhat later, a fashion arose to carry asymptotic expansions to very high order, thus obtaining accurate answers to small Reynolds number expansions for values of Re as large as 100;[23] but such procedures are essentially limited to linear problems, although the calculation to high order of the (convergent) expansion for Stokes waves (see below) has also been done.

The Not-So-Simple Pendulum[24]

As a postscript, we might add that the weakly damped pendulum, in the form

$$l\ddot{\theta} + k\dot{\theta} + g\sin\theta = 0, \tag{3}$$

is commonly used as a pedagogical exercise to discuss perturbation methods, including the Poincaré–Lindstedt method, the method of multiple scales, and the method of Kuzmak–Luke. Implicitly it is based on the idea that the viscous damping term is given by Stokes's law. But what actually *is* a realistic value for k, and is it small? Assuming Stokes drag and weak damping, the maximum velocity of a pendulum with angular amplitude θ_M is $v = \theta_M\sqrt{gl}$, so that for a spherical bob of radius a, the Reynolds number is

$$Re = \frac{\rho_a\sqrt{gl}\,\theta_M a}{\mu_a}, \tag{4}$$

where ρ_a is air density and μ_a is air viscosity. Further, for Stokes drag we have

$$k = \frac{9\pi\mu_a l}{2\rho_s a^2}, \tag{5}$$

where ρ_s is the bob density, and if (3) is made dimensionless as

$$\ddot{\theta} + \beta\dot{\theta} + \sin\theta = 0, \tag{6}$$

then the definition of β is

$$\beta = \frac{9\pi\mu_a}{2\rho_s a^2}\sqrt{\frac{l}{g}}. \tag{7}$$

Using values for air, and also $\rho_s = 3\times10^3$ kg m^{-3}, $a = 4$ cm, $l = 1$ m, $\theta_M = 0.1$, we find $\beta \sim 1.7\times10^{-5}$, definitely small, but $Re \sim 835$! The weakly damped pendulum is indeed weakly damped, but the damping should be more nearly quadratic!

The Hele-Shaw Cell

In a report to the British Association, Stokes derives Laplace's equation for the stream function, or equivalently the pressure, in a Hele-Shaw cell, using the combined assumptions of slow flow and a narrow gap.[25] His concern was with the streamlines of flow past an obstacle, but another experiment which has been of great interest is that in which one fluid displaces another in a Hele-Shaw cell. Because the governing equations are also those of flow in a porous medium, this experiment is also of great interest to the oil industry, since oil wells commonly function by the injection of water to push out the oil in the porous sedimentary basin. When a less viscous fluid displaces a more viscous one (such as water displacing oil, or air displacing water) a uniformly advancing front is unstable (this is the *Saffman–Taylor* instability). In a relatively

Fig. 11.1 Digitized image of the formation of a dendritic fingering pattern in a Hele-Shaw cell air (black) injected centrally into oil (white). From Mathiesen *et al.*, 'The Universality Class of Diffusion-Limited Aggregation and Viscous Fingering', *Europhysics Letters* 76 (2006), figure 1, reproduced with permission.

narrow channel, the interface evolves to form a long finger-like tendril, whereas for a more extended interface a number of fingers are produced; the phenomenon is thus referred to as viscous fingering.[26] Wave number selection is enabled by surface tension, which acts as a stabilising effect at high wave numbers. As can be seen in Figure 11.1, the fingers formed from a central injection source break up and form a dendritic pattern as the interface evolves.

There are a number of interesting questions which this phenomenon raises. Perhaps the first is what width a single finger in a narrow channel will have. Saffman and Taylor found an exact solution for a steadily symmetric travelling finger of the less viscous fluid,[27] which took the form of a one parameter family, the parameter being the ratio λ of the finger to the channel width. In this solution, the surface tension was taken to be zero. In experiments, λ took a value close to one half. The question then arises, what selects this value?

This kind of question is one that permeates pattern selection problems. Perhaps the most well-known example is the travelling wave solution of the Fisher equation, which exists for all wave speeds $c \geq 2$. In practice, $c = 2$ is selected, since solutions with $c > 2$ require very specific asymptotic decay rates at large distances. A similar notion occurs in the Ivantsov dendrite solution (which is discussed further below), where an exact solution for a steadily growing needle crystal from a supercooled liquid has a velocity of propagation related to the tip radius of curvature, but neither is determined uniquely.

A different kind of example occurs in the analysis of roll waves, such as those seen in Figure 11.2. A classic analysis, originally due to Dressler,[28] shows that the waves can be described as a one parameter family of periodic travelling shock waves,[29] where the wavelength can be taken to be the free parameter. What determines the wavelength? It seems it should be associated with a second upstream boundary condition, but the answer is not known. Other problems where wavelength selection is an issue are in thermal convection in a large pan of fluid heated from below, whose lateral extent is much greater than the depth. Sometimes ideas of 'optimum transport' (for example) are used to define a constraint on such systems, but as Howard comments,

> One sometimes gets the impression, especially from reading general books about physics, that many people regard a variational formulation as an essential component of a true and deep understanding of the fundamental character of almost anything. This idealistic but rather narrow-minded attitude, a bit akin to the once-popular view that planetary orbits must obviously be circles, probably limits scientific progress somewhat.[30]

Fig. 11.2 Roll waves propagating down a road in Clonlara, Co. Clare.

There is a fundamental philosophical issue with the mechanism of wavelength selection; a particularly fraught example is the use of reaction-diffusion theory in morphogenesis to explain such things as teeth or finger formation. While there are occasional exceptions, we generally have five digits on hands and feet; how can a pattern selection theory be so robust?

Moving Contact Lines

The front of a glacier is one example of a place where the viscous ice, the much less viscous air, and the solid substrate, all meet. The edge of a raindrop on a windscreen is another. More generally, the line on a surface which separates two fluids in contact is called the *contact line*, and is illustrated in Figure 11.3. The Navier–Stokes equations run into a severe difficulty at a

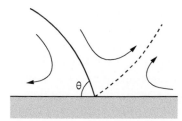

Fig. 11.3 The contact line at the interface of two fluids and a solid surface.

contact line, as was pointed out by Dussan V. and Davis.[31] It is not the equations themselves which are at fault, but the assumption of a no-slip condition. The difficulty is a purely kinematic one, and can be understood with reference to Figure 11.3. We suppose the contact line moves as a droplet advances, but we take a moving coordinate system, such that the contact line is fixed in this moving frame. As indicated in Figure 11.3, the resultant velocities (assuming no slip) of the advancing fluid (to the left) and the displaced fluid (to the right) involve rolling motions. For the advancing fluid, there is not really a problem; but for the displaced fluid, there is a dividing streamline (dashed in the figure) and, as shown, the fluid approaches the contact line along both interfaces, so that it must be ejected along the dashed streamline. Consequently there is a discontinuity in the displaced fluid velocity at the contact line. In more detail, the local flow near the contact line is governed by Stokes's equation for slow flow, and in two dimensions and polar coordinates (r, ϕ) with ϕ measured from the dotted streamline in Figure 11.3, the stream function in the displaced fluid is

$$\psi = \frac{U(\phi \cos \phi \sin \gamma - \gamma \cos \gamma \sin \phi)}{\gamma - \sin \gamma \cos \gamma}, \tag{8}$$

where $2\gamma = \pi - \theta$. The shear stress in the wall is then found to be

$$\tau_{r\phi} = \frac{2\mu U}{r} \left(\frac{\sin \gamma \sin \frac{1}{2}\gamma}{\gamma - \sin \gamma \cos \gamma} \right), \tag{9}$$

where μ is viscosity, and this becomes infinite as $r \to 0$. This in itself is not a problem, it is the fact that the force on the wall is also infinite (because the singularity is not integrable) that renders the solution unphysical.

Two resolutions of this conundrum have been proposed; the first is to resort to a Navier slip condition, as in (1). And actually, there is some theoretical justification for this, since the boundary condition which one derives for the Boltzmann equation does allow for such slip. The other is to avoid the problem by allowing a 'precursor' film to exist in front of the contact line, so that in effect there is no contact line at all. Such a film is associated with thermodynamic effects of 'pre-wetting'. Nevertheless, finite contact angles are commonly measured, and these panaceas are not that satisfactory.

The Highest Water Wave

Stokes treated nonlinear water waves by means of expansions in powers of the amplitude,[32,33] incidentally inventing the Poincaré–Lindstedt method. He was concerned with travelling waves, and famously ascertained that the highest wave would have a crest angle of 120°. This particular theme was continued with some vigour in the twentieth century, particularly with the work of Michael Longuet-Higgins,[34] but the whole issue of nonlinear waves in general became big business, at least partly on account of John Scott Russell's observation of the great wave of translation. The solitary wave led eventually to the formulation of the Korteweg–de Vries equation, the discovery of the soliton, and the subsequent explosion of research in the 1960s and beyond on integrable nonlinear partial differential equations, which continues, sometimes in abstruse forms, to this day.

One circuitous route to some of these nonlinear *amplitude equations* lies through the development of multiple scales techniques. These became very fashionable in the 1970s, following the equally popular development of applications of Hopf bifurcation, which itself arose from developments of the perturbation methods introduced by Poincaré, for example (except actually by Stokes, as mentioned above). Indeed, it is now well understood that the Korteweg–de Vries equation can formally be derived by an asymptotic expansion procedure describing long waves in shallow water, and other such equations can be derived in a similar fashion; for example the non-linear Schrödinger equation describes the evolution of a weakly nonlinear wave amplitude in a dispersive system.

It is worth illustrating how such equations arise from general principles. The presentation here follows that of Newell.[35] Suppose we have a nonlinear evolution equation for a variable $u(x,t)$ which can be written in the form

$$L\left(\frac{\partial}{\partial t}, \frac{\partial}{\partial x}\right)u = N_2(u,u) + N_3(u) + \dots, \qquad (10)$$

where L is a linear operator, N_2 is quadratically nonlinear (and symmetric and bilinear, i.e. linear in each argument), N_3 is cubic, etc. We have $Le^{i(kx-\omega t)} = L(-i\omega, ik)e^{i(kx-\omega t)}$, and thus travelling waves of wavenumber k and frequency ω exist if

$$L(-i\omega, ik) = 0, \qquad (11)$$

and this defines the *dispersion relation* for frequency as a function of wavenumber, $\omega = \omega(k)$. Differentiating this with respect to k, we find successively

$$L_2 = \omega' L_1,$$
$$\frac{1}{2}\omega'^2 L_{11} - \omega' L_{12} + \frac{1}{2}L_{22} = -\frac{1}{2}i\omega'' L_1, \qquad (12)$$

where $L = L(\xi, \eta)$, $L_1 = \partial L / \partial \xi$, $L_{11} = \partial^2 L / \partial \xi^2$, etc. Next, for small amplitude waves, we suppose $\varepsilon << 1$, and write

$$u \sim \varepsilon u_0 + \varepsilon^2 u_1 + \dots, \qquad (13)$$

and we anticipate the use of multiple scales by defining

$$X = \varepsilon x, \quad T = \varepsilon t, \quad \tau = \varepsilon^2 t, \qquad (14)$$

and taking the functions $u_k = u_k(x,t,X,T,\tau)$. Equating powers of ε in the resulting expansion, we find at leading order $Lu_0 = 0$, thus $u_0 = A(X,T,\tau)e^{i(kx-\omega t)} + (cc)$ (cc denotes the complex conjugate), and at second order

$$Lu_1 = -\left(L_1\frac{\partial}{\partial T} + L_2\frac{\partial}{\partial X}\right)u_0 + N_2(u_0,u_0). \qquad (15)$$

Secular terms on the right are those proportional to $e^{i(kx-\omega t)}$, and there are none in the quadratic term, which contains terms proportional to $A^2 e^{2i(kx-\omega t)}$ and $|A|^2$; in some cases the latter can be secular, if $u = 1$ is in the null space of L; we will assume not for simplicity. The first term on the right of (15) is secular, and in view of (12), can be written as $(A_T + \omega' A_X)L_1 e^{i(kx-\omega t)}$, and thus we require

$$A = A(\xi, \tau), \quad \xi = X - \omega'T: \tag{16}$$

$\omega'(k)$ is the *group velocity*. The solution for u_2 then consists of a particular solution consisting of parts proportional to $A^2 e^{2i(kx-\omega t)}$ (and complex conjugate) and $|A|^2$ and a free wave which can be absorbed into u_0.

At second order, we have

$$Lu_2 = \left(L_1 \frac{\partial}{\partial T} + L_2 \frac{\partial}{\partial X} \right) u_1$$
$$- \left(L_1 \frac{\partial}{\partial \tau} + \frac{1}{2} L_{11} \frac{\partial^2}{\partial T^2} + L_{12} \frac{\partial^2}{\partial X \partial T} + \frac{1}{2} L_{22} \frac{\partial^2}{\partial X^2} \right) u_0 + 2N_2(u_0, u_1) + N_3(u_0). \tag{17}$$

Ignoring the free wave for u_1 which we absorb into u_0, there are no secular terms in the first term on the right. Using (12) and (16), the second term can be written as $-\left(A_\tau - \frac{1}{2} i\omega'' A_{\xi\xi} \right) L_1 e^{i(kx-\omega t)}$, which is secular, and the two nonlinear terms produce secular terms which we can write in the form $i\beta |A|^2 AL_1 e^{i(kx-\omega t)}$. Removal of the secular terms yields the amplitude equation

$$iA_\tau + \frac{1}{2} \omega'' A_{\xi\xi} + \beta |A|^2 A = 0, \tag{18}$$

and this is the nonlinear Schrödinger (NLS) equation; for our assumed dispersive system β is real.

An interesting application of this theory takes us back to Stokes and his monochromatic waves. First note that the Stokes wave is the exact solution $A = A_0 e^{i\beta A_0^2 \tau}$ of (18). It turns out that this monochromatic wave is unstable if $\beta\omega'' > 0$, and for water waves in water of depth h, this is the case for deep water, specifically if $kh > 1.36$; this is the famous Benjamin–Feir modulational instability, and the integrability properties of the NLS equation show that the wave train then breaks up into a collection of (envelope) solitons.

It has been suggested that the modulational instability may be a causative factor in the occurrence of *rogue waves* in the ocean. These waves, occasionally observed by mariners (but usually without precise measurement) occur in stormy seas when the *significant wave height* (essentially a statistical measure of mean wave height) is of the order of ten metres. Every so often, an enormous wave of height perhaps twenty metres will suddenly appear.

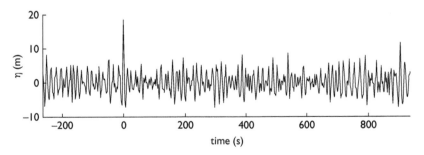

Fig. 11.4 Time series of the Draupner wave recorded in the North Sea on 1 January 1995. Figure reproduced from Adcock *et al.*, 'Did the Draupner wave occur in a crossing sea?', *Proceedings of the. Royal Society of. London* A 467 (2011), pp. 3004–21, with permission. I have resisted the temptation to include a more dramatic representation.

There are various mechanisms which can produce such exceptional wave heights, and they can be used to generate them also in wave tanks; one such theory relies on the modulational instability theory, but the issue is one of current debate. For a recent discussion, see Fedele et al.[36]

Arising out of the Mist

It is in his letter to the editor of the *Acta Mathematica* that Stokes makes his famous remarks on asymptotic series.[37] The letter summarizes his earlier papers on the asymptotic expansion of certain functions,[38, 39, 40] including what we now call the Airy function. The issue is that if an analytic function $f(z)$ has an asymptotic expansion

$$f \sim a_0 + \frac{a_1}{z} + \dots \tag{19}$$

as $z \to \infty$ in the complex plane which is uniformly valid for all $\arg z$, then the function must in fact be analytic at ∞; if the function is singular at ∞, then asymptotic expansions are typically only valid in finite sectors of the complex plane: this is *Stokes's phenomenon*. When computing an asymptotic expansion for the Airy function, defined for example by the integral

$$\mathrm{Ai}(z) = \frac{1}{2\pi} \int_{-\infty}^{\infty} e^{i\left(zs + \frac{1}{3}s^3\right)} ds = \frac{1}{2\pi i} \int_{C:[\infty e^{-2\pi i/3}, \infty e^{2\pi i/3}]} e^{zt - \frac{1}{3}t^3} dt, \tag{20}$$

as $z \to \infty$, one uses the method of steepest descents to deform the contour through an appropriate saddle (this is the one at $t = -\sqrt{z}$ if $\arg z = 0$). As $\arg z$ changes, the deformed contour changes also, but can still pass directly from $\infty e^{-2\pi/3}$ to $\infty e^{2\pi/3}$. This changes for $\arg z > \frac{2}{3}\pi$, where the steepest descent contour through $-\sqrt{z}$ now goes to $+\infty$, and so passage to $\infty e^{2\pi/3}$ requires an addition of a further steepest descent contour from $+\infty$ to $\infty e^{2\pi/3}$ through the other saddle at $t = \sqrt{z}$. The situation is illustrated by Hinch,[41] although one has to draw in one's own steepest descent contours. This introduces a sudden second (sub-dominant) expansion at this *Stokes line*, and is perplexing as it introduces an apparent discontinuity in the asymptotic expansions despite the analyticity of the function. Specifically, the asymptotic expansions obtained take the form

$$\mathrm{Ai}(z) \sim \frac{z^{-1/4}}{2\sqrt{\pi}} \left[e^{-\zeta} \sum_0^{\infty} \frac{(-1)^k c_k}{\zeta^k} \left\{ \pm i e^{\zeta} \sum_0^{\infty} \frac{c_k}{\zeta^k} \right\} \right], \quad \zeta = \frac{2}{3} z^{3/2}, \tag{21}$$

where[42]

$$c_k = \frac{\Gamma\left(k + \frac{5}{6}\right)\Gamma\left(k + \frac{1}{6}\right)}{2^{k+1}\pi k!}, \tag{22}$$

and the expression in curly brackets arises from the second steepest descent contour from ∞ to $\infty e^{2\pi i/3}$, and is only present when $|\arg z| > \frac{2}{3}\pi$ (the upper sign applies for $\arg z > \frac{2}{3}\pi$, and the lower for $\arg z < -\frac{2}{3}\pi$). This is where, in his letter to the editor of *Acta*, Stokes's famous description occurs: 'As θ passes through the critical value, the inferior term enters as it were into a mist, is hidden for a little from view, and comes out with its coefficient changed.'

Exponential Asymptotics

The resolution of this paradox, the apparent discontinuity in the describing asymptotic expansions, is associated with the emergence of the subject of exponential asymptotics, which, roughly speaking, deals with the provenance of exponentially small terms in asymptotic expansions that in the normal way of things can be formally ignored to all algebraic orders (thus a term e^{-z} can be neglected to all orders in (19)). Much of this recent work is inspired by the book by Dingle,[43] and has been prominently expounded by Berry.[44, 45] The 1988 paper in particular is a very elegant and historically informed summary of the basics of what Berry calls 'hyperasymptotics', also known as 'asymptotics beyond all orders'.

It is worth recounting the ideas of the method. We know that in asymptotic expansions of functions defined by integrals such as that in (20), the terms first decrease and then increase (this is why the series is asymptotic and not convergent). The asymptotic expansion for the Airy function (based on the contour C in (20)) is given by the expression in (21) with the term in curly brackets omitted. Using Stirling's formula, we can show that for large k, $(k-r)! \sim k^{k-r} e^{-k} \sqrt{2\pi k}$, and thus

$$c_k \sim \frac{(k-1)!}{2^{k+1}\pi},\tag{23}$$

and the k-th term in the expansion in (21) is a minimum when $k \approx 2|\zeta|$ (and is exponentially small).

The idea of the method is then to replace the asymptotic terms beyond this minimum term with their asymptotic form, and then 're-form' the diverging series using Borel summation. In practice, we take $K \approx |2\zeta|$, and then

$$\begin{aligned}
\mathrm{Ai}(z) &\sim \frac{1}{2\sqrt{\pi}} z^{-1/4} e^{-\zeta} \left[\sum_0^K \frac{(-1)^k c_k}{\zeta^k} + \sum_{K+1}^\infty \frac{(-1)^k c_k}{\zeta^k} \right] \\
&\sim \frac{1}{2\sqrt{\pi}} z^{-1/4} \left[e^{-\zeta} \sum_0^K \frac{(-1)^k c_k}{\zeta^k} + i e^{\zeta} \left\{ \frac{e^{-2\zeta}}{2\pi i} \sum_{K+1}^\infty \frac{(k-1)!}{(-2\zeta)^k} \right\} \right].
\end{aligned}\tag{24}$$

We write $F = -2\zeta$ and define the Stokes multiplier to be

$$S(F) \sim \frac{e^F}{2\pi i} \sum_{K+1}^\infty \frac{(k-1)!}{F^k}.\tag{25}$$

Noting that the Stokes lines $\arg z = \pm\frac{2}{3}\pi$ correspond to $\arg\zeta = \pm\pi$, we see that the discontinuity in the asymptotic expansions is caused by the apparent discontinuity of S across the positive F axis, which 'switches on' the sub-dominant term in curly brackets in (24).

To see this explicitly, we use the integral formula for the factorial to reconstitute $S(F)$ as the Borel sum

$$S = \frac{e^F}{2\pi i} \int_0^\infty e^{-t} \sum_K^\infty \left(\frac{t}{F}\right)^j \frac{dt}{F} = \frac{1}{2\pi i} \int_0^\infty \left(\frac{t}{F}\right)^K \frac{e^{F-t} dt}{F-t}.\tag{26}$$

Presuming that we increase $\arg z$ through $\frac{2}{3}\pi$, and thus $\arg F$ through zero, it is appropriate to define $F = F^* e^{i\alpha}$ and to examine the approximate form of (26) when F^* is large and α is small. We then have $F \approx F^*(1 + i\alpha - \frac{1}{2}\alpha^2 ...)$, and in particular

$$\sigma = \frac{ImF}{\sqrt{2ReF}} \approx \sqrt{\frac{F^*}{2}}\alpha. \tag{27}$$

We put $t = F^*\left(1+\sqrt{\frac{2}{F^*}}s\right)$, $K = F^*$ in the integral and expand for large F^* and small $\alpha \sim 1/\sqrt{F^*}$

(thus $\sigma \sim O(1)$), to find, at leading order,

$$S \approx \frac{e^{-\sigma^2}}{2\pi i}\int_{-\infty}^{\infty}\frac{e^{-s^2}\,ds}{i\sigma - s}. \tag{28}$$

This is a known function, the plasma dispersion function,[43] and is related to the error function.[42] If $\sigma < 0$, then

$$S \approx S_-(\sigma) = \frac{1}{2}[1 + \mathrm{erf}\,\sigma], \tag{29}$$

whereas for $\sigma > 0$, $S = S_+ = S_- - 1$. However, when $\sigma > 0$, the term in curly brackets in (21) must be included, and this has the effect of adding one to S. It follows that (29) provides a uniformly accurate description of the jump across the Stokes line.

Dendrite Growth

There are a number of applications and analyses of exponential asymptotics discussed in the book edited by Segur, Tanveer and Levine.[46] Among these is the problem of dendritic crystal growth. In the explosion of applications of applied mathematics that occurred in the 1960s and thereafter, interest in crystal growth grew, at least partly because of its importance in alloy solidification in materials science. When a casting is made, the liquid mixture (such as lead and tin) is solidified inwards from the exterior wall, and in so doing, a planar solidification front is in practice never found. It is unstable due to *constitutional supercooling*, and the interface becomes at first wavy, but then rapidly columnar, and finally *dendritic*. These dendrites form as growing spears, with secondary and even tertiary branches. The same phenomenon occurs in supercooled pure liquids, and was extensively studied experimentally by Marty Glicksman.[47] An example is shown in Figure 11.5. Particularly in the right hand figure, we can see that initially growth is almost parabolic, and indeed there is a famous exact solution of the relevant free boundary diffusion problem due to Ivantsov, but the growth velocity was arbitrary. Later the idea arose that inclusion of (small) surface tension might provide the key to fixing the velocity.

The problem is complicated, and in their way, physicists have introduced a simpler 'toy' model, which aims to inform the issue. The 'geometric' model of crystal growth is this:

$$\varepsilon^2\theta''' + \theta' = \cos\theta, \tag{30}$$

where $\theta' = d\theta/ds$, θ is the tangent angle of the interface (oriented as in Figure 11.5) and s is arc length measured anti-clockwise from the dendrite tip. The boundary conditions for the 'needle' crystal are that

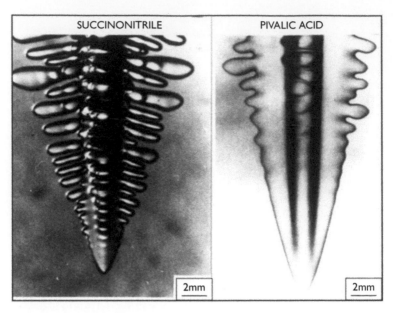

Fig. 11.5 One of Marty Glicksman's iconic images of dendrite growth; this is from Martin E. Glicksman, 'Mechanism of Dendritic Branching', *Metallurgical and Materials Transactions* A 43, pp. 391–404 (2012), used with permission.

$$\theta \to \pm\frac{1}{2}\pi \text{ as } s \to \pm\infty, \tag{31}$$

and the small parameter ε represents the effects of surface tension. The question to be asked here is, can we select ε to ensure that a solution to this boundary problem exists?

First, as is intuitive, any such solution has to be odd, and thus we would require $\theta''(0) = 0$. It turns out that if we select a solution with $\theta(-\infty) = -\frac{1}{2}\pi$, then for small ε, $\theta''(0) \neq 0$: it is in fact exponentially small. This is related to the fact that as $\theta \to \pm\frac{1}{2}\pi$, the equation allows free oscillations $\propto e^{\pm is/\varepsilon}$, and exponentially small such terms are excited as s increases. Equivalently, if we start at $s = 0$ seeking an odd solution in which $\theta(0) = \theta''(0) = 0$, then as $s \to \pm\infty$, exponentially small oscillations are excited. It is *most* tempting to associate these with the secondary dendrites which can be seen in Figure 11.5, though I do not know whether that association has been made.

The Slow Manifold

Because the preceding example is nonlinear, it is difficult to provide detailed analytic results, though with some effort it is possible. A much simpler and linear example occurs in a problem associated with meteorology. As is implied by our dendrite discussion, exponentially small terms beyond asymptotic expansions can be found directly from perturbation expansions of ordinary differential equations. There is a close relation between such asymptotic expansions and the presence of singularities of analytic functions in the complex plane, which somehow informs the approximation process. A particularly simple example arises in the analysis of the 'slow manifold' of solutions in a model of atmospheric dynamics.[48] The problem to be considered is the forced ordinary differential equation

$$\ddot{\xi} + \xi = R(\tau), \quad \tau = \varepsilon t, \quad \varepsilon = 1, \tag{32}$$

where the overdots indicate differentiation with respect to the fast time t. We suppose $R(\pm\infty) = 0$, and that $\xi \to 0$ as $t \to -\infty$.

The asymptotic solution is readily constructed, and takes the form

$$\xi = (1 + \varepsilon^2 D^2)^{-1} R \sim \sum_0^\infty (-\varepsilon^2)^n R^{(2n)}, \tag{33}$$

where $D = d / d\tau$, and Borel summation of this yields

$$\xi \sim \mathrm{Re} \int_0^\infty R(\tau - i\varepsilon s) e^{-s} ds. \tag{34}$$

Thus the asymptotic solution is slowly varying; but fast oscillations $\propto e^{\pm it}$ are free solutions of (32), and it turns out that such components occur in the exact solution, although of exponentially small amplitude.

The exact solution of (32) is simply

$$\xi = \frac{1}{\varepsilon} \mathrm{Im} \int_{-\infty}^\tau R(U) \exp\left[\frac{i(\tau - U)}{\varepsilon} \right] dU, \tag{35}$$

and this can be transformed to

$$\xi = \mathrm{Re} \int_0^{-i\infty} R(\tau - i\varepsilon s) e^{-s} ds. \tag{36}$$

Simply by transforming the contour, we see that the exact solution can be written as

$$\xi = \mathrm{Re}\left[\int_0^\infty R(\tau - i\varepsilon s) e^{-s} ds + \frac{2\pi}{\varepsilon} \sum_j R_j e^{-s_j/\varepsilon} e^{i(t-t_j)} \right], \tag{47}$$

where $R(\tau)$ has poles at $\tau_j - is_j$, $\tau_j = \varepsilon t_j$, and the summation is over all the poles such that $t_j < t$. Thus in a magical way, the solution picks up exponentially small fast oscillations as time cruises past the complex singularities of R in the landscape.

Stokes and Diffraction

Stokes, like many of his contemporaries, was very much involved with ongoing investigations of the natural world, and in particular the basic nature of physics, subjects which we now take for granted. Particularly, he was very interested in optics, the understanding of light, how it propagates, and in such phenomena as interference fringes or diffraction patterns. It is important to realize the historical context of his investigations. In his long article on diffraction,[49] for example, concerned with diffraction patterns formed by light passing through a small aperture, the wave-particle duality is still an issue, Maxwell's equations do not exist, but the ether does! His paper is largely concerned with solutions of the wave equation.

Intuitively, we understand how light is transmitted: it propagates undisturbed as *rays* and is reflected or refracted at surfaces. If we shine a unidirectional beam of light at a cylinder of radius a, for example, as shown in Figure 11.6, then the rays are straight lines which are reflected at the surface. The same processes of reflection and diffraction also occur for other

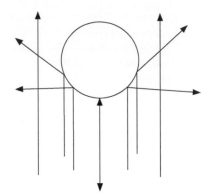

Fig 11.6 Light rays reflected from a cylinder.

kinds of wave propagation, for example for water waves or sound waves. For an incident two-dimensional monochromatic wave of frequency ω and intensity Φ, such that Φ satisfies the wave equation $\Phi_{tt} = c^2 \nabla^2 \Phi$, we can write $\Phi = \phi(x, y)e^{-i\omega t}$, and then ϕ satisfies the Helmholtz equation

$$\nabla^2 \phi + k^2 \phi = 0, \tag{48}$$

where $k = \omega / c$ is the wavenumber, c being the wave speed (for light waves this is the speed of light). The incident wave is represented by $\phi = e^{iky}$, and then suitable boundary conditions for the situation shown in Figure 11.6 are to take

$$\frac{\partial \phi}{\partial r} = 0 \quad \text{on} \quad r = a,$$

$$\sqrt{r}\left(\frac{\partial}{\partial r} - ik\right)(\phi - e^{iky}) \to 0 \quad \text{as} \quad r \to \infty. \tag{49}$$

The boundary conditions on the cylinder depend on the physical system: for light waves there are different boundary conditions for the electric and magnetic fields. The far field boundary condition is Sommerfeld's radiation condition, to indicate that the reflected wave is outgoing. The reflected wave is determined by the boundary condition at the cylinder surface. Since, using polar coordinates r and θ, $e^{iky} = e^{ikr\sin\theta} = \sum_m e^{im\theta} J_m(kr)$, the exact solution can be written in the form

$$\phi = \sum_m e^{im\theta}\left[J_m(kr) + A_m H_m^{(1)}(kr)\right] = e^{iky} + \sum_m A_m e^{im\theta} H_m^{(1)}(kr), \tag{50}$$

where the coefficients A_m are chosen to satisfy the boundary condition on the cylinder: the Hankel function of the first kind is that corresponding to an outgoing wave. This solution was given by Rayleigh.[50]

Ray theory underpins its geometrical description by using an assumption that the wavelength is very small, as is the case for light, or for water waves entering a harbour, or for sound transmission past buildings, so that when suitably non-dimensionalised, k is large. Asymptotic solutions of (48) then take the WKB form

$$\phi \sim e^{iky} \sum_0^\infty \frac{i^n A_n}{k^n}, \tag{51}$$

where ψ satisfies the eikonal equation

$$|\nabla \psi| = 1, \tag{52}$$

and

$$2\nabla \psi . \nabla A_n + A_n \nabla^2 \psi = \nabla^2 A_{n-1} \tag{53}$$

(and $A_{-1} = 0$). Because the Helmholtz equation is linear, there are actually two such solutions ϕ_1 and ϕ_2 which are superposed (as evident in (50)). As is well known, the eikonal equation can be solved using Charpit's method; the characteristics are straight lines, and these are the rays.[51] The incoming rays (corresponding to the solution $\phi_1 = e^{iky}$) are reflected at the boundary (the reflected rays are the characteristics for the WKB solution ϕ_2) so as to satisfy the boundary condition there, but no characteristics reach the shadow region behind the cylinder, which is therefore dark.

The jump from the illuminated region to the shadow is facilitated by the process of diffraction, described for example by Keller.[52] The ray solution becomes singular where the rays are tangent to the boundary, and the reflected and incident ray become coincident there. The consequence is the formation of a boundary layer which can be described by writing ϕ_1, for example, as $\phi_1 = \chi e^{iky}$, rescaling x near the right hand tangent ray (at $x = a$, $y = 0$) as $x = a + X / \sqrt{2k}$, so that to leading order

$$\chi_{XX} + 2i\chi_y = 0, \tag{54}$$

with the (similarity) solution being a Fresnel integral; this gives the commonly seen diffraction pattern of light and dark stripes. Further exploration of the asymptotic solutions at high wavenumber brings us back to exponential asymptotics and Stokes lines.[51]

Conclusions

As Kelvin (1903) put it in his scientific assessment, 'Stokes ranged over the whole domain of natural philosophy in his work and thought'.[53] (He excluded only electricity from this assessment.) In the nineteenth century, mathematical physics (as we might now say) was still a distillation of the earlier concept of natural philosophy, the knowledge of all things. The modern concept of applied mathematics had not separated from its parents, and indeed many of the techniques we now take for granted did not exist. As Berry comments, in constructing his asymptotic approximations for the Airy integral, Stokes discovered the WKB approximation, and invented the method of stationary phase more or less in passing.[44]

As a subject evolves, so does its language and even its name. In writing the Navier–Stokes equations, for example, Stokes does not have the partial derivative sign available to him. And we talk about mathematical physics, but more recently theoretical mechanics, and now just applied mathematics, or even physical applied mathematics. In the same way, geology became geophysics, subsequently Earth science, or even Earth systems science. Since Stokes's time, the subjects of applications have broadened as well. Kelvin in 1903 mentions hydrodynamics and elasticity, sound, light, radiant heat and chemistry. To these we can later add aerodynamics, and then meteorology and oceanography, and now materials science, industry, finance, biology, and a host more. Mathematics has grown as well: numerical analysis, functional analysis, stochastic processes, perturbation theory have become the instruments of the present day applied mathematician. But a direct lineage can be traced back to the scientific giants of the nineteenth century, and Stokes's place among these is secure and certain.

• •

ACKNOWLEDGEMENTS

It was most certainly an honour and a privilege to write this commentary, and an invitation to do so was one I could hardly refuse, and I thank Alastair Wood and his co-editors for this opportunity. At the same time, I recognize that I am hardly the most suitable person to undertake such a task. I am most grateful to my colleagues and friends for their help, notably John Gibbon for information on the existence problem for the Navier–Stokes equations, Jon Chapman for his potted guide to ray theory, and Chris Farmer for his insights into the relationship of Kelvin to Stokes. I acknowledge the support of the Mathematics Applications Consortium for Science and Industry (www.macsi.ul.ie) funded by the Science Foundation Ireland mathematics grant 12/IA/1683.

ENDNOTES

References to the following works are abbreviated as shown:

MPP   George Gabriel Stokes, *Mathematical and Physical Papers*, 5 vols (Cambridge: Cambridge University Press, 1880–1905, repr. 2009)

MSC   Joseph Larmor (ed.), *Memoir and Scientific Correspondence of the late Sir George Gabriel Stokes, Bart.*, 2 vols (Cambridge: Cambridge University Press, 1907), reprinted 2010

CSK   David B. Wilson (ed.), *The Correspondence between Sir George Gabriel Stokes and Sir William Thomson, Baron Kelvin of Largs*, 2 vols (Cambridge: Cambridge University Press, 1990)

CHAPTER 1

1. Joseph Larmor (ed.), *Memoir and Scientific Correspondence of the late Sir George Gabriel Stokes, Bart.* (Cambridge: Cambridge University Press, 1907).
2. J. Heueston, personal communication, 2003.
3. Alex D. D. Craik, *Mr. Hopkins' Men: Cambridge Reform and British Mathematics in the 19th Century* (London: Springer-Verlag, 2008).
4. David B. Wilson, 'Arbiters of Victorian Science: George Gabriel Stokes and Joshua King', in Kevin C. Knox and Richard Noakes (eds), *From Newton to Hawking: A History of Cambridge University's Lucasian Professors of Mathematics* (Cambridge: Cambridge University Press, 2003), pp. 295–342.
5. Michael V. Berry and Christopher J. Howls, 'Divergent Series: Taming the Tails', in Nicholas J. Higham (ed.), *The Princeton Companion to Applied Mathematics* (Princeton, NJ: Princeton University Press, 2015).
6. David B. Wilson (ed.), *The Correspondence between Sir George Gabriel Stokes and Sir William Thomson, Baron Kelvin of Largs*, 2 vols (Cambridge: Cambridge University Press, 1990). [Several extracts from the correspondence, along with their dates, are quoted in this chapter. Because the correspondence has been arranged in chronological order by Wilson, individual letters are easily located and the page number of each extract is not given.]
7. George Gabriel Stokes, *Mathematical and Physical Papers*, 5 vols (Cambridge: Cambridge University Press, 1880–1905).
8. Denis Weaire, 'Rich Silence: Some of the Contributions of G. G. Stokes to Physics', *Mathematics Today* 32 (1996).
9. Crosbie Smith and M. Norton Wise, *Energy and Empire: A Biographical Study of Lord Kelvin* (Cambridge: Cambridge University Press, 1989).
10. Michael C. W. Sandford, personal communication, 2017.

CHAPTER 2

1. Joseph Larmor (ed.), *Memoir and Scientific Correspondence of the Late Sir George Gabriel Stokes, Bart.* (Cambridge: Cambridge University Press, 1907).
2. Hugh Montgomery-Massingberd (ed.), *Burke's Irish Family Records* (5th edn, London: Burke's Peerage, 1976).
3. David B. Wilson, *Catalogue of the Manuscript Collections of Sir George Gabriel Stokes and Sir William Thomson, Baron Kelvin of Largs in Cambridge University Library* (Cambridge: Cambridge University Library, 1976).
4. David B. Wilson, *Kelvin and Stokes: A Comparative Study in Victorian Physics* (Bristol: Adam Hilger, 1987).
5. A.G. Stokes, *A Stokes Family of Dublin* (Australia: The Author, 1986). Privately published work located in the National Library of Ireland; catalogue entry http://catalogue.nli.ie/Record/vtls000182381.
6. http://www.britishsurnames.co.uk/surname/stokes/stats.
7. Teresa Stokes, Gabriel Stokes of Dublin (1681–1768) Descendants, https://www.ancestry.co.uk/family-tree/tree/53697794/family. An invitation to view is obtainable through the author.
8. Alan Boswell, private communication, 2018.
9. Andrew Pritchard, *English Patents: Being a Register of All Those Granted for Inventions in the Arts, Manufactures, Chemistry, Agriculture etc etc.* (London, 1847).
10. Index to the Act or Grant Books, and to Original Wills, of the Diocese of Dublin to the year 1800, in *Reports of the Deputy Keeper of the Public Records and Keeper of the State Papers in Ireland.*
11. Nicholas Carlisle, *A Topographical Dictionary of Ireland* (London: William Miller, 1810).
12. John C. Bowmer (ed.), 'Irish Notes', *Proceedings of the Wesley Historical Society* 36 (October 1968), p. 179.
13. J. A. Bennett, '(John) Thomas Romney Robinson (1793–1882)', *Oxford Dictionary of National Biography* (Oxford: Oxford University Press, 2004), https://doi.org/10.1093/ref:odnb/23883.
14. *Report of the Twenty-Second Meeting of the British Association for the Advancement of Science; held at Belfast in September 1852* (London: John Murray, 1853).
15. Peter Bryan and Nick Wise, 'Cambridge New Town—A Victorian Microcosm', *Proceedings of the Cambridge Antiquarian Society* 94 (2005), pp. 199–216.
16. *The Carthusian* 12 No 396 (December 1916).
17. *Northern Echo* (28 August 1893) and *Manchester Times* (1 September 1893).
18. A. D. Morrison-Low, 'Gabriel Stokes (bap. 1682, d.1768)', *Oxford Dictionary of National Biography* (Oxford: Oxford University Press, 2004), https://doi.org/10.1093/ref:odnb/69095.
19. *Records of the Cathedral of the Holy Trinity commonly called Christ Church, Dublin, 12th–20th Cent.*, p. 56. In Representative Church Body Library, Dublin.
20. John E. Burnett and Alison D. Morrison-Low, 'Scientific instrument making in Dublin, 1700–1830', in *Vulgar and Mechanick: The Scientific Instrument Trade in Ireland, 1650–1921* (Dublin: Royal Dublin Society, 1989).
21. W. B. S. Taylor, *History of the University of Dublin* (London: T. Cadell, 1845), p. 424.
22. J. Walsh, *Sketches of Ireland 60 Years Ago* (Dublin: James McGlashan, 1849), p. 133.
23. L. Perry Curtis jun., 'William Stokes (1804–1878)', *Oxford Dictionary of National Biography*, https://doi.org/10.1093/ref:odnb/26561.
24. William Stokes, *William Stokes: His Life and Work* (London: T. Fisher Unwin, 1898).
25. William Stokes, *The Life and Labours in Art and Archæology of George Petrie* (London: Longmans, Green, 1868).
26. Nollaig Ó Muraíle, ''Whitley Stokes (1830–1909), *Oxford Dictionary of National Biography*, https://doi.org/10.1093/ref:odnb/36315.
27. Elizabeth Boyle and Paul Russell (eds), *The Tripartite Life of Whitley Stokes (1830–1909)* (Dublin: Four Courts Press, 2011).

28. D'A. Power, rev. J. B. Lyons, 'Sir William Stokes (1839–1900)', *Oxford Dictionary of National Biography*, https://doi.org/10.1093/ref:odnb/26562.

29. C. L. Falkiner, rev. Marie-Louise Legg, 'Margaret M'Nair Stokes (1832–1900)', *Oxford Dictionary of National Biography*, https://doi.org/10.1093/ref:odnb/26558.

30. A. F. Hurst, rev. Mary E. Gibson, 'Adrian Stokes (1887–1927)', *Oxford Dictionary of National Biography*, https://doi.org/10.1093/ref:odnb/36311.

31. William Betham, Sketch Pedigrees, Mss, National Library of Ireland Call numbers GO MSS 261–76.

CHAPTER 3

1. Alex D. D. Craik, *Mr Hopkins' Men: Cambridge Reform and British Mathematics in the 19th Century* (London: Springer-Verlag, 2008), p. 45.

2. Wranglers were students who had gained first-class honours in the Mathematical Tripos. Romilly, who was writing after attending the meeting where the order of merit had been decided, had been Fourth Wrangler in 1813 and he 'long remembered his sufferings when he sat for the examination' (which was partly oral in his time) so was well qualified to judge: see Denys Arthur Winstanley, *Early Victorian Cambridge* (Cambridge: Cambridge University Press, 1940), p. 152.

3. Isabella Lucy Humphry, 'Notes and Recollections', in *MSC* vol. I, pp. 1–49, on pp. 3–5. Most of what is known about Stokes's early education comes from this memoir written by his daughter, Mrs Laurence Humphry, written after his death.

4. Elias Voster, *Elimatus or the Rules of Arithmetick Digested and Explained in Mercantile Course* (Dublin: Isaac Jackson, 1768).

5. Humphry, 'Notes and Recollections', p. 5.

6. Elias Voster set up a school in Cork for teaching arithmetic and book-keeping. His text was first published in 1768 and ran into multiple editions during the eighteenth and nineteenth centuries. It is not known which edition was used by Stokes.

7. *Dublin University Magazine Advertiser* (July 1838) (Dublin: William Curry, Jun. and Company), p. 6.

8. Humphry, 'Notes and Recollections', p. 5.

9. Senior Optimes were students who had gained second-class honours in the Mathematical Tripos. The Classical Tripos, established in 1822, was a voluntary examination open only to students who had already gained honours in the Mathematical Tripos. The Classical Tripos became independent of the Mathematical Tripos in 1850.

10. Humphry, 'Notes and Recollections', p. 6.

11. Today George Birch Jerrard is best remembered for his mistaken belief that he had found a formula for solving equations of the fifth degree. In 1824, Niels Henrik Abel had proved that such a formula (based on only fundamental algebraic operations) could not exist but Jerrard refused to accept it, publishing erroneous results and persisting in his belief to the end of his life: see R. A. Bryce, 'George Birch Jerrard (1804–1863)', *Oxford Dictionary of National Biography* (Oxford: Oxford University Press, 2004), https://doi.org/10.1093/ref:odnb/14788.

12. *Bristol Mercury* (2 March 1833), p. 3.

13. The founding of the college is described in detail by John Latimer, former editor of the liberal leaning *Bristol Mercury*, and author of the highly regarded four-volume *Annals of Bristol*: John Latimer, *The Annals of Bristol in the Nineteenth Century* (Bristol: W. & F. Morgan, 1887), pp. 140–2.

14. Council of Bristol College, *Outline of the Plan of Education to be pursued in the Bristol College* (Bristol: John Taylor, 1830).

15. Craik, *Mr Hopkins' Men*, pp. 90–1.

16. William Whewell, *Thoughts on the Study of Mathematics as a Part of a Liberal Education: The second edition, to which is added a letter to the Editor of the Edinburgh Review, etc.* (London: John W. Parker; Cambridge: J. and J. J. Deighton, 1836).

17. Council of Bristol College, *Outline*, pp. 6, 9–10.
18. *Bristol Mercury* (1 December 1829), Letter to the Editor, p. 3.
19. One famous pupil was the writer Walter Bagehot who subsequently studied mathematics at University College London with Augustus De Morgan.
20. In 1836 the see of Bristol was united with that of Gloucester. In 1897 the two were again separated
21. Latimer, *Annals of Bristol*, p. 141.
22. Humphry, 'Notes and Recollections', p. 6.
23. Humphry, 'Notes and Recollections', p. 6.
24. Francis William Newman, *The Theorems of Taylor and Maclaurin in a Finite Form* (Oxford: Vincent, 1830).
25. Francis William Newman, *The Difficulties of Elementary Geometry, especially those which concern the straight line, the plane and the theory of parallels* (London: William Ball & Co.,1841).
26. In 1851 Newman taught elementary geometry at the Ladies' College (later Bedford College) in succession to Augustus De Morgan; among those who attended his classes was the novelist George Eliot: see Derek Ball, 'Mathematics in George Eliot's Novels', PhD thesis, University of Leicester, 2016, p. 77.
27. Lord Rayleigh, 'Sir George Gabriel Stokes, Bart. 1819–1903', *Proceedings of the Royal Society* 75 (1905), pp. 199–216, on p. 200;. repr. in *MPP* vol. V pp. ix–xxv, on p. ix.
28. *Bristol Mercury* (9 July 1836), p. 3.
29. *Bristol Mercury* (8 July 1837), p. 4.
30. Humphry, 'Notes and Recollections', p. 6.
31. David B. Wilson, 'Arbiters of Victorian Science: George Gabriel Stokes and Joshua King', in Kevin C. Knox and Richard Noakes (eds), *From Newton to Hawking: A History of Cambridge University's Lucasian Professors of Mathematics* (Cambridge: Cambridge University's Press, 2003), pp. 295–342, on p. 300.
32. Humphry, 'Notes and Recollections', p. 7.
33. John Sykes ended up as Third Wrangler to Stokes. He later examined for the Mathematical Tripos several times. He went on to have a long career in education, including examining for The Society of Arts.
34. The term 'coach' for a private tutor is thought to have originated in Oxford in around 1830 and to have come into use in Cambridge in the early 1830s: see Andrew Warwick, *Masters of Theory: Cambridge and the Rise of Mathematical Physics* (Chicago: University of Chicago Press, 2003), p. 89.
35. Alex Craik in his comprehensive study of Hopkins has found no evidence of Sykes being coached by Hopkins: see (1: p. 4).
36. June Barrow-Green, 'A Correction to the Spirit of Too Exclusively Pure Mathematics: Robert Smith (1689–1768) and his Prizes at Cambridge University', *Annals of Science* 56 (1999), pp. 271–316.
37. Robert Leslie Ellis who was the Senior Wrangler a year ahead of Stokes attended Peacock's lectures, so it is likely that Stokes did also. See Harvey Goodwin, 'Biographical Memoir of Robert Leslie Ellis, M.A.', in *The Mathematical and Other Writings of Robert Leslie Ellis, M.A.*, ed. William Walton (Cambridge: Deighton, Bell, and Co., 1863), pp. ix–xxxvi, on p. xv.
38. James Challis, *Syllabus of a Course of Experimental Lectures on the Equilibrium and Motion of Fluids and on Optics* (London: John W. Parker; Cambridge: J. and J.J. Deighton, 1838).
39. Royal Society Election Certificate, EC/1838/30. Hymers' proposers for Fellowship included George Peacock and William Whewell.
40. R. E. Anderson, rev. Maria Panteki, 'John Hymers (1803–1887)', *Oxford Dictionary of National Biography* (Oxford: Oxford University Press, 2004), https://doi.org/10.1093/ref:odnb/14340.
41. Warwick, *Masters of Theory*, pp. 146–7.
42. Martha Somerville, *Personal Recollections, from Early Life to Old Age of Mary Somerville* (London: John Murray, 1873), p. 172.
43. Warwick, *Masters of Theory*, pp. 84–5.
44. Warwick, *Masters of Theory*, p. 84.

45. *The Gentleman's Magazine*, 1866, p. 706, quoted in Tony Crilly, *Arthur Cayley: Mathematician Laureate of the Victorian Age* (Baltimore, MD: Johns Hopkins University Press, 2006), p. 40.

46. Crosby Smith and M. Norton Wise, *Energy and Empire: A Biographical Study of Lord Kelvin* (Cambridge: Cambridge University Press, 1989), p. 76.

47. Letter from Francis Galton to his father, 11 November 1841, quoted in Warwick, *Masters of Theory*, p. 85.

48. Warwick, *Masters of Theory*, p. 183.

49. Diary of Robert Leslie Ellis, 3 December 1838. Trinity College Cambridge, Add Ms a.219.1. For further comments about Ellis's dislike of the Mathematical Tripos, see (34: pp. 184–5).

50. George Gabriel Stokes, (1838–40). MSS Stokes Collection PA 2-24, Manuscripts and University Archives, Cambridge University Library. Wood notes that Stokes's 'writing was so bad that he eventually became one of the first people in Britain to make regular use of a typewriter': Alastair Wood, 'George Gabriel Stokes 1819–1903: An Irish Mathematical Physicist', *IMS Bulletin* 35 (1995), pp. 49–58, on p. 53.

51. The other courses in Stokes's notes are on an example of tracing a curve and differential equations, constrained motion of a point, impulsive force and impact, motion of two or three bodies attracting one another and lunar inequalities, and astronomical instruments.

52. A generation later, Isaac Todhunter would choose the opposite path, taking up textbook writing in preference to teaching, and thereby amassed a small fortune. See June Barrow-Green, '"The Advantage of Proceeding from an Author of Some Scientific Reputation": Isaac Todhunter and his Mathematical Textbooks', in Jonathan Smith and Christopher Stray (eds), *Teaching and Learning in Nineteenth-Century Cambridge* (Cambridge: Boydell Press and Cambridge University Library, 2001), pp. 177–203.

53. W. Hopkins, *Remarks on Certain Proposed Regulations Respecting the Studies of the University and the Period of Conferring the Degree of B.A.* (Cambridge: J. & J. J. Deighton, 1841), p. 10.

54. Wilson, 'Arbiters of Victorian Science', p. 302.

55. George Peacock, *Observations on the Statutes of the University of Cambridge* (London: John W. Parker; Cambridge: J. and J. J. Deighton, 1841), p. 153.

56. George Peacock, *Observations on the Statutes of the University of Cambridge*, p. 156.

57. Warwick, *Masters of Theory*, p. 87.

58. Hopkins, *Remarks on Certain Proposed Regulations*, p. 153.

59. Craik, *Mr Hopkins' Men*, p. 100.

60. Hopkins was thus earning something in region of £750 a year through his coaching, a substantial amount and a sum much higher than the stipend of most of the university professors.

61. Humphry, 'Notes and Recollections', p. 9.

62. Warwick, *Masters of Theory*, p. 156.

63. Until 1850 students who had not obtained honours in the Mathematical Tripos were not allowed to sit the Classical Tripos. It was therefore possible for high-achieving classics students to fail to get a degree.

64. Goodwin, 'Robert Leslie Ellis', p. xix.

65. Robert Leslie Ellis said of Bristed: 'I rather like him though he is vain and a little American: decidedly clever.' Diary of Robert Leslie Ellis, 25 December 1843. Trinity College Cambridge, Add Ms a.218.41.

66. Charles Astor Bristed, *Five Years in an English University*, 2 vols (New York: Putnam & Co., 1852), p. 331.

67. Humphry, 'Notes and Recollections', p. 7.

68. Bristed, *Five Years in an English University*, p. 126.

69. Bristed, *Five Years in an English University*, p. 163.

70. Perhaps surprisingly and unlike many Tripos students, Goodeve was not put off mathematics by the experience. He went on to have a solid career teaching mathematics at a number of institutions including the Royal Military College Woolwich, Kings College London, and the Royal College of Science.

71. The papers were designated pure mathematics, natural philosophy or problems. See the Appendix for the 'Plan of the Examination' and copies of the first and the last papers. All the papers, together with the order of merit, are reproduced in the *Cambridge University Magazine*, vol. II (Cambridge: W. P. Grant; 1843), pp. 189–208). The solutions to the three designated 'Problems' papers are given in John Adams Coombe, *Solutions of the Cambridge Problems for the Years 1840, 1841* (Cambridge: John W. Parker, 1841).

72. Newton, when investigating orbital motion in the *Principia*, proved propositions involving force by considering accelerations: see Bruce Pourcia, 'The Preliminary Mathematical Lemmas of Newton's Principia', *Archive for History of Exact Sciences* 52/3 (1998), pp. 279–95, on p. 291.

73. Diary of Robert Leslie Ellis, 15 January 1840, Trinity College Cambridge, Add Ms a.218.41. Thomas Burcham was the Father (today Praelector) of Trinity College. For further details of the ceremony, see (34: pp. 207–8).

74. *The Hull Packet* (29 January 1841).

75. *The Times*, 25 January 1841, p. 5, and 29 January 1841, p. 5; the former supporting the status quo as 'judicious and right' and the latter vehemently opposing it.

76. Bristed, *Five Years in an English University*, p. 85.

77. George Peacock, *Observations on the Statutes of the University of Cambridge*, p. 153.

78. The relevant statute stated: 'That there be not contained in any paper more Questions than Students well prepared have generally been found able to answer within the time allowed for that Paper.' The problems with the examination—both with its form and its content—generated much discussion which culminated in major reforms in 1848. Not only was the examination revised but a formal syllabus of mathematical studies was also introduced. For details see (34: pp. 101–3).

79. For details of the history of the Smith's Prize together with a list of winners, see Barrow-Green, 'Robert Smith and His Prizes'.

80. Famously, in 1854 Stokes himself set a Smith's Prize question which included the first appearance in print of what would later become known as Stokes's theorem.

81. The second prize was won by the Second Wrangler, Henry Cadman Jones of Trinity, who afterwards made a career at the Bar. Cadman was present at the Stokes's Jubilee in Cambridge in 1899 and the fact that he was able to come and offer his congratulations was considered one of the 'most pleasing features' of the occasion: [Anon.], 'The Jubilee of Sir George Gabriel Stokes', *Nature* 60 (1899), pp. 125–9, on p. 125.

82. When William Thomson won the first Smith's Prize having come a disappointing second in the Senate-House, William Hopkins wrote to Thomson's father 'The examination, as you are probably aware, is altogether of a higher character than that of the Senate-House, being, in fact, intended to furnish a higher test of the merits of the first men', see Barrow-Green, 'Robert Smith and His Prizes', p. 280.

83. Letter from N. C. Fenwick to Stokes, 1 February 1841. Stokes Manuscripts. Add. MS 7656. F69.

CHAPTER 4

1. Cf. Edmund Whittaker, *A History of the Theories of Aether and Electricity: From the Age of Descartes to the Close of the Nineteenth Century* (London: Longmans, Green, 1910), Chapter 4. Jed Buchwald, *The Rise of the Wave Theory of Light: Optical Theory and Experiment in the Early Nineteenth Century* (Chicago: University of Chicago Press, 1989). André Chappert, *L'édification au XIXᵉ siècle d'une science des phénomènes lumineux* (Paris: Vrin, 2004). Olivier Darrigol, *A History of Optics from Greek Antiquity to the Nineteenth Century* (Oxford: Oxford University Press, 2012), Chapter 5.

2. George Biddell Airy, *Mathematical Tracts on the Lunar and Planetary Theories, the Figure of the Earth, Precession and Nutation, the Calculus of Variations, and the Undulatory Theory of Optics. Designed for the use of students in the university* (2nd edn, Cambridge: Deighton, 1831). On the British reception in general, cf. Geoffrey Cantor, *Optics after Newton: Theories of Light in Britain and Ireland, 1704–1840* (Manchester: Manchester University Press, 1983), pp. 159–65; Buchwald, *Rise of*

the *Wave Theory of Light*, Chapter 12. Xiang Chen, *Instrumental Traditions and Theories of Light: The Uses of Instruments in the Optical Revolution* (Dordrecht: Kluwer, 2000), Chapters 1–2. On the reception in Cambridge, cf. Cantor, *Optics after Newton*. David B. Wilson, 'Arbiters of Victorian Science: George Gabriel Stokes and Joshua King', in Kevin C. Knox and Richard Noakes (eds), *From Newton to Hawking: A History of Cambridge University's Lucasian Professors of Mathematics* (Cambridge: Cambridge University Press, 2003), pp. 295–342, on p. 302.

3. On the diversity of British optics, cf. Chen, *Instrumental Traditions*. On the persistence of Newtonian colour problems, cf. Alan E. Shapiro, *Fits, Passions, and Paroxysms: Physics, Method, and Chemistry and Newton's Theories of Colored Bodies and Fits of Easy Reflection* (Cambridge: Cambridge University Press, 1993), Chapter 9. It should be remembered that Newton wanted most of his optics to be independent of the deeper nature of light.

4. Cf. David B. Wilson, *Kelvin and Stokes: A Comparative Study in Victorian Physics* (Bristol: Adam Hilger, 1987), pp. 27, 34.

5. Stokes to Thomson, 12 February 1858, *CSK*, pp. 230–2.

6. The best informed reviews of Stokes's works in optics are found in Lord Rayleigh [John William Strutt], 'Sir George Gabriel Stokes, Bart. 1819–1903', *Proceedings of the Royal Society* 75 (1905), pp. xiii–xix, xx–xxi; repr. in *MPP* vol. V, pp. ix–xxv, and in Jed Buchwald, 'Why Stokes Never Wrote a Treatise in Optics', in Peter Harman and Alan Shapiro (eds), *An Investigation of Difficult Things: Essays on Newton and the History of Exact Sciences* (Cambridge: Cambridge University Press, 1992), pp. 451–76.

7. G. G. Stokes, 'On the Theories of the Internal Friction of Fluids in Motion, and of the Equilibrium and Motion of Elastic Solids' [read 14 April 1845], *Transactions of the Cambridge Philosophical Society* 8 (1849), pp. 287–319; repr. in *MPP* vol. I, pp. 75–129. Cf. Olivier Darrigol, *Worlds of Flow: A History of Hydrodynamics from the Bernoullis to Prandtl* (Oxford: Oxford University Press, 2005), Chapter 3. As explained by Fresnel and Saint-Venant, a continuous medium whose points act in pairs through central forces cannot be rigid because a shear in this kind of medium does not imply any stress. In modern terms, the symmetry of the medium is such that a translation of a layer over another does not imply any change in the potential energy of the molecular forces. It does so in the case of a lattice of point-like molecules.

8. Cf. Buchwald, *Rise of the Wave Theory of Light*, Chapter 11. The Laplacian model of an elastic solid as a lattice of molecules interacting through central forces leads to one elastic constant only in the isotropic case (unless the solid is compressed in its original state). In contrast, a linear isotropic stress–strain relation generally has two constants.

9. Stokes, 'Theories of the Internal Friction of Fluids', pp. 121 (jelly), 124–9. George Green, 'On the Laws of Reflexion and Refraction of Light at the Common Surface of Two Non-crystallized Media' (1838), *Transactions of the Cambridge Philosophical Society* 7 (1838–42), pp. 1–24.

10. G. G. Stokes, 'On the Constitution of the Luminiferous Aether', *Philosophical Magazine* 22 (1848), p. 12; repr. in *MPP* vol. II, pp. 8–13. On Stokes's ether, his theory of aberration, and the ensuing polemic with Challis, cf. David B. Wilson, 'George Gabriel Stokes on Stellar Aberration and the Luminiferous Ether', *British Journal for the History of Science*, 6 (1972), pp. 57–72; and Wilson, *Kelvin and Stokes*, Chapter 6.

11. G. G. Stokes, 'On the Aberration of Light' [a], *Report of the Fifteenth Meeting of the British Association for the Advancement of Science; held at Cambridge in June 1845* (London: John Murray, 1846), Notices and Abstracts p. 9; G. G. Stokes, 'On the Aberration of Light' [b], *Philosophical Magazine* 27 (1845), pp. 9–15; repr. in *MPP* vol. I, pp. 134–40. James Challis, 'On the Aberration of Light', *Report of the Fifteenth Meeting of the British Association for the Advancement of Science; held at Cambridge in June 1845* (London: John Murray, 1846), Notices and Abstracts p. 9; James Challis, 'A Theoretical Explanation of the Aberration of Light', *Philosophical Magazine* 27 (1845), pp. 321–7.

12. James Bradley, 'A letter from the Reverend Mr. James Bradley Savilian Professor of astronomy at Oxford, and F. R. S. to Dr. Edmund Halley Astronom. Reg. &c giving an account of a new discovered motion of the fix'd stars', *Philosophical Transactions of the Royal Society of London* 35 (1727–8) [read

and published 1729], pp. 637–61. Leonhard Euler, 'Mémoire sur l'effet de la propagation successive de la lumiere dans l'apparition tant des planètes que des comètes', *Mémoires de l'Académie Royale des sciences et des belles-lettres de Berlin* 2 (1746), pp. 141–81. Thomas Young, 'The Bakerian Lecture: Experiments and Calculations Relative to Physical Optics', *Philosophical Transactions of the Royal Society of London* 94 (1804); repr. in *Miscellaneous Works*, 3 vols (London: John Murray, 1855), vol.1, pp. 179–91, on p. 188. Cf. Jean Eisenstaedt, *Avant Einstein: Relativité, lumière, gravitation* (Paris: Le Seuil, 2005), pp. 74–84, Darrigol, *History of Optics*, pp. 129, 258–60.

13. Stokes, 'On the Aberration of Light' [b], *MSS* vol. I, pp. 134–40. The reasoning is here given in the simplified version of *MPP* I, pp. 139–40.

14. Stokes, 'On the Aberration of Light' [a].

15. Challis, 'On the Aberration of Light', p. 323. On the entire debate, cf. Baden Powell, 'Remarks on Some Points of the Reasoning in the Recent Discussions on the Theory of the Aberration of Light' *Philosophical Magazine* 29 (1846), pp. 425–40.

16. Challis, 'On the Aberration of Light', p. 324.

17. G. G. Stokes, 'Remarks on Professor Challis's Theoretical Explanation of the Aberration of Light', *Philosophical Magazine* 28 (1846), pp. 15–17.

18. James Challis, 'On the Aberration of Light, in Reply to Mr. Stokes', *Philosophical Magazine* 28 (1846), pp. 90–3, pp. 92,93; James Challis, 'On the Principles to be Applied in Explaining the Aberration of Light', *Philosophical Magazine* 28 (1846), pp. 176–7.

19. G. G. Stokes, 'On the Aberration of Light', *Philosophical Magazine* 28 (1846), pp. 335–6.

20. James Challis, 'On the Aberration of light, in Reply to Mr. Stokes', *Philosophical Magazine* 28 (1846), pp. 393–4; G. G. Stokes, 'On the Constitution of the Luminiferous Aether, viewed with Reference to the Phænomenon of the Aberration of Light', *Philosophical Magazine* 29 (1846), pp. 6–10, on pp. 9–10.

21. Stokes, 'On the Constitution of the Luminiferous Aether, viewed with Reference to the Phænomenon of the Aberration of Light'.

22. By analogy with ordinary fluids of small viscosity, Stokes expected the irrotational motion around the sphere to be unstable. But his ether differed from an ordinary fluid by quickly dissipating any shear stress that would appear when the velocity of the sphere gradually increased from zero to its actual value. See also Stokes, 'On the Constitution of the Luminiferous Aether' [1848].

23. This conjecture was not too far-fetched in the 1840s, because the modern kinetic theory of gases, according to which the molecules of a gas occupy a negligible fraction of its volume, was still marginal in Britain.

24. Stokes, 'On the Aberration of Light' [1846], pp. 7–9.

25. Augustin Fresnel, 'Sur l'influence du mouvement terrestre dans quelques phénomènes optiques', *Annales de chimie et de physique* 9 (1818), pp. 57–66, 286. Stokes, 'On the Aberration of Light' [1846], p. 81. Cf. Darrigol, *History of Optics*, pp. 258–61 and further references there.

26. Hippolyte Fizeau, 'Sur les hypothèses relatives à l'éther lumineux, et sur une expérience qui paraît démontrer que le mouvement des corps change la vitesse avec laquelle la lumière se propage dans leur intérieur', *Comptes rendus hebdomadaires des séances de l'Académie des sciences* 33 (1851), pp. 349–55. James Clerk Maxwell, 'On the Influence of the Motion of the Heavenly Bodies on the Index of Refraction of Light': letter to William Huggins, 10 June 1867, quoted in Huggins, 'Further Observations on the Spectra of Some of the Stars and Nebulae, with an Attempt to Determine therefrom whether these Bodies are moving towards or from the Earth, also Observations on the Spectra of the Sun and of Comet II., 1868', *Philosophical Transactions of the Royal Society of London* 158 (1868), pp. 532–5, on p. 535. James Clerk Maxwell, 'Ether', *Encyclopaedia Britannica*, 9th edn, vol. 8 (1878), pp. 568–72, on p. 570. Cf. Michel Janssen and John Stachel, 'The Optics and Electrodynamics of Moving Bodies', Preprint 265, Max Planck Institut für Wissenschaftsgeschichte, Berlin, 2004. Chappert, *L'édification au XIXᵉ siècle d'une science des phénomènes lumineux*, Chapter 10.

27. H. A. Lorentz, 'Over den invloed, dien de beweging der aarde op de lichtverschijnselen uitofent', *Verslagen en Medeelingen der Koninklijke Akademie van Wetenschappen* (Amsterdam: Johannes Müller, 1886), pp. 297–372; transl. as 'De l'influence du mouvement de la terre sur les phénomènes lumineux' in *Archives Néerlandaises des sciences exactes et naturelles* 21 (1887), pp. 103–76; repr. in *Collected Papers*, vol. IV (The Hague : Martinus Nijhoff, 1937), pp. 153–214. Cf. Jed Buchwald, 'The Michelson Experiment in the Light of Electromagnetic Theory before 1900', in Stanley Goldberg and Roger H. Stuewer (eds), *The Michelson Era in American Science 1870-1930* (New York: American Institute of Physics, 1988), pp. 55–70.

28. George Biddell Airy, 'The Bakerian Lecture: On the Theoretical Explanation of an Apparent New Polarity in Light', *Philosophical Transactions of the Royal Society of London* 130 (1840), pp. 225–44, on p. 226.

29. David Brewster, 'On a New Property of Light', *Report of the Seventh Meeting of the British Association for the Advancement of Science; held at Liverpool in September 1837* (London: John Murray, 1838), Notices and Abstracts pp. 12–13. David Brewster, 'On a New Kind of Polarity in Homogeneous Light', *Report of the Eighth Meeting of the British Association for the Advancement of Science; held at Newcastle in August 1838* (London: John Murray, 1839), Notices and Abstracts pp. 12–13. George Biddell Airy, 'Theoretical Explanation of an Apparent New Polarity in Light'; George Biddell Airy, 'Supplement to a Paper "On the Theoretical Explanation of an Apparent New Polarity in Light"', *Philosophical Transactions of the Royal Society of London* 131 (1841), pp. 1–10. The bands are most commonly known as the 'Talbot bands' as they were reported in William Henry Fox Talbot, 'An Experiment on the Interference of Light', *Philosophical Magazine* 10 (1837), pp. 364 (without the polarity discovered by Brewster). On these bands, Brewster's polarity, and their explanation by Airy and Stokes, cf. Chen, *Instrumental Traditions*, Chapter 5.

30. Airy's explanation did not satisfy Brewster: see David Brewster, 'On a New Polarity of Light, with an Examination of Mr. Airy's Explanation of it on the Undulatory Theory', *Report of the Fifteenth Meeting of the British Association for the Advancement of Science; held at Cambridge in June 1845* (London: John Murray, 1846), Notices and Abstracts pp. 7–8.

31. Baden Powell, 'On a New Case of the Interference of Light', *Philosophical Transactions of the Royal Society of London* 138 (1848), pp. 213–26. G. G. Stokes, 'On the Theory of Certain Bands Seen in the Spectrum', *Philosophical Transactions of the Royal Society of London* 138 (1848); repr. in *MPP* vol, II, pp. 14–35. Cf. Stokes to Kelvin [October 1847], in *CSK*, p. 33, which announces this memoir as well as the later memoir on the central spot of Newton's rings (see note 35).

32. The characteristic function of a domain is the function that takes the value one in this domain and vanishes outside this domain.

33. Instead of appealing to the properties of the Fourier transform, Stokes integrates the explicit trigonometric expressions of the diffraction amplitudes.

34. Augustin Fresnel, 'Quelques observations sur les principales objections de Newton contre le système des vibrations lumineuses et sur les difficultés que présente son hypothèse des accès', *Bibliothèque universelle des sciences, belles-lettres, et arts* 22 (1823), pp. 73–91, on p. 87.

35. Humphrey Lloyd, 'Report on the Progress and Present State of Physical Optics', *Report of the Fourth Meeting of the British Association for the Advancement of Science; held at Edinburgh in 1834* (London: John Murray, 1835), pp. 295–414, on p. 310. Matthew O'Brien, 'On the Reflexion and Refraction of Light at the Surface of an Uncrystallized Body', *Transactions of the Cambridge Philosophical Society* 8 (1849), pp. 7–30. G. G. Stokes, 'On the Formation of the Central Spot of Newton's Rings beyond the Critical Angle' [read 11 December 1848], *Transactions of the Cambridge Philosophical Society* 8 (1849), pp. 642–58; repr. in *MPP* vol. II, pp. 56–81. G. G. Stokes, 'On the Mode of Disappearance of Newton's Rings in passing the Angle of Total Internal Reflexion', *Report of the Twentieth Meeting of the British Association for the Advancement of Science; held at Edinburgh in July and August 1850* (London: John Murray, 1851), Notices and Abstracts p. 19; repr. in *MPP* vol. II, pp. 358–9.

36. Lloyd, 'Progress and Present State of Physical Optics', pp. 344–5. Siméon Denis Poisson, 'Sur le phénomène des anneaux colorés', *Annales de chimie et de physique* 22 (1823), pp. 337–47.

37. G. G. Stokes, 'On the Perfect Blackness of the Central Spot in Newton's Rings, and on the Verification
of Fresnel's Formulae for the Intensities of Reflected and Refracted Rays', *Cambridge and Dublin
Mathematical Journal* 4 (1849); repr. in *MPP* vol. II, pp. 89–103. Cf. Wilson, *Kelvin and Stokes*,
pp. 110–11.

38. G. G. Stokes, 'On the Colours of Thick Plates' [read 19 May 1851], *Transactions of the Cambridge
Philosophical Society* 9 (1856), Part II pp. 147–76; repr. in *MPP* vol. III, pp. 155–96, on p. 160. On
earlier studies, cf. Shapiro, *Fits, Passions, and Paroxysms*, pp. 150–70; Darrigol, *History of Optics*,
pp. 100, 176.

39. G. G. Stokes, 'On the Dynamical Theory of Diffraction' [read 26 November 1849], *Transactions of the
Cambridge Philosophical Society* 9 (1856), pp. 1–62; repr. in *MPP* vol. II, pp. 243–328. Cf. Buchwald,
'Why Stokes Never Wrote a Treatise in Optics', pp. 456–9; Darrigol, *History of Optics*, pp. 273–6.

40. Siméon Denis Poisson, 'Mémoire sur l'intégration de quelques équations linéaires aux différences
partielles, et particulièrement de l'équation générale du mouvement des fluides élastiques', *Mémoires
de l'Académie des sciences*, 3 (1819), pp. 121–76, on pp. 133–4.

41. Stokes, 'On the Dynamical Theory of Diffraction', p. 31n (footnote in which Stokes indicates how the
formula for sound differs from his formula for transverse diffracted waves; there is an erroneous factor cos θ in the resulting formula).This expression gives a non-zero backward radiation for a single
surface element but a negligible backward radiation when integrated over a surface large compared
with the wavelength.

42. Stokes, 'On the Dynamical Theory of Diffraction', pp. 4–5, 35–61. Cf. Whittaker, *History of the
Theories of Aether and Electricity*, pp. 168–70.

43. Cf. Whittaker, *History of the Theories of Aether and Electricity*, Chapter 5; Kenneth Schaffner,
Nineteenth-Century Aether Theories (Oxford: Oxford University Press, 1972). Darrigol, *History of
Optics*, pp. 227–39.

44. Stokes, 'On the Dynamical Theory of Diffraction', pp. 323–6.

45. G. G. Stokes, 'On the Change of Refrangibility of Light', *Philosophical Transactions of the Royal Society
of London* 142 (1852), pp. 463–562; repr. in *MPP* vol. III, pp. 266–413, on pp. 361–2.

46. Carl Alexander Holtzmann, 'Das polarisirte Licht schwingt in der Polarisationsebene', *Annalen der
Physik* 99 (1856), pp. 446–51. G. G. Stokes, 'On the Polarization of Diffracted Light', *Philosophical
Magazine* 13 (1857), pp. 159–61; repr. in *MPP* vol. IV, pp. 74–6.

47. Ludvig Valentin Lorenz, 'Bestimmung der Schwingungsrichtung des Lichtäthers durch die Polarisation
des gebeugten Lichtes', *Annalen der Physik* 111 (1860), pp. 315–28. Cf. Helge Kragh, *Ludvig Lorenz:
A Nineteenth-Century Theoretical Physicist* (Copenhagen: Det Kongelige Danske Videnskabernes
Selskab, 2018), pp. 63–5. Stokes, 'On the Change of Refrangibility of Light', *MPP* vol. III, pp. 327–8 (see
also Stokes to Strutt, 10 August 1872, in *MSC* vol. II pp. 101–2). Louis Georges Gouy, 'Recherches
expérimentales sur la diffraction', *Annales de chimie et de physique* 52 (1886), pp. 145–92.

48. Cf. Olivier Darrigol, 'Poincaré's Light', in Bertrand Duplantier and Vincent Rivasseau (eds), *Henri
Poincaré, 1912-2012*, Progress in Mathematical Physics 67 (Basel: Birkhäuser, 2015), pp. 1–50.

49. Cf. Darrigol, *History of Optics*, pp. 274–5.

50. G. G. Stokes, 'On the Composition and Resolution of Streams of Different Light from Different
Sources' [read 16 February and 15 March 1852], *Transactions of the Cambridge Philosophical Society*
9 (1856); repr. in *MPP* vol. III, pp. 233–58, on pp. 256–7 (on Airy), 234 (citation). For a modern take
on Stokes's parameters, cf. William McMaster, 'Polarization and the Stokes Parameters', *American
Journal of Physics* 22 (1954), pp. 351–62.

51. Introducing the Pauli matrices $\sigma_1 = \begin{pmatrix} 1 & 0 \\ 0 & -1 \end{pmatrix}$, $\sigma_2 = \begin{pmatrix} 0 & 1 \\ 1 & 0 \end{pmatrix}$, $\sigma_3 = \begin{pmatrix} 0 & -i \\ i & 0 \end{pmatrix}$, we have $\rho = \frac{1}{2}(A1 + B\sigma_1 + C\sigma_2 + D\sigma_3)$.

52. Stokes, 'Composition and Resolution of Streams of Different Light', pp. 233–4.

53. G. G. Stokes, 'Report on Double Refraction', *Report of the Thirty-Second Meeting of the British Association for the Advancement of Science; held at Cambridge in October 1862* (London: John Murray, 1863), pp 253–82; repr. in *MPP* vol. IV, pp. 157–202. Cf. Buchwald, 'Why Stokes Never Wrote a Treatise in Optics', pp. 463–7. On the competing dynamical theories, cf. Whittaker, *History of the Theories of Aether and Electricity*, pp. 125–30, 143–4, 161–6, 171–2; Schaffner, *Nineteenth-Century Aether Theories*, Chapter 4; Darrigol, *History of Optics*, pp. 214–20, 229–35.

54. Augustin Cauchy, 'Mémoire sur la théorie de la lumière', *Mémoires de l'Académie des sciences* 10 (1831), pp. 293–316; 'Notes de M. Cauchy sur l'optique, adressées à M. Libri: Première note', *Comptes Rendus hebdomadaires des séances de l'Académie des sciences* 2 (1836), pp. 341–9; 'Mémoire sur la polarisation rectiligne et la double réfraction', *Mémoires de l'Académie des sciences* 18 (1839), pp. 153–216. Stokes, 'Report on Double Refraction', *MPP* vol. IV, pp. 158–68.

55. George Green, 'On the Propagation of Light in Crystallized Media', *Transactions of the Cambridge Philosophical Society* 7 (1842), pp. 121–40. Stokes, 'Report on Double Refraction', *MPP* vol. IV, pp. 168–76.

56. Cf. Whittaker, *History of the Theories of Aether and Electricity*, pp. 153–4.

57. Even though Cauchy still relied on a molecular medium, he could have two elastic constants by assuming an originally compressed state of the medium.

58. Augustin Cauchy, 'Mémoire sur la polarisation des rayons réfléchis ou réfractés par la surface de séparation de deux corps isophanes et transparents', *Comptes Rendus hebdomadaires des séances de l'Académie des sciences* 9 (1839), pp. 676–91. William Thomson [Lord Kelvin], 'On the Reflexion and Refraction of Light', *Philosophical Magazine* 26 (1888), pp. 414–25, 500–1. Cf. Darrigol, *History of Optics*, pp. 235–6.

59. James MacCullagh, 1839. 'An Essay towards the Dynamical Theory of Crystalline Reflexion and Refraction', *Transactions of the Royal Irish Academy* 21(1846), pp. 17–50.

60. Stokes, 'Report on Double Refraction', *MPP* vol. IV, pp. 178–80, 196–7. Cf. Darrigol, *History of Optics*, pp. 243–4.

61. Cf. Olivier Darrigol, *Electrodynamics from Ampère to Einstein* (Oxford: Oxford University Press, 2000), pp. 190, 323.

62. Stokes, 'Report on Double Refraction', *MPP* vol. IV, p. 197.

63. Stokes, 'Report on Double Refraction', *MPP* vol. IV, pp. 180–1.

64. G. G. Stokes, 'On Some Cases of Fluid Motion' [read 20 May 1843], *Transactions of the Cambridge Philosophical Society* 8 (1849); repr. in *MPP* vol. I, pp. 17–68, on pp. 27–8. Stokes to Thomson, 12 Feb 1858, *CSK*, pp. 230–2. Cf. Buchwald 'Why Stokes Never Wrote a Treatise in Optics', pp. 467–8. The electromagnetic theory gives the same result as Fresnel's, with the magnetic permeability μ instead of ρ and the inverse $[\varepsilon]^{-1}$ of the anisotropic dielectric permittivity instead of $[K]$. Stokes's theory would correctly describe waves in a magnetically anisotropic and dielectrically isotropic medium.

65. G. G. Stokes, 'On the Foci of Lines seen through a Crystalline Plate', *Proceedings of the Royal Society of London* 26 (1877); repr. in *MPP* vol. V, pp. 6–23, on p. 22. Stokes, 'Report on Double Refraction', *MPP* vol. IV, pp. 157–202, on pp. 182–3 (allusion), 187–9 (new method of measurement); G. G. Stokes, 'On the Law of Extraordinary Refraction in Iceland Spar', *Proceedings of the Royal Society of London* 20 (1872) ; repr. in *MPP* vol. IV, p. 336. Stokes to Thomson, 1 September 1866, *CSK*, p. 325. William Rankine in 1849 and Lord Rayleigh in 1871 independently proposed the same theory: see Stokes to Strutt, 23 March 1871, in *MSC* vol. II pp. 99–100. As Richard Glazebrook showed in 1888, isotropic inertia agrees with Fresnel's predictions when combined with the labile-ether theory: cf. Buchwald, 'Why Stokes Never Wrote a Treatise in Optics', pp. 472–5; Darrigol, *History of Optics*, p. 236n.

66. Stokes to Thomson, 12 and 13 December 1848, *CSK*, pp. 54–8; James MacCullagh, 'On the Laws of Double Refraction in Quartz' [read 22 February 1836], *Transactions of the Royal Irish Academy* 17 (1837), pp. 461–9. George Biddell Airy, 'On the Equations applying to Light under the Action of Magnetism', *Philosophical Magazine* 28 (1846), pp. 469–77. James Clerk Maxwell, 'On Physical Lines

of Force' [1861–62], *Philosophical Magazine* 21 (1861), pp. 161–75, 281–91, 338–48; 23 (1862), pp. 12–24, 85–95; on pp. 85–95 (1862).

67. Thomson to Stokes, 28 January 1856, *CSK*, p. 209; Stokes to Thomson, 4 February 1856, *CSK*, pp. 210–11.

68. William Thomson [Lord Kelvin], 'Dynamical Illustrations of the Magnetic and Helicoidal Rotary Effects of Transparent Bodies on Polarized Light', *Proceedings of the Royal Society of London* 8 (1856), pp. 150–8. Thomson to Stokes, 17 June 1857, *CSK*, pp. 225–6; 23 December 1857, *CSK*, pp. 227–30, on p. 229 ('unanswerable'); Stokes to Thomson, 12 February 1858, *CSK*, pp. 230–2, on p. 232. Cf. Ole Knudsen, 'The Faraday Effect and Physical Theory', *Archive for History of Exact Sciences*, 15 (1976), pp. 235–81.

CHAPTER 5

1. John Herschel, 'Άμόρφωτα, No. I.—On a Case of Superficial Colour presented by a Homogeneous Liquid Internally Colourless', *Philosophical Transactions of the Royal Society of London* 135 (1845), pp. 143–5. John Herschel, 'Άμόρφωτα No. II.—On the Epipolic Dispersion of Light, being a Supplement to a Paper entitled, "On a Case of Superficial Colour Presented by a Homogeneous Liquid Internally Colourless"', *Philosophical Transactions of the Royal Society of London* 135 (1845), pp. 147–53. On the colours of bodies, cf. Alan Shapiro, *Fits, Passions, and Paroxysms: Physics, Method, and Chemistry and Newton's Theories of Colored Bodies and Fits of Easy Reflection* (Cambridge: Cambridge University Press, 1993), Chapter 9.

2. On the history of fluorescence, cf. Edmund Newton Harvey, *A History of Luminescence from the Earliest Times until 1900* (Philadelphia, PA: American Philosophical Society, 1957), Chapter 9.

3. Frederick William Lewis Fischer, then professor of natural and experimental philosophy at St Andrews.

4. Stokes to Thomson, 14 August 1851, *CSK* pp. 120–1; G. G. Stokes, 'On the Change of Refrangibility of Light' [read 27 May 1852], *Philosophical Transactions of the Royal Society of London* 142 (1852), pp. 463–562; repr. in *MPP* vol. III, pp. 266–413, pp. 271–2. On Stokes's discovery, cf. Frank A. J. L. James, 'The Conservation of Energy, Theories of Absorption and Resonating Molecules, 1851–1854: G. G. Stokes, A. J. Ångström and W. Thomson', *Notes and Records of the Royal Society*, 38 (1983), pp. 79–107; Xiang Chen, *Instrumental Traditions and Theories of Light: The Uses of Instruments in the Optical Revolution* (Dordrecht: Kluwer, 2000), pp. 137–41.

5. Stokes, 'On the Change of Refrangibility of Light', *MPP* vol. III, pp. 272–3 (experiment with filter), pp. 261–2 (experiment with prism), p. 265 (citation). Later in the century, exceptions to Stokes's law were debated and confirmed: cf. Marjorie Malley, 'A Heated Controversy on Cold Light', *Archive for History of Exact Sciences* 42 (1991), pp. 173–86. Contrary to Malley's opinion (p. 182), Stokes never questioned this law and instead suggested a plausible mechanism for seeming exceptions: George Gabriel Stokes, *Burnett Lectures on Light in Three Courses, delivered at Aberdeen in November, 1883, December, 1884, and November, 1885* (London: Macmillan, 1887), pp. 264–5. Also, Stokes never downgraded fluorescence as being 'too much of a specialty as to be of very general interest': he used these words (p. 155) to qualify the complex variations of the fluorescence spectrum with the frequency of the incoming light.

6. Stokes, 'On the Change of Refrangibility of Light', *MPP* vol. III, p. 288.

7. Stokes, 'On the Change of Refrangibility of Light', *MPP* vol. III, pp. 289n (citation), 386 (phosphorescence). In a sequel to this memoir (G. G. Stokes, 'On the Change of Refrangibility of Light.—II' [read 16 June 1853], *Transactions of the Cambridge Philosophical Society* 9 (1856); repr. in *MPP* vol. IV, pp. 1–17, p. 4n), Stokes expressed his preference for the theory-neutral 'fluorescence' and his belief in the novelty of fluorescence with respect to phosphorescence. Edmond Becquerel soon demonstrated a small delay in the emission of fluorescence light and used this finding to deny any fundamental difference with phosphorescence: cf. Edmond Becquerel, *La lumière, ses causes et ses effets*, 2 vols (Paris: Firmin Didot, 1867), vol. 1 pp. 320–1.

8. In 1842 John Herschel had observed a much widened, ultra-violet-inclusive spectrum by means of turmeric paper. He regarded this observation as a proof of the visibility of ultra-violet light, ascribing its normal lack of lightening power to surface absorption. Cf. Stokes, 'On the Change of Refrangibility of Light', *MPP* vol. III, p. 323.

9. Stokes, 'On the Change of Refrangibility of Light', *MPP* vol. III, pp. 312, 335.

10. Stokes, 'On the Change of Refrangibility of Light.—II', *MPP* vol. IV, p. 1.

11. G. G. Stokes, 'On the Existence of a Second Crystallizable Fluorescent Substance (Paviin) in the Bark of the Horse-Chestnut', *Quarterly Journal of the Chemical Society of London* 11 (1859), pp. 17-21; repr. in *MPP* vol. IV, pp. 113–16; G. G. Stokes, 'Optical Characters of Purpurine and Alizarine', Appendix to Edward Schunck, 'On the Colouring Matters of Madder', *Quarterly Journal of the Chemical Society of London* 12 (1860), pp. 198–221, on pp. 219–21; repr. in *MPP* vol. IV, pp. 122–6; G. G. Stokes, 'On the Supposed Identity of Biliverdin with Chlorophyll, with Remarks on the Constitution of Chlorophyll', *Proceedings of the Royal Society of London* 13 (1864), pp. 144–5; repr. in *MPP* vol. IV, pp. 236–7; G. G. Stokes, 'On the Discrimination of Organic Bodies by their Optical Properties' [1864], *Proceedings of the Royal Institution of Great Britain* 4 (1862–6), pp. 223–31; repr. in *MPP* vol. IV, pp. 238–48; G. G. Stokes, 'On the Application of the Optical Properties of Bodies to the Detection and Discrimination of Organic Substances', *Journal of the Chemical Society* 17 (1864), pp. 304–18; repr. in *MPP* vol. IV, pp. 249–63; G. G. Stokes, 'On the Reduction and Oxidation of the Colouring Matter of the Blood', *Proceedings of the Royal Society of London* 13 (1864), pp. 355–64; in *MPP* vol. IV, pp. 264–76.

12. G. G. Stokes, 'On the Change of Refrangibility of Light, and the Exhibition thereby of Chemical Rays', *Proceedings of the Royal Institution of Great Britain* 1 (1853), pp. 259–64; repr. in *MPP* vol. IV, pp. 22–9. G. G. Stokes, 'On the Long Spectrum of Electric Light', *Philosophical Transactions of the Royal Society of London* 152 (1862), pp. 599–619; repr. in *MPP* vol. IV, pp. 203–35.

13. Stokes, 'On the Change of Refrangibility of Light', *MPP* vol. III, pp. 391–2 , 410–13 (recollection of a model imagined before 1867).

14. Stokes, 'On the Change of Refrangibility of Light', *MPP* vol. III, pp. 394–6. On Talbot and Herschel, cf. William McGucken, *Nineteenth-Century Spectroscopy: Development of the Understanding of Spectra 1802–1897* (Baltimore, MD: Johns Hopkins Press, 1969), pp. 18–19.

15. Thomson to Stokes, 20 February 1854, *CSK*, p. 134; Joseph Fraunhofer, 'Bestimmung des Brechungs- und Strahlungsvertreuungs-Vermögens verschiedener Glasarten, in Bezug auf die Vervollkommnung achromatischer Fernrohre', *Annalen der Physik* 56 (1817), pp. 264–313; pp. 278–83 (sun), 290 (stars), 269 (D in flames), 312 (D in sunlight). Cf. McGucken, *Nineteenth-Century Spectroscopy*, pp. 3–4; Myles Jackson, *Spectrum of Belief: Joseph von Fraunhofer and the Craft of Precision Optics* (Cambridge, MA: MIT Press, 2000).

16. Stokes to Thomson, 24 February 1854, *CSK*, pp. 135–6; François Napoléon Marie (abbé) Moigno, *Répertoire d'optique moderne ou analyse complète des travaux modernes relatifs aux phénomènes de la lumière*, 4 vols (Paris: Franck, 1850), vol. III, p. 1237.

17. Thomson to Stokes, 2 March 1854, *CSK*, pp. 137–9; Stokes to Thomson, 7 March 1854, *CSK*, pp. 140–1, and 26 November 1855, *CSK*, p. 204. On this story, cf. Stokes to Thomson, 5 July 1871, *CSK*, pp. 356–9; McGucken, *Nineteenth-Century Spectroscopy*, pp. 22–4.

18. Cf. McGucken, *Nineteenth-Century Spectroscopy*, pp. 23–34; Christa Jungnickel and Russell McCormmach, *Intellectual Mastery of Nature: Theoretical Physics from Ohm to Einstein*, vol. 1: *The Torch of Mathematics, 1800–1870*, vol. 2: *The Now Mighty Theoretical Physics, 1870–1925* (Chicago: University of Chicago Press, 1986), vol. 1, pp. 297–302.

19. Stokes, in *MPP* vol. IV, p. 130.

20. For instance, Peter Guthrie Tait, *Lectures on Some Recent Advances in Physical Science* (London: Macmillan, 1876), p. 191.

21. William Thomson, 'Physical Considerations regarding the Possible Age of the Sun's Heat', *Philosophical Magazine* 23 (1862), pp. 158–60, on pp. 158–9. Stokes to Roscoe, 7 February 1862, in

MSC Vol. II, p. 83; Stokes to Thomson, 11 January 1876, in *CSK*, pp. 410–11; G. G. Stokes, 'On the Early History of Spectrum Analysis' [Stokes to Whitmell, 23 December 1875], *Nature* 12 (1876); repr. in *MPP* vol. IV, pp. 133–6.

22. Thomson to Stokes, 27 January 1884, *CSK*, p. 550; Stokes to Thomson, 30 January 1884, *CSK*, pp. 551–2. On the history of anomalous dispersion, cf. Olivier Darrigol, *A History of Optics from Greek Antiquity to the Nineteenth Century* (Oxford: Oxford University Press, 2012), pp. 250–2.

23. Werner Haidinger, 'Über das direkte Erkennen des polarisirten Lichtes und der Lage der Polarizationsebene', *Annalen der Physik* 63 (1844), pp. 29–39. G. G. Stokes, 'On Haidinger's Brushes', *Report of the Twentieth Meeting of the British Association for the Advancement of Science; held at Edinburgh in July and August 1850* (London: John Murray, 1851), Notices and Abstracts pp. 20–1; repr. in *MPP* vol. II, pp. 362–4; G. G. Stokes, 'Einige neuere Ansichten über die Natur der Polarisations-büschel' [Stokes to Haidinger, 9 February 1854], *Annalen der Physik* 96 (1855); repr. in *MPP* vol. IV, p. 60. Hermann Helmholtz, *Handbuch der physiologischen Optik* (Leipzig: Voss 1867), pp. 421–3.

24. G. G. Stokes, 'On the Optical Properties of a recently discovered Salt of Quinine', *Report of the Twenty-Second Meeting of the British Association for the Advancement of Science; held at Belfast in September 1852* (London: John Murray, 1853), pp. 15–16; repr. in *MPP* vol. IV, pp. 18–21. G. G. Stokes, 'Die Richtung der Schwingungen des Lichtäthers im polarisirten Lichte: Mittheilung aus einem Schreiben des Hrn. Prof. Stokes, nebst Bemerkungen von W. Haidinger' [Stokes to Haidinger, 9 February 1854], *Annalen der Physik* 96 (1855); repr. in *MPP* vol. IV, pp. 50–4. On the history of the Polaroïd, cf. Edwin Herbert Land, 'Some Aspects on the Development of Sheet Polarizers', *Journal of the Optical Society of America* 41 (1951), pp. 957–63.

25. G. G. Stokes, 'On the Cause of the Occurrence of Abnormal Figures in Photographic Impressions of Polarized Rings', *Philosophical Magazine* 6 (1853), pp. 107–13; repr. in *MPP* vol. IV, pp. 30–7; G. G. Stokes, 'Mittheilung aus einem Schreiben des Hrn. Prof. Stokes, über das optische Schachbrettmuster' [Stokes to Haidinger, 9 February 1854], *Annelen der Physik* 96 (1855); repr. in *MPP* vol. IV, pp. 55–60. G. G. Stokes, 'On a Phenomenon of Metallic Reflexion', *Report of the Forty-Sixth Meeting of the British Association for the Advancement of Science; held at Glasgow in September 1876* (London: John Murray, 1877), pp. 41–2; repr. in *MPP* vol. IV, pp. 361–4. G. G. Stokes, 'On the Cause of the Light Border frequently noticed in Photographs just outside the Outline of a Dark Body seen against the Sky: with Some Introductory Remarks on Phosphorescence', *Proceedings of the Royal Society of London*, 34 (1882); repr. in *MPP* vol. V, pp. 117–24.

26. G. G. Stokes, 'On a Remarkable Phenomenon of Crystalline Reflection', *Proceedings of the Royal Society of London* 38 (1885); repr. in *MPP* vol. V, pp. 164–79. Lord Rayleigh [John William Strutt], 'Sir George Gabriel Stokes, Bart. 1819–1903', *Proceedings of the Royal Society* 75 (1905), pp. 199–216; repr. in *MPP* vol. V, pp. ix–xxv, on p. xxi. Lord Rayleigh [John William Strutt], 'On the Reflexion of Light at a Twin Plane of a Crystal', *Philosophical Magazine* 26 (1888), pp. 256–65.

27. G. G. Stokes, 'On the Nature of the Röntgen Rays' [1896]. *Proceedings of the Cambridge Philosophical Society* 9 (1895–8), pp. 215–16; repr. in *MPP* vol. V, pp. 254–5. G. G. Stokes, 'On the Nature of the Röntgen Rays' [1897], *Memoirs and Proceedings of the Manchester Literary and Philosophical Society* 41 (1896–7), pp. 1–28; repr. in *MPP* vol. V, pp. 256–77. Cf. David B. Wilson, *Kelvin and Stokes: A Comparative Study in Victorian Physics* (Bristol: Adam Hilger, 1987), pp. 203–9.

28. G. G. Stokes, 'On a Mode of Measuring the Astigmatism of a Defective Eye', *Report of the Nineteenth Meeting of the British Association for the Advancement of Science; held at Birmingham in September 1849* (London: John Murray, 1850), Notices and Abstracts pp. 10–11; repr. in *MPP* vol. II, pp. 172–5; G. G. Stokes, 'On the Determination of the Wave Length corresponding with any Point of the Spectrum', *Report of the Nineteenth Meeting of the British Association for the Advancement of Science; held at Birmingham in September 1849* (London: John Murray, 1850), Notices and Abstracts p. 11; repr. in *MPP* vol. II, pp. 176–7. G. G. Stokes, 'On Metallic Reflexion', *Report of the Twentieth Meeting of the British Association for the Advancement of Science; held at Edinburgh in July and August 1850* (London: John Murray, 1851), Notices and Abstracts pp. 19–20; repr. in *MPP* vol. II, pp. 360.

G. G. Stokes, 'On a New Elliptic Analyser', *Report of the Twenty-First Meeting of the British Association for the Advancement of Science; held at Ipswich in July 1851* (London: John Murray, 1852), Notices and Abstracts pp. 14–15; repr. in *MPP* vol. III, pp. 197–202. G. G. Stokes, 'On the Achromatism of a Double Object-glass', *Report of the Twenty-Fifth Meeting of the British Association for the Advancement of Science; held at Glasgow in September 1855* (London: John Murray, 1856), Notices and Abstracts pp. 14–15; repr. in *MPP* vol. IV, pp. 63–4. G. G. Stokes, 'On the Principles of the Chemical Correction of Object-Glasses', *Photographic Journal* (1873); repr. in *MPP* vol. IV, pp. 344–54. G. G. Stokes, 'On the Intensity of the Light Reflected from or Transmitted through a Pile of Plates', *Proceedings of the Royal Society of London* 11 (1862), pp. 545–56; repr. in *MPP* vol. IV, pp. 145–56. G. G. Stokes, 'On a Formula for Determining the Optical Constants of Doubly Refracting Crystals', *Cambridge and Dublin Mathematical Journal* 1 (1846), pp. 183–7; repr. in *MPP* vol. I, pp. 148–52. G. G. Stokes, 'Report on Double Refraction', *Report of the Thirty-Second Meeting of the British Association for the Advancement of Science; held at Cambridge in October 1862* (London: John Murray, 1863), pp 253–82; repr. in *MPP* vol. IV, pp. 157–202, on pp. 187–9. G. G. Stokes, 'On the Law of Extraordinary Refraction in Iceland Spar', *Proceedings of the Royal Society of London* 20 (1872) ; repr. in *MPP* vol. IV, p. 336. G. G. Stokes, 'On the Foci of Lines Seen through a Crystalline Plate', *Proceedings of the Royal Society of London* 26 (1877); repr. in *MPP* vol. V, pp. 6–23.

29. G. G. Stokes, 'Notice of the Researches of the late Rev. W. Vernon Harcourt on the Conditions of Transparency in Glass and the Connexion between the Chemical Constitution and Optical Properties of Different Glasses', *Report of the Forty-First Meeting of the British Association for the Advancement of Science; held at Edinburgh in August 1871* (London: John Murray, 1872), Notices and Abstracts pp 38–41; repr. in *MPP* vol. IV, pp. 339–43, on p. 340 (citation); *MSC* vol. I, pp. 201 (Harcourt), 204 (Grubb); *MSC* vol. II, p. 87 (Grubb); G. G. Stokes and J. Hopkinson, 'On the Optical Properties of a Titano-Silicic Glass', *Report of the Forty-Fifth Meeting of the British Association for the Advancement of Science; held at Bristol in August 1875* (London: John Murray, 1876), pp. 26–7; repr. in *MPP* vol. IV, pp. 358–60 (titanic glass); G. G. Stokes, 'On the Question of a Theoretical Limit to the Apertures of Microscopic Objectives', *Journal of the Royal Microscopical Society* 1 (1878); repr. in *MPP* vol. V, pp. 36–9. G. G. Stokes, 'On an Easy and at the Same Time Accurate Method of Determining the Ratio of the Dispersions of Glasses intended for Objectives', *Proceedings of the Royal Society of London* 27 (1878), pp. 485–94; repr. in *MPP* vol. V, pp. 40–51. On Harcourt and Stokes, cf. Wilson, *Kelvin and Stokes*, p. 189. On the Zeiss success, cf. Stuart Feffer, '*Microscopes to Munitions: Ernst Abbe, Carl Zeiss, and the Transformation of Technical Optics, 1850–1914*, PhD dissertation, University of California at Berkeley, 1994.

30. Thomson, William [Lord Kelvin], 'Copley Medal. Sir G. Gabriel Stokes, Bart., F.R.S.', *Proceedings of the Royal Society of London*, 54 (1893), pp. 389–91; cited in *MSC* vol. I, pp. 268–70, on p. 269.

31. Cf. Wilson, *Kelvin and Stokes*, pp. 41, 44–5; David B. Wilson, 'Arbiters of Victorian Science: George Gabriel Stokes and Joshua King', in Kevin C. Knox and Richard Noakes (eds), *From Newton to Hawking: A History of Cambridge University's Lucasian Professors of Mathematics* (Cambridge: Cambridge University Press, 2003), pp. 295–342.

32. William Thomson [Lord Kelvin], 'The Scientific Work of Sir George Stokes', *Nature* 67 (1903), pp. 337–8; cited in *MSC* vol. I, pp. 307–9, on p. 309: 'I wish some of the students who have followed his Lucasian lectures could publish to the world his *Opticae Lectiones*; it would be a fitting sequel to the "Opticae Lectiones" of his predecessor in the Lucasian chair, Newton.'

33. Cf. Rayleigh, 'George Gabriel Stokes', pp. xx–xxi; Jed Buchwald, 'Why Stokes Never Wrote a Treatise in Optics', in Peter Harman and Alan Shapiro (eds), *An Investigation of Difficult Things: Essays on Newton and the History of Exact Sciences* (Cambridge: Cambridge University Press, 1992), pp. 451–76, on pp. 451–2.

34. Stokes, *Burnett Lectures on Light*. Cf. Buchwald, 'Why Stokes Never Wrote a Treatise in Optics', p. 453.

35. Stokes, *Burnett Lectures on Light*, p. 169.

36. On Stokes's theological concerns, cf. Wilson, *Kelvin and Stokes*, Chapter 4.

37. G. G. Stokes, 'On the Numerical Calculation of a Class of Definite Integrals and Infinite Series', *Transactions of the Cambridge Philosophical Society* 9 (1850); repr. in *MPP* vol. II, pp. 329–57. G. G. Stokes, 'Note on Internal Reflexion', *Proceedings of the Royal Society of London* 11 (1861); repr. in *MPP* vol. IV, pp. 137–44 (property of wave surface required for extension of Stewart–Kirchhoff law when transparent medium between the absorbers is anisotropic). Thomson, 'The Scientific Work of Sir George Stokes', *MSC* vol. I, p. 308.

38. Joseph John Thomson, 'The Stokes Jubilee', *Cambridge Review* 20 (1899), pp. 370–1, on p. 370.

39. G. G. Stokes, 'On the Perfect Blackness of the Central Spot in Newton's Rings, and on the Verification of Fresnel's Formulae for the Intensities of Reflected and Refracted Rays', *Cambridge and Dublin Mathematical Journal* 4 (1849); repr. in *MPP* vol. II, pp. 89–103, on p. 97.

40. Rayleigh, 'George Gabriel Stokes', p. xix: ' The reflection suggests itself that scientific men should be kept to scientific work, and should not be tempted to assume heavy administrative duties, at any rate until such time as they have delivered their more important messages to the world'; Buchwald 'Why Stokes never wrote a treatise in optics'. On the rise of microphysical approaches, cf. Jed Buchwald, *From Maxwell to Microphysics: Aspects of Electromagnetic Theory in the Last Quarter of the Nineteenth Century* (Chicago: University of Chicago Press, 1985).

41. Thomson, 'Copley Medal', *MSC* vol. I, p. 269.

42. Lord Rayleigh [John William Strutt] [Allocution at the Stokes memorial ceremony of 7 July 1904], *MSC* vol. I, pp. 318–20, on p. 319; Rayleigh, 'George Gabriel Stokes', p. xix; Richard Claverhouse Jebb, 'To Sir George Gabriel Stokes', *Cambridge Review* 20 (1899), p. 370.

CHAPTER 6

1. George Gabriel Stokes, *Mathematical and Physical Papers*, 5 vols (Cambridge: Cambridge University Press, 1880–1905).

2. G. G. Stokes, 'On Some Cases of Fluid Motion' [read on 20 May 1843], *Transactions of the Cambridge Philosophical Society* 8 (1849); repr. in *MPP* vol. I, pp. 17–68, on p. 54.

3. G. G. Stokes, 'On the Theories of the Internal Friction of Fluids in Motion, and of the Equilibrium and Motion of Elastic Solids' [read 14 April 1845], *Transactions of the Cambridge Philosophical Society* 8 (1849), pp. 287–319; repr. in *MPP* vol. I, pp. 75–129.

4. G. G. Stokes, 'On the Theory of Oscillatory Waves', *Transactions of the Cambridge Philosophical Society* 8 (1847), pp. 441–55; repr. in *MPP* vol. I, pp. 197–229.

5. G. G. Stokes, 'On the Effect of the Internal Friction of Fluids on the Motion of Pendulums', *Transactions of the Cambridge Philosophical Society* 9 (1850), Part II pp. 8–106; repr.in *MPP* vol. III, pp. 1–141.

6. G. G. Stokes, 'On the Highest Wave of Uniform Propagation (Preliminary Notice)', *Proceedings of the Cambridge Philosophical Society* 4 (1883), pp. 361–65.

7. Olivier Darrigol, 'Between Hydrodynamics and Elasticity Theory: The First Five Births of the Navier–Stokes Equation', *Archive for History of Exact Sciences* 56 (2002), pp. 95–150.

8. Alex D. D. Craik, 'The Origins of Water Wave Theory', *Annual Review of Fluid Mechanics* 36 (2004), pp. 1–28.

9. J. Challis, 'On the Principles of Hydrodynamics', *Philosophical Magazine* 1/4 (1851), pp. 26–38.

10. Pijush K. Kundu and Ira M. Cohen, *Fluid Mechanics* (2nd edn, San Diego, CA: Academic Press, 2002).

11. Peter Bauer, Alan Thorpe and Gilbert Brunet, 'The Quiet Revolution of Numerical Weather Prediction', *Nature* 525 (2015), pp. 47–55.

12. Olivier Darrigol, 'The Spirited Horse, The Engineer and the Mathematician: Water Waves in Nineteenth-Century Hydrodynamics', *Archive for History of Exact Sciences* 58 (2003), pp. 21–95.

13. George Biddell Airy, 1841: 'Tides and Waves', *Encyclopaedia Metropolitana* (1817–45), pp. 241–396.

14. D. J. Korteweg and G. de Vries, 'On the Change of Form of Long Waves Advancing in a Rectangular Canal, and on a New Type of Long Stationary Waves', *Philosophical Magazine* 39/5 (1895), pp. 422–43.
15. Horace Lamb, *Hydrodynamics* (6th edn, Cambridge: Cambridge University Press, 1932), p. 425.
16. Ali Hasan Nayfeh, *Perturbation Methods* (New York: John Wiley & Sons, 2000).
17. G. G. Stokes, 'Supplement to a Paper on the Theory of Oscillatory Waves', *MPP* vol. I, pp. 314–26.
18. Darrigol, 'Between Hydrodynamics and Elasticity Theory', p. 62.
19. Alex D. D. Craik, 'George Gabriel Stokes on Water Wave Theory', *Annual Review of Fluid Mechanics* 37 (2005), pp. 23–42, on p. 30.
20. Lamb, *Hydrodynamics*, p. 417 eqn (3).
21. G. B. Whitham, *Linear and Nonlinear Waves* (New York: John Wiley & Sons, 1999), p. 12 eqn (1.33).
22. Lamb, *Hydrodynamics*, p. 419 eqn (17).
23. Lamb, *Hydrodynamics*, p. 381.

CHAPTER 7

1. The chapter is based in part on R. B. Paris, 'The Mathematical Work of G. G. Stokes', *Mathematics Today* 32 (1996), pp. 43–6.
2. G. H. Hardy, 'Sir George Stokes and the Concept of Uniform Convergence', *Proceedings of the Cambridge Philosophical Society* 19 (1918), pp. 149–56.
3. G. G. Stokes, 'On the Critical Values of the Sum of Periodic Series', *Transactions of the Cambridge Philosophical Society* ? (1847), pp. 533–583; repr. in *MPP* vol. I, pp. 236–313.
4. P. G. Dirichlet, 'Sur la convergence des séries trigonométriques qui servent à représenter une fonction arbitraire entre des limites données', *Journal für die reine und angewandte Mathematik* 4 (1829), pp. 157–69.
5. *MPP* vol. I, p. 251.
6. E. T. Whittaker and G. N. Watson, *A Course of Modern Analysis* (4th edn, Cambridge: Cambridge University Press, 1962), Section 9.3.
7. Bernhard Riemann, 'Ueber die Darstellbarkeit einer Function durch eine trigonometrische Reihe' (1854), repr. in *The Collected Works of Bernhard Riemann* (New York: Dover, 1953), pp. 227–64.
8. William Rowan Hamilton, 'On Fluctuating Functions', *Transactions of the Royal Irish Academy* 19 (1843), pp. 264–321.
9. *MPP* vol. II, p. 178.
10. *MPP* vol. II, p. 329.
11. George Biddell Airy, 'On the Intensity of Light in the Neighbourhood of a Caustic', *Transactions of the Cambridge Philosophical Society* 6 (1838), pp. 379–402.
12. *MPP* vol. II, p. 335.
13. Bernhard Riemann, 'Sullo svolgimento del quoziente di due series ipergeometriche in frazione continua infinita' (1863), repr. in *The Collected Works of Bernhard Riemann* (New York: Dover, 1953), pp. 424–30.
14. *MPP* vol. II, p. 341.
15. *MPP* vol. IV, p. 77.
16. *MPP* vol. IV, p. 283.
17. *MPP* vol. V, p. 283.
18. H. Poincaré, 'Sur les intégrales irrégulières des équations linéaires', *Acta Mathematica* 8 (1886), pp. 295–344.
19. T.-J. Stieltjes, 'Recherches sur quelques séries semi-convergentes', *Annales scientifiques de l'École Normale Supérieure* 3 (1886), pp. 201–58.
20. M. V. Berry, 'Uniform Asymptotic Smoothing of Stokes's Discontinuities', *Proceedings of the Royal Society of London* A 422 (1989), pp. 7–21.

21. Michael V. Berry and Christopher J. Howls, 'Divergent Series: Taming the Tails', in Nicholas J. Higham (ed.), *The Princeton Companion to Applied Mathematics* (Princeton, NJ: Princeton University Press, 2015).

22. R. B. Paris and A. D. Wood, 'Stokes Phenomenon Demystified', *Bulletin of the Institute of Mathematics and Its Applications* 31 (1995), pp. 21–8.

23. *MPP* vol. V, p. 221.

24. R. B. Paris, 'The Discrete Analogue of Laplace's Method', *Computers and Mathematics with Applications* 61 (2011), pp. 3024–3034.

CHAPTER 8

1. Mark McCartney, 'William Thomson: An Introductory Biography', in Raymond Flood, Mark McCartney and Andrew Whitaker (eds), *Kelvin: Life, Labours and Legacy* (Oxford: Oxford University Press, 2008), pp. 1–22, on p. 13.

2. Duke of Sussex, 'Address Delivered before the Royal Society', *Abstracts of Papers Printed in the Philosophical Transactions of the Royal Society of London* 3 (1830–37), pp. 140–55, on p. 141. For more on the evolution of the peer review process of the Royal Society, see Noah Moxham and Aileen Fyfe, 'The Royal Society and the Prehistory of Peer Review, 1665–1965', *The Historical Journal*, 61/4 (2018), pp. 863–89 and Sloan Despeaux, 'Fit to Print? Referee Reports on Mathematics for the Nineteenth-Century Journals for the Royal Society of London', *Notes and Records of the Royal Society of London* 65 (2011), pp. 233–52.

3. Michael Foster, 'Appreciation', in Joseph Larmor (ed.), *Memoir and Scientific Correspondence of the Late Sir George Gabriel Stokes*, vol. 1 (Cambridge: Cambridge University Press, 1880), pp. 97–103, on p. 98.

4. Foster, 'Appreciation', p. 99.

5. Melinda Baldwin, 'Tyndall and Stokes: Correspondence, Referee Reports and the Physical Sciences in Victorian Britain', in Bernard Lightman and Michael S. Reidy (eds), *The Age of Scientific Naturalism: Tyndall and His Contemporaries* (London: Routledge, 2016), pp. 171–86, on p. 176.

6. Andrew John Harrison, 'Scientific Naturalists and the Government of the Royal Society, 1850–1900', PhD thesis, Open University, 1988, pp. 115–16.

7. Mrs. [Isabella Lucy] Laurence Humphry, 'Notes and Recollections', in Larmor, *Memoir*, pp. 1–49, on pp. 37.

8. Harrison, 'Scientific Naturalists', p. 97.

9. Harrison, 'Scientific Naturalists', pp. 97–8.

10. Baldwin, 'Tyndall and Stokes', p. 175.

11. For more on the issue of referees as Council members versus Fellows, see Despeaux, 'Fit to Print?, p. 244.

12. Harrison, 'Scientific Naturalists', p. 85.

13. Moxham and Fyfe, 'The Royal Society', p. 17.

14. Harrison, 'Scientific Naturalists', p. 92.

15. Baldwin, 'Tyndall and Stokes', p. 180.

16. Baldwin, 'Tyndall and Stokes', p. 174.

17. N. L. Biggs, 'T.P. Kirkman, Mathematician', *Bulletin of the London Mathematical Society* 13 (1981): pp. 97–120, on pp. 97–8.

18. Grand Prix de Mathématique question for 1861; quoted in Biggs, 'T.P. Kirkman', p. 105: 'Perfectionner en quelque point important la théorie géométrique des polyèdres.'

19. Biggs, 'T.P. Kirkman', pp. 106–7.

20. Thomas P. Kirkman to Stokes, 13 April 1861, Stokes Papers, Cambridge University Library, Add 7656, RS242.

21. Alexander MacFarlane related that Kirkman preferred to write 'polyedron' instead of 'polyhedron', because the spelling of 'periodic' is not 'perihodic'. Alexander MacFarlane, *Lectures on Ten British Mathematicians of the Nineteenth Century* (New York: John Wiley & Sons, 1916), p. 126.

22. Referee's Report, Thomas Archer Hirst, 23 June 1861, Royal Society Archives, London, R.R.4.147.

23. Referee's Report, William Spottiswoode to Stokes, 4 July 1861, Royal Society Archives, London, R.R.4.148.

24. Referee's Report, Arthur Cayley, 1861 [day and month unknown], Royal Society Archives, London, R.R.4.150.

25. Thomas P. Kirkman to Stokes, 14 December 1861, Stokes Papers, Cambridge University Library, Add 7656, RS284.

26. Tony Crilly, *Arthur Cayley: Mathematician Laureate of the Victorian Age* (Baltimore, MD: Johns Hopkins University Press, 2006), p. 143. Indeed, while Kirkman 'was an isolated figure who remained outside the metropolitan fold, Cayley respected him and responded to his research.' p. 144.

27. Thomas P. Kirkman to Stokes, 21 December 1861, Stokes Papers, Cambridge University Library, Add 7656, RS287. Kirkman's emphasis.

28. Referee's Report, Arthur Cayley, 17 February 1862, Royal Society Archives, London, R.R.4.150.

29. Thomas P. Kirkman, 'On the Theory of the Polyedra', *Philosophical Transactions of the Royal Society of London* 152 (1862), pp. 121–65.

30. Michael Foster, 'A Note on the History of the Statutes of the Society', *Proceedings of the Royal Society of London* 50 (1891–92), pp. 501–15, on p. 512. This practice remained in effect until 1984. Despeaux, 'Fit to Print?, p. 248. In a letter to Thomas Huxley, Joseph Hooker explained the practice: 'I have long thought that the retention of rejected papers was a course that had its awkward side; it is so often regarded, however unreasonably, as "suppression" of the papers, which, added to rejection, piles the horrors. We must be unfettered in our power of rejection and we must keep the originals as our pièces justificatives, and I see no middle course but that of offering copies to be made at the author's expense.' Hooker to Huxley, in Harrison, 'Scientific Naturalists', p. 111.

31. Thomas P. Kirkman, 'Applications of the Theory of the Polyedra to the Enumeration and Registration of Results', *Proceedings of the Royal Society of London* 12 (1862–3), pp. 341–80.

32. For more on the evolution of the *Proceedings*, see Despeaux, 'Fit to Print?, pp. 241–4.

33. Moxham and Fyfe, 'The Royal Socicty', p. 17.

34. Thomas P. Kirkman to Stokes, 2 Jan. 1862, Stokes Papers, Cambridge University Library, Add 7656, RS291.

35. Mary Boas Hall, *All Scientists Now: The Royal Society in the Nineteenth Century* (Cambridge: Cambridge University Press, 1984), p. 136.

36. Sir Henry George Lyons, *The Royal Society 1660–1940: A History of Its Administration under Its Charters* (New York: Greenwood Press, 1968), p. 270.

37. Lyons, *The Royal Society*, p. 270.

38. See Chapter 1 of this volume.

39. William Huggins, 'Appreciation', in Larmor, *Memoir*, p. 103.

40. Howard Grubb in Larmor, *Memoir*, pp. 346–7. For more on this consultancy work, see Chapter 1, this volume.

41. Baldwin, 'Tyndall and Stokes', p. 174.

42. Baldwin, 'Tyndall and Stokes', p. 174.

43. Hall, *All Scientists Now*, p. 136.

44. Stokes quoted in Huggins, p. 104.

45. Hall, *All Scientists Now*, p. 139.

46. P. G. Tait, 'Scientific Worthies: V. George Gabriel Stokes', *Nature* **12** (15 July 1875), pp. 201–3, on pp. 201–2.

47. Thomson to Stokes, November 1884 in Alastair Wood, 'Fifty-Eight Years of Friendship: Kelvin and Stokes', in *Kelvin: Life, Labours, and Legacy*, pp. 64–85, on p. 80. Stokes did not apply for the chair in order to not, in his words, 'block the way of promotion to younger men who might be reasonably expected to rise in their profession'. Ibid.

48. Michael Foster to Lady Rayleigh, 1884 in Robert John Strutt, *Life of John William Strutt* (London: Edward Arnold & Co., 1924), p. 168.
49. Hall, *All Scientists Now*, p. 139.
50. Stokes to Thomas Romney Robinson, 1 December 1877, in Larmor, *Memoir*, pp. 39–41, on p. 40.
51. Baldwin, 'Tyndall and Stokes', p. 172.
52. Humphry, 'Notes and Recollections', p. 24.
53. Humphry, 'Notes and Recollections', pp. 36–7.
54. Baldwin, 'Tyndall and Stokes', p. 172.
55. For example, Foster to Huxley, 4 December 1884, Huxley Papers, Imperial College London, 4.233 in Harrison, 'Scientific Naturalists', p. 223.
56. Baldwin, 'Tyndall and Stokes', p. 173. For more on scientific naturalism and especially its role in mathematics and physics, see Bernard Lightman and Michael S. Reidy (eds), *The Age of Scientific Naturalism: Tyndall and His Contemporaries* (London: Routledge, 2016).
57. William White (ed.), *The Journals of Walter White* (London: Chapman and Hall, 1898), p. 236.
58. Harrison, p. 317.
59. A. J. Meadows, *Science and Controversy: A Biography of Sir Normal Lockyer, Founder Editor of* Nature (2nd edn, London: Macmillan, 2008), p. 226.
60. [Thomas Huxley], 'M.P., P.R.S.' *Nature* 37 (17 November 1887): 49–50.
61. Ibid.
62. W.T. Thiselton Dyer, 'Politics and the Presidency of the Royal Society', *Nature* **37** (1 December 1887): 103–4.
63. Balfour Stewart, 'Politics and the Presidency of the Royal Society', *Nature* **37** (24 November 1887): 76.
64. George Gabriel Stokes, 'President's Address', *Proceedings of the Royal Society of London* **43** (1887–8), pp. 184–210, on pp. 189–90.
65. Foster, 'Appreciation', p. 102.
66. Huxley to Stokes, 1 December 1887, in *MSC*, p. 178.
67. Huxley to Stokes, 1 December 1887, in *MSC*, p. 178, p. 179.
68. Foster, 'Appreciation', p. 101.
69. Humphry, 'Notes and Recollections', p. 44. Stokes actually spoke three times in five years. For more on Stokes in Parliament, see Chapter 1, this volume.
70. George Gabriel Stokes, 'President's Address', *Proceedings of the Royal Society of London* 48 (1890), pp. 465–75, on p. 472.
71. Humphry, 'Notes and Recollections', p. 44.
72. Humphry, 'Notes and Recollections', p. 45.
73. David B. Wilson, *Kelvin and Stokes: A Comparative Study in Victorian Physics* (Bristol: Adam Hilger, 1987), p. 208.
74. Stokes's long tenure as a Royal Society officer was not that unusual during the nineteenth century. Sharpey and Foster served terms of seventeen years and twenty-two years respectively as 'B Secretaries', and John Evans was Treasurer for twenty years. Lyons, *The Royal Society*, p. 274.

CHAPTER 9
1. David Deutsch, *The Fabric of Reality* (London: Allen Lane, 1997).
2. Paul Sibly, 'The Prediction of Structural Failures', PhD thesis, University College London, 1977.
3. T. M. Charlton, *A History of Theory of Structures in the Nineteenth Century* (Cambridge: Cambridge University Press, 1982).
4. P. G. Sibly and A. C. Walker, 'Structural Accidents and their Causes', *Proceedings of the Institution of Civil Engineers* 62 (1977), pp. 191–208.
5. Henry Petroski, *Design Paradigms: Case Histories of Error and Judgment in Engineering* (Cambridge: Cambridge University Press, 1994).

6. N. Subramanian, 'I-35W Mississipi Bridge Failure—Is it a Wake Up Call?', *Indian Concrete Journal* (February 2008), 29–38.

7. L. T. C. Rolt, *Red for Danger: A History of Railway Accidents and Railway Safety Precautions* (London: Bodley Head, 1955).

8. Benjamin Baker, *Long-Span Railway Bridges* (London: E. & F. N. Spon, 1873).

9. Arnold Koerte, *Two Railway Bridges of an Era: Firth of Forth and Firth of Tay* (Basel: Birkhaüser, 1992).

10. David McCullough, *The Great Bridge* (New York: Simon and Schuster, 1972).

11. D. W. Smith, 'Bridge Failure', *Proceedings of the Institution of Civil Engineers* 60 (1960), pp. 367–82.

12. Andrew Lally, 'Steel Box Girder Bridges', *Engineering Journal*, American Institute of Steel Construction, 10 (1973), pp. 117–30.

13. Christopher Alexander, *Notes on the Synthesis of Form* (Cambridge, MA: Harvard University Press, 1964).

14. 'Angers Bridge Disaster', https://www.youtube.com/watch?v=OskvclB2DNU [amateur video].

15. J. C. Jeaffreson, *The Life of Robert Stephenson, F.R.S.* (London: Longman, Green, 1864).

16. L. T. C. Rolt, *George and Robert Stephenson: The Railway Revolution* (London: Longmans, 1960).

17. J. G. H. Warren, *A Century of Locomotive Building by Robert Stephenson and Co., 1823–1923* (Newcastle upon Tyne: Andrew Reid, 1923; repr. Newton Abbot, David & Charles, 1970).

18. Reports on the collapse of the Dee Bridge, http://www.railwaysarchive.co.uk/documents/BoT_DeeBridge1847.pdf.

19. Peter R. Lewis and Colin Gagg, 'Aesthetics versus Function: The Fall of the Dee Bridge, 1847', *Interdisciplinary Science Reviews* 29 (2004), pp. 177–91.

20. Hector Velez, 'Design Complacency: The Dee Bridge Collapse'.

21. 'Obituary of Sir Charles William Pasley,. *Proceedings of the Royal Society* 12 (1862–3), pp. xx–xxv.

22. *Report of the Commissioners Appointed to Inquire into the Application of Iron to Railway Structures* (London: William Clowes and Sons for Her Majesty's Stationery Office, 1849).

23. G. G. Stokes, 'Discussion of a Differential Equation relating to the Breaking of Railway Bridges', *Cambridge Philosophical Transactions* 8 (1849), pp. 707–35; repr. in *Mathematical and Physical Papers.*, vol. 2 (Cambridge: Cambridge University Press, 1883), pp. 178–220.

24. Isaac Todhunter, *A History of the Theory of Elasticity and of the Strength of Materials from Galilei to the Present Time*, ed. Karl Pearson, 2 vols (Cambridge: Cambridge University Press, 1886–93).

25. Homersham Cox, 'The Dynamical Deflection and Strain of Railway Girders', *Civil Engineer and Architect's Journal* 11 (1849), pp. 258–64.

26. S. P. Timoshenko, 'On the Forced Vibrations of Bridges', *Philosophical Magazine* 43 (1922), pp. 1018–19.

27. Yeong-Bin Yang and Jong-Dar Yau, 'Vehicle-Bridge Interaction Element for Dynamics Analysis', *Journal of Structural Engineering* 123 (1997), pp. 1512–18.

28. T. Arvidsson and R. Karoumi, 'Train-Bridge Interaction— A Review and Discussion of Key Model Parameters', *International Journal of Rail Transportation* 2 (2014), pp. 147–86.

29. A. Clebsch,*Théorie de l'Elasticité des Corps Solides*, trans. A. J. C. Barré de Saint-Venant and A. Flamant, with additional notes by A. J. C. Barré de Saint-Venant (Paris: Dunod, 1883); German orig. *Theorie der Elastizitat fester Körper* (Leipzig: Teubner, 1862).

30. John Prebble, *The High Girders* (London: Secker and Warburg, 1956).

31. John Thomas, *The Tay Bridge Disaster—New Light on the 1879 Tragedy* (Newton Abbot: David and Charles, 1972).

32. *Report of the Court of Inquiry and Report of Mr Rothery, upon the Circumstances Attending the Fall of a Portion of the Tay Bridge on 28th December 1878* (London: Eyre and Spottiswoode for Her Majesty's Stationery Office, 1880).

33. 'Story of the Tay Bridges', *Wonders of World Engineering*, http://wondersofworldengineering.com/tay-bridges.html.

34. *Report of the Wind Pressure (Railway Structures) Commission* (20 May 1881), http://www.railwaysarchive.co.uk/documents/WindComm_Tay1880.pdf.

35. Rolt Hammond, *The Forth Bridge and its Builders* (London: Eyre & Spottiswoode, 1964).

36. *Appendix to Report and Evidence taken upon the Court of Inquiry into the Tay Bridge Disaster* (London: Eyre and Spottiswoode for Her Majesty's Stationery Office, 1880).

37. G. G. Stokes, 'Discussion of the Results of some Experiments with Whirled Anemometers', *Proceedings of the Royal Society* 32 (1881), 170–88; repr. in *Mathematical and Physical Papers*, vol. 5 (Cambridge: Cambridge University Press, 1905), pp. 73–94.

38. G. G. Stokes, 'On the Determination of the Constants of the Cup Anemometer by Experiments with a Whirling Machine', *Philosophical Transactions of the Royal Society of London* 5 (1878), 818–21; repr. in *Mathematical and Physical Papers*, vol. 5 (Cambridge: Cambridge University Press, 1905), pp. 95–9.

CHAPTER 10

1. Horace Mann (ed.), *Census of Great Britain, 1851: Religious Worship in England and Wales* (London: George Routledge, 1854).

2. Stephen J. Gould, 'Fall in the house of Ussher', *Natural History* 100 (1991), pp. 12–21.

3. Ronald L. Numbers, *The Creationists: From Scientific Creationism to Intelligent Design* (2nd edn, Cambridge, MA: Harvard University Press, 2006).

4. Peter J. Bowler, *Evolution: The History of An Idea* (4th edn, Berkeley, CA: University of California Press, 2009).

5. David N. Livingstone, *Darwin's Forgotten Defenders: The Encounter between Evangelical Theology and Evolutionary Thought* (Vancouver: Regent College, 1987).

6. Frederick Temple, Rowland Williams, Baden Powell, Henry Bristow Wilson, C. W. Goodwin, Mark Pattison and Benjamin Jowett, *Essays and Reviews* (London: John W. Parker and Son, 1860).

7. J. W. Burrow, *The Crisis of Reason: European Thought, 1848–1914* (New Haven, CT: Yale University Press, 2000).

8. Stewart J. Brown, *Providence and Empire: Religion, Politics and Society in the United Kingdom, 1815–1914* (Harlow: Pearson, 2008).

9. Gerald Parsons, 'Biblical Criticism in Victorian Britain: From Controversy to Acceptance?', in Gerald Parsons (ed.), *Religion in Victorian Britain*, Volume II: *Controversies* (Manchester: Manchester University Press, 1988), pp 238–57.

10. Howard R. Murphy, 'The Ethical Revolt against Christian Orthodoxy in Early Victorian England', *American Historical Review* 60 (1955), pp. 800–17.

11. Dominic Erdozain, *The Soul of Doubt: The Religious Roots of Unbelief from Luther to Marx* (New York: Oxford University Press, 2016).

12. Charles Darwin, *The Autobiography of Charles Darwin, 1809–1882: with original omissions restored* (London: Collins, 1958), p. 87.

13. James R. Moore, *The Post-Darwinian Controversies: A Study of the Protestant Struggle to Come to Terms with Darwin in Great Britain and America, 1870–1900* (Cambridge: Cambridge University Press, 1979).

14. Frank M. Turner, 'The Victorian Conflict between Science and Religion: A Professional Dimension', *Isis* 69 (1978), pp. 356–76.

15. David N. Livingstone, 'Myth 17. That Huxley Defeated Wilberforce in Their Debate over Evolution and Religion', in Ronald L. Numbers (ed.), *Galileo Goes to Jail and Other Myths about Science and Religion* (Cambridge, MA: Harvard University Press, 2009), pp. 152–60.

16. Joseph Larmor (ed.), *Memoir and Scientific Correspondence of the Late Sir George Gabriel Stokes*, vol. 1 (Cambridge: Cambridge University Press; 1907), pp. 2–3.

17. Larmor, *Memoir*, p. 46.

18. David B. Wilson, 'A Physicist's Alternative to Materialism: The Religious Thought of George Gabriel Stokes', *Victorian Studies* 28/1 (1984), p. 76.

19. Larmor, *Memoir*, p. 47.
20. David B. Wilson (ed.), *The Correspondence between Sir George Gabriel Stokes and Sir William Thomson, Baron Kelvin of Largs*, vol. 1: *1849–1869* (Cambridge: Cambridge University Press, 2011), p. 60.
21. Wilson, *The Correspondence between Sir George Gabriel Stokes and Sir William Thomson*, vol. 1, p. 61.
22. George Gabriel Stokes, *Conditional Immortality: A Help to Sceptics* (London: James Nisbet, 1897), pp. 28–9.
23. D. W. Bebbington, *Evangelicalism in Modern Britain: A History from the 1730s to the 1980s* (London: Unwin Hyman, 1989).
24. Richard Whately, *A View of the Scripture Revelations Concerning a Future State* (London: Fellowes, 1829).
25. J. W. Colenso, *St Paul's Epistle to the Romans: Newly Translated and Explained from a Missionary Point of View* (Cambridge: Macmillan, 1861).
26. Brown, *Providence and Empire*, pp 238–9.
27. F. W. Farrar, *Mercy and Judgment* (London: Macmillan, 1881), p. 14.
28. Wilson, 'A Physicist's Alternative to Materialism', p. 80.
29. Edward White, *Life in Christ: Four Discoveries upon the Scripture Doctrine that Immortality is the Peculiar Privilege of the Regenerate* (London: Jackson and Walford, 1846), quotation from p. 286.
30. Dale A. Johnson, 'Popular Apologetics in Late Victorian England: The Work of the Christian Evidence Society', *Journal of Religious History* 11/4 (1981), p. 558.
31. *Report of a Conference on Conditional Immortality* (London: Elliot Stock, 1876).
32. P. Barker to Stokes, 18 May 1877, Cambridge University Library Add. MS 7656, C391.
33. Stokes, *Conditional Immortality*, p. 9.
34. Stokes, *Conditional Immortality*, p. 10.
35. Stokes, *Conditional Immortality*, pp. 10–11.
36. Stokes, *Conditional Immortality*, p. 12.
37. Stokes, *Conditional Immortality*, p. 12. Here Stokes quotes 1 Corinthians 15:22.
38. Stokes, *Conditional Immortality*, p. 14. Stokes again quotes from 1 Corinthians 15:22.
39. Stokes, *Conditional Immortality*, p. 14.
40. Stokes, *Conditional Immortality*, p. 15.
41. Stokes, *Conditional Immortality*, p. 16.
42. Stokes, *Conditional Immortality*, p. 18.
43. Stokes, *Conditional Immortality*, p. 19.
44. Stokes, *Conditional Immortality*, p. 23.
45. Stokes, *Conditional Immortality*, p. 24.
46. Stokes, *Conditional Immortality*, p. 25.
47. Brown, *Providence and Empire*, pp. 98–9.
48. Alastair Wood, 'Fifty-eight Years of Friendship: Kelvin and Stokes', in Raymond Flood, Mark McCartney and Andrew Whitaker, (eds), *Kelvin: Life, Labours and Legacy* (Oxford: Oxford University Press, 2008), pp. 69, 73–4.
49. Aileen Fyfe, 'The Reception of William Paley's "Natural Theology" in the University of Cambridge', *British Journal for the History of Science* 30/3 (1997), pp. 321–35.
50. Darwin, *Autobiography*, p. 59.
51. G. G. Stokes, 'On the Bearings of the Study of Natural Science, and of the Contemplation of the Discoveries to which that Study Leads, on our religious ideas', *Journal of the Transactions of the Victoria Institute* 14 (1881), p. 247.
52. William Paley, *Natural Theology: Or, Evidences of the Existence and Attributes of the Deity* (12th edn, London: R. Faulder, 1809), p. 3.
53. Robert M. Young, *Darwin's Metaphor: Nature's Place in Victorian Culture* (Cambridge: Cambridge University Press, 1985), p. 127.

54. Jonathan R. Topham, 'Beyond the "Common Context": The Production and Reading of the Bridgewater Treatises', *Isis* 89 (1998), pp. 233–62.

55. Stokes, 'On the Bearings of the Study of Natural Science', p. 228.

56. George Gabriel Stokes, *Burnett Lectures on Light: First Course, on the Nature of Light, Delivered at Aberdeen in November*, 1883 (London: Macmillan; 1884), pp. viii, 88.

57. George Gabriel Stokes, *Natural Theology: The Gifford Lectures Delivered before the University of Edinburgh in 1891* (London and Edinburgh: Adam and Charles Black, 1891), p. 1.

58. Stokes, *Natural Theology 1891*, p. 270.

59. Stokes, *Natural Theology 1891*, p. 231.

60. David N. Livingstone, *Dealing with Darwin: Place, Politics, and Rhetoric in Religious Engagements with Evolution* (Baltimore, MD: Johns Hopkins University Press, 2014), p. 67.

61. George Gabriel Stokes, *Natural Theology: The Gifford Lectures Delivered before the University of Edinburgh in 1893* (London: Adam and Charles Black, 1893), p. 47.

62. Stokes, *Natural Theology 1893*, p. 46.

63. [James Reddie], *Scientia scientiarum: being some account of the origin and objects of the Victoria Institute, or, the Philosophical Society of Great Britain* (3rd edn, London: Robert Hardwicke, 1866), p. 26.

64. W. H. Brock and R. M. Macleod, 'The Scientists' Declaration: Reflexions on Science and Belief in the Wake of *Essays and Reviews*, 1864–5', *British Journal for the History of Science* 9/1(1976), pp. 39–66.

65. Reddie, *Scientia scientiarum*, p. 28.

66. 1 Timothy 6:20, Holy Bible, King James Version.

67. F. W. H. Petrie to Stokes, 12 August 1871, Cambridge University Library Add. MS 7656, CM04984, V57.

68. F. W. H. Petrie, to Stokes, 9 December 1871, Cambridge University Library Add. MS 7656, CM04984, V58.

69. F. W. H. Petrie to Stokes, 11 March 1872, Cambridge University Library Add. MS 7656, CM04984, V59.

70. Philip L. Marston, 'Maxwell, Faith and Physics', in Raymond Flood, Mark McCartney, and Andrew Whitaker (eds), *James Clerk Maxwell: Perspectives on His Life and Work* (Oxford: Oxford University Press, 2014), p. 278.

71. F. W. H. Petrie to Stokes, Cambridge University Library Add. MS 7656, CM04984, V61, V62, V63, V65, V68, V69.

72. George Gabriel Stokes, 'Absence of Real Opposition between Science and Religion', *Journal of the Transactions of the Victoria Institute* 17 (1884), pp. 195–203.

73. Marie Boas Hall, *All Scientists Now: The Royal Society in the Nineteenth Century* (Cambridge: Cambridge University Press, 1984), p. 136.

74. T. C. F. Stunt, 'The Victoria Institute: The First Hundred Years', *Faith and Thought* 94/3 (1965), p. 174.

75. George Gabriel Stokes, 'The Perception of Light', *Journal of the Transactions of the Victoria Institute* 29 (1897), p. 21.

CHAPTER 11

1. An arresting description of this explosion is described in Ray Bradbury's short story 'The Sound of Thunder', which appears in his collection *The Golden Apples of the Sun*, originally published in 1953; the story is noteworthy as being a literary precursor of 'the butterfly effect' commonly referred to in descriptions of chaos theory.

2. Lord Rayleigh, 'Sir George Gabriel Stokes, Bart. 1819–1903', *Proceedings of the Royal Society* 75 (1905), pp. 199–216;. repr. in *MPP* vol. V pp. ix–xxv.

3. As a young academic, one learns to appreciate this sixth sense. An example is Stokes's derivation of the equations describing the Hele-Shaw cell, discussed below. A particular favourite of mine is the derivation of the Taylor's dispersion coefficient, produced more or less by black magic, although the

result is a secure consequence of perturbation theory, which does not appear in Taylor's account: see G. I. Taylor, 'Dispersion of Soluble Matter in a Solvent Flowing Slowly through a Tube', *Proceedings of the Royal Society of London* A 219 (1953), pp. 186–203.

4. G. G. Stokes, 'On the Theories of the Internal Friction of Fluids in Motion, and of the Equilibrium and Motion of Elastic Solids' [read 14 April 1845], *Transactions of the Cambridge Philosophical Society* 8 (3) (1849), pp. 287–319; repr. in *MPP* vol. I, pp. 75–129, section 1, article 4.

5. G. K. Batchelor, *An Introduction to Fluid Dynamics* (Cambridge: Cambridge University Press, 1967).

6. e.g. C. Truesdell, *A First Course in Rational Continuum Mechanics*, vol. 1: *General Concepts* (New York: Academic Press, 1977).

7. Kolumban Hutter, *Theoretical Glaciology* (Dordrecht: Reidel, 1983).

8. Jacob Bear, and Yehuda Bachmat, *Introduction to Modeling of Transport Phenomena in Porous Media* (Dordrecht: Kluwer, 1990).

9. R. N. Hills, D. E. Loper and P. H. Roberts, 'A Thermodynamically Consistent Model of a Mushy Zone', *Quarterly Journal of Mechanics and Applied Mathematics* 36 (1983), pp. 505–39.

10. D. A. Drew and R. T. Wood, 'Overview and Taxonomy of Models and Methods', presented at the International Workshop on Two-Phase Flow Fundamentals', National Bureau of Standards, Gaithersburg, MD, 22–27 September 1985.

11. Jean Leray, 'Sur le mouvement d'un liquide visqueux emplissant l'espace', *Acta Mathematica* 63 (1934), pp. 193–248.

12. C. Truesdell, *Rational Thermodynamics* (New York: McGraw-Hill, 1969).

13. See, for example, Carlo Cercignani, *The Boltzmann Equation and its Applications* (New York: Springer-Verlag, 1988).

14. See Sydney Chapman and T. G. Cowling, *The Mathematical Theory of Non-Uniform Gases*, 3rd edn (Cambridge: Cambridge University Press, 1970).

15. G. G. Stokes, 'Theories of the Internal Friction of Fluids in Motion', op. cit. (note 4), section 1, article 6.

16. Daniel Hillel, *Fundamentals of Soil Physics* (New York: Academic Press, 1980).

17. S. Goldstein, *Modern Developments in Fluid Dynamics*, vol. 2 (Oxford: Clarendon Press, 1938).

18. Alfred Wegener, *The Origin of Continents and Oceans*, 4th edn, trans. John Biram (New York: Dover, 1966).

19. See Arthur Holmes, *Principles of Physical Geology*, 3rd edn, rev. Doris Holmes (New York: John Wiley, 1978) or Geoffrey F. Davies, *Dynamic Earth: Plates, Plumes and Mantle Convection* (Cambridge: Cambridge University Press, 1999).

20. K. M. Cuffey and W. S. B. Paterson, *The Physics of Glaciers*, 4th edn (Amsterdam: Elsevier, 2010).

21. G. G. Stokes, 'On the Effect of the Internal Friction of Fluids on the Motion of Pendulums', [read 9 December 1850], *Transactions of the Cambridge Philosophical Society* 9 (2), Part II pp. [8]–[106]; repr. in *MPP* vol. III, pp. 1–141.

22. Ian Proudman and J. R. A. Pearson, 'Expansions at Small Reynolds Numbers for the Flow past a Sphere and a Circular Cylinder', *Journal of Fluid Mechanics* 2 (1957), pp. 237–62.

23. Milton Van Dyke, *Perturbation Methods in Fluid Mechanics*, annotated edn (Stanford, CA: Parabolic Press, 1975).

24. With apologies to David Acheson, see *From Calculus to Chaos: An Introduction to Dynamics* (Oxford: Oxford University Press, 1997), Chapter 12.

25. G. G. Stokes, 'Mathematical Proof of the Identity of the Stream Lines Obtained by Means of a Viscous Film with those of a Perfect Fluid Moving in Two Dimensions', *Report of the Sixty-Eighth Meeting of the British Association for the Advancement of Science; held at Bristol in September 1898* (London: John Murray, 1899), pp. 143–4; repr. in *MPP* vol. V, pp. 278–82.

26. J. Mathiesen, I. Procaccia, H. L. Swinney and M. Thrasher, 'The Universality Class of Diffusion-Limited Aggregation and Viscous Fingering', *Europhysics Letters* 76 (2) (2006), pp. 257–63.

27. P. G. Saffman and Geoffrey Taylor, 'The Penetration of a Fluid into a Porous Medium or Hele-Shaw Cell Containing a More Viscous Liquid', *Proceedings of the Royal Society of London* A 245 (1958), pp. 312–29.

28. R. F. Dressler, 'Mathematical Solution of the Problem of Roll Waves in Inclined Open Channels', *Communications on Pure and Applied Mathematics* 2 (1949), pp. 149–94.

29. Dressler had two parameters, but did not take account of a prescribed upstream mass flow.

30. Louis N. Howard, 'Bounds on Flow Quantities', *Annual Review of Fluid Mechanics* 4 (1972), pp. 473–94.

31. Elizabeth B. Dussan V. and S. H. Davis, 'On the Motion of a Fluid–Fluid Interface along a Solid Surface', *Journal of Fluid Mechanics* 65 (1974), pp. 71–95.

32. G. G. Stokes, 'On the Theory of Oscillatory Waves', [read 1 March 1847], *Transactions of the Cambridge Philosophical Society* 8 (4) (1847), pp. 441–55; repr. in *MPP* vol. I, pp. 197–229.

33. G. G. Stokes, 'Supplement to a Paper on the Theory of Oscillatory Waves'. *MPP* vol. I, pp. 314–26.

34. See for example M. S. Longuet-Higgins and M. J. H. Fox, 'Theory of the Almost-Highest Wave, Part 2: Matching and Analytic Extension', *Journal of Fluid Mechanics* 85/4 (1978), pp. 769–86.

35. Alan C. Newell, *Solitons in Mathematics and Physics* (Philadelphia, PA: Society for Industrial and Applied Mathematics, 1985).

36. Francesco Fedele, Joseph Brennan, Sonia Ponce de León, John Dudley and Frédéric Dias, 'Real World Ocean Rogue Waves Explained without the Modulational Instability', *Nature Scientific Reports* 6 (2016), art. 27715.

37. G. G. Stokes, 'On the Discontinuity of Arbitrary Constants that Appear as Multipliers of Semi-Convergent Series (A letter to the Editor)', *Acta Mathematica* 26 (1902), pp. 393–7; repr. in *MPP* vol. V, pp. 278–82.

38. G. G. Stokes, 'On the Numerical Calculation of a Class of Definite Integrals and Infinite Series' [read 11 March 1850], *Transactions of the Cambridge Philosophical Society* 9 (1) (1856), pp. 166–187; repr. in *MPP* vol. II, pp. 329–57.

39. G. G. Stokes, 'On the Discontinuity of Arbitrary Constants which Appear in Divergent Developments' [read 11 May 1857], *Transactions of the Cambridge Philosophical Society* 10 (1) (1864), pp. 105–28; repr. in *MPP* vol. IV, pp. 77–109.

40. G. G. Stokes, 'Supplement to a Paper on the Discontinuity of Arbitrary Constants which Appear in Divergent Developments' [read 25 May 1868], *Transactions of the Cambridge Philosophical Society* 11 (2) (1871), pp. 412–25; repr. in *MPP* vol. IV, pp. 283–98.

41. E. J. Hinch, *Perturbation Methods* (Cambridge: Cambridge University Press, 1991), Figure 3. 3.

42. The simplest derivation of equation (21) is to use the coefficients c_k is to use the method of steepest descents; alternatively one can derive the same expression (without the normalization factor being determined) directly from the differential equation, though it is more laborious. Incidentally, the expression in Milton Abramowitz and Irene A. Stegun (eds), *Handbook of Mathematical Functions* (Washington, DC: National Bureau of Standards, 1964) appears different but is equivalent, using the triplication formula for the gamma function.

43. R. B. Dingle, *Asymptotic Expansions: Their Derivation and Interpretation* (London: Academic Press, 1973).

44. M. V. Berry, 'Stokes' Phenomenon: Smoothing a Victorian Discontinuity', *Publications mathématiques de l'I. H. É. S.* 68 (1988), pp. 211–21.

45. M. V. Berry, 'Uniform Asymptotic Smoothing of Stokes's Discontinuities', *Proceedings of the Royal Society of London* A 422 (1989), pp. 7–21.

46. Harvey Segur, Saleh Tanveer and Herbert Levine (eds), *Asymptotics beyond All Orders*, NATO ASI series B, vol. 284 (New York: Plenum Press, 1991).

47. Martin E. Glicksman, 'Mechanism of Dendritic Branching', *Metallurgical and Materials Transactions* A 43 (2011), pp. 391–404.

48. A. C. Fowler and G. Kember, 'On the Lorenz–Krishnamurthy Slow Manifold', *Journal of the Atmospheric Sciences* 53 (1996), pp. 1433–7.

49. G. G. Stokes, 'On the Dynamical Theory of Diffraction' [read 26 November 1849], *Transactions of the Cambridge Philosophical Society* 9 (1) (1856), pp. 1–62; repr. in *MPP* vol. II, pp. 243–328.

50. Lord Rayleigh, 'On the Electromagnetic Theory of Light', *Philosophical Magazine* (5th set.) 12 (73) (1881), pp. 81–101.
51. e.g. S. J. Chapman, J. M. H. Lawry, J. R. Ockendon and R. H. Tew, 'On the Theory of Complex Rays', *SIAM Review* 41 (3) (1999), pp. 417–509.
52. Joseph B. Keller, 'Geometrical Theory of Diffraction', *Journal of the Optical Society of America* 52 (2) (1962), pp. 116–30.
53. Lord Kelvin, 'The Scientific Work of Sir George Stokes', *Nature* 67 (1903), pp. 337–8; repr. in *Mathematical and Physical Papers*, Lord Kelvin, ed. Joseph Larmor, vol. VI (Cambridge: Cambridge University Press, 1911, repr. 2011), pp. 339–44.

NOTES ON CONTRIBUTORS

June Barrow-Green is Professor of History of Mathematics at the Open University.

Michael Berry is Melville Wills Professor of Physics (Emeritus) at the University of Bristol.

Olivier Darrigol is a Research Director at the SPHere team of CNRS in Paris, and a Research Scholar at UC-Berkeley's OHST. He is the author of several books on the history of quantum mechanics, electro-dynamics, fluid mechanics, optics, and Boltzmann's statistical mechanics.

Sloan Evans Despeaux is Professor of Mathematics at Western Carolina University in Cullowhee, NC, USA. She is currently Secretary of the International Commission for the History of Mathematics.

Andrew Fowler is Stokes Professor at the University of Limerick and a Senior Research Fellow at Corpus Christi College, Oxford.

Peter Lynch is an Emeritus Professor in the School of Mathematical Sciences, University College Dublin. His main research is in numerical weather prediction and dynamic meteorology. He writes popular articles on mathematics and blogs at www.thatsmaths.com.

Stuart Mathieson is a Postgraduate Research student at Queen's University, Belfast.

Mark McCartney is Senior Lecturer in Mathematics at the University of Ulster. He is currently President of the British Society for the History of Mathematics.

Richard B. Paris is Emeritus Reader in Mathematics at Abertay University. He was formerly employed by Euratom at the French Atomic Energy Commission.

Michael C. W. Sandford, Retired from RAL Space in 2003 as head of Space Instrumentation and Future Projects. Now as a descendant of Gabriel Stokes he measures marathons and researches genealogy.

Andrew Whitaker is Emeritus and Visiting Professor of Physics at Queen's University Belfast and works on the foundations of quantum theory and the history of physics.

Alastair Wood is Emeritus Professor of Mathematics at Dublin City University. Earlier research in asymptotic analysis led to an investigation of Stokes's wider legacy. He currently lives in the Pyrenées-Orientales.

INDEX